e/10

A lovely, Intelligent, and adventurous young lady, who is also a 'free Spirit'...
...just like me.

Hope you enjoy the read, and I look forward to hearing your comments...
...be they 😊 good or bad?

TEL (0034) 674 886 442
e-mail: shaun.donovan77@yahoo.co.uk

BAT OUT OF HELL

THE MOVIE

SHAUN DONOVAN

authorHOUSE®

AuthorHouse™ UK
1663 Liberty Drive
Bloomington, IN 47403 USA
www.authorhouse.co.uk
Phone: 0800.197.4150

Published by AuthorHouse 05/01/2019

ISBN: 978-1-7283-8684-3 (sc)
ISBN: 978-1-7283-8685-0 (e)

Print information available on the last page.

Any people depicted in stock imagery provided by Getty Images are models, and such images are being used for illustrative purposes only.
Certain stock imagery © Getty Images.

This book is printed on acid-free paper.

Because of the dynamic nature of the Internet, any web addresses or links contained in this book may have changed since publication and may no longer be valid. The views expressed in this work are solely those of the author and do not necessarily reflect the views of the publisher, and the publisher hereby disclaims any responsibility for them.

Other titles in this series:

BATTLE OF THE GREYHOUNDS

PART I - AMERICA

From scorching deserts to snow-capped mountains, through forest fires and gangland war-zones, Shaun Donovan rides the Greyhound Bus to the four corners of America…. and beyond.

Having survived an unbelievable white water rafting trip in Colorado, a death-defying sky-dive in Las Vegas and an unforgettable swim with sharks in Florida, Shaun journeys on, scuba-diving for shipwrecks and coral reefs in the Florida Keys, before riding several roller-coasters for pleasure as he travels the continent.

Lucky escapes, a pilgrimage and a dream fulfilled as his 20,000km (12,000 miles) trek through forty-eight states, Canada and the Bahamas are completed and a promise to his children is finally kept.

An epic voyage of discovery, combining thrills and excitement, disappointment and despair, as each normal days ride is turned into the adventure of a lifetime.

To see the American journey in full colour pictures, just go to Shaun's web-site: www.taffys-travels.com

On the web-site there are also links to all of Shaun's other books —which include lots more amazing and exciting journeys around the world.

Other titles in this series:

BATTLE OF THE GREYHOUNDS

PART II - AUSTRALIA

After successfully completing his 12,000 mile bus ride around America, Shaun Donovan now embarks on his latest and greatest challenge -to circumnavigate the coastline of Australia. Apart from using the Australian Greyhound Bus Company to traverse the majority of the continent, Shaun and his fiancée, Sally, also ride the famous 'Indian Pacific' and 'Ghan' trains from west to east and south to north of this vast continent, as part of their 20,000km (12,000 miles) journey -which must be completed in less than 50 days.

In a compelling story of determination, desperation -and in some cases deprivation, read how Shaun and Sally are roasted alive in sizzling heat and stifling humidity, get drowned-out in tropical storms and flash-flooding -and survive an onslaught of electrifying lightning bolts, devastating cyclones and bone-shaking earthquakes before their journey is finally at an end.

Add to this a lethal concoction of shark, stingray, and snake attacks they stumble upon during their trip -and mix it up with crocodile, box jelly-fish, blue-ringed octopus and deadly stone-fish encounters -and you've got yourself one hell of a story-line to tell.

Like his American voyage of discovery, many things were learned and achieved, but there was also despair as his dreams of scuba diving on the Great Barrier Reef and visiting Ayres Rock were shattered due to weather conditions and personal circumstances, both of which he swears *will* be done on his return to one of the most diverse and exciting continents on the planet.

To see Shaun and Sally's circumnavigation of Australian in full colour pictures, just go to Shaun's web-site: www.taffys-travels.com

On the web-site there are also links to all of Shaun's other books —which include lots more amazing and exciting journeys around the world.

TIMESHARE – A JOURNEY INTO THE UNKNOWN

'Timeshare – a journey into the unknown' is a very frank account of the ten years Shaun Donovan spent working as a sales representative and a manager in the industry, both in the Canary Islands, and also on the island of Cyprus. During that time he closed over three million pounds worth of personal sales, along with training hundreds of new recruits to become 'timeshare professionals'.

With over 3,000 *tours* to his name and around 700 sales under his belt, Shaun has made many friendships in the business, not only with his fellow colleagues, but also with many of his clients, who kept coming back year after year to see him, (often to spend more money), after he had introduced them to the wonderful world of luxurious holidays.

Inside the book he also describes in detail how he broke all the ethics of his profession, by mixing business with pleasure, and running off with his client! Read how their *Shirley Valentine* romance eventually turns into a living nightmare, as everything goes tragically wrong for them and their world of dreams is systematically torn apart.

Apart from all the heartache and despair, there are also some wonderful holidays, which Shaun and his family enjoyed together, along with loads of great travel stories, which include two unbelievable bus journey's across America and Australia, a ferry-hop around all seven Canary Islands – and two unforgettable cruises to the Greek Islands and Egypt.

Shaun's manuscript is a compelling catalogue of anecdotes, which has all the ingredients of love, hate and compassion, violence, drugs and embezzlement - along with one of the best insights into the world of timeshare ever produced.

Combine all this with the unparalleled passion of one man, who truly believes that his product is the best thing since sliced bread, and you've got yourself a book which may possibly change the way you think about one of the most lucrative and volatile industries in the world today.

To see Shaun's timeshare story, along with his worldwide travels, in full colour pictures, just go to Shaun's web-site: www.taffys-travels.com

On the web-site there are also links to all of Shaun's other books —which include lots more amazing and exciting journeys around the world.

CHRONICLES OF A BIKER

'Chronicles of a Biker' is a compilation of anecdotes which Shaun Donovan wrote during his 12 years of riding (and *racing*) motor-cycles between 1975 and 1987. Now, in 2012, some 36 years after writing his first *memoirs of a misspent youth*, Shaun has encapsulated dozens of these truly memorable and remarkable (and in some cases *unbelievable*) stories into one book. In the first few chapters, read how Shaun survives an onslaught of terrifying crashes, each one leaving him with several broken bones -or huge chunks of flesh hanging from his battered body, as he lives-on to fight another day. Also included are the tales of his *unlawful misdemeanours*, as he is continually chased —and caught by the police, culminating in several endorsements, along with the loss of his driving licence on no fewer than two occasions. The second-half of the book is dedicated to three amazing journeys which he undertook between July 1981 and August 1983. Spanning a distance of nearly 15,000km on the road (and a further 2,000 nautical miles on ferries), Shaun traverses 13 countries, 1 principality and several islands in Western Europe, before crossing continents into North Africa, to say his first *'Hello'* to the countries of Morocco and Tunisia.

Apart from crashing in Spain, dodging psychopathic drivers in Italy, and having to put-up with various punctures and breakdowns along the way, Shaun has a whale of a time with various friends and partners, as he crosses the Pyrenees Mountains, glides over the Austrian Alps, and blasts his way up and down the motorways of western Europe, discovering a handful of paradise beaches, dozens of quaint little villages -and several wonders of the world, before his *triple-adventure* is finally at an end. Just like Shaun's first two books, *'Battle of the Greyhounds, Part I - America, and 'Battle of the Greyhounds, Part II – Australia'*, which tell the stories of his epic journeys, as he circumnavigates these vast continents by Greyhound Bus, Shaun writes every *biking* tale with such conviction and an unprecedented passion, that one could truly believe they were riding pillion-passenger with him from start to finish.

To see an outline of Shaun's epic journeys around Europe and North Africa in full colour pictures, just go to Shaun's web-site: www.taffys-travels.com

On the web-site there are also links to all of Shaun's other books—which include lots more amazing and exciting journeys around the world.

ASIA —ME & MY GIRL

After living apart for over 18 years, Shaun Donovan and his teenage daughter, Hayley, decide to *make-up for lost time*, by embarking on a 60,000km (40,000 miles) *voyage of discovery*; a journey that would inevitably take them half-way around the world. 'Asia —Me & My Girl' tells the amazing true story of the first half of their incredible adventure, as they travel 20,000km (12,000 miles) *overland* from their hometown of Cardiff, in South Wales, to the Far Eastern shores of Singapore.

Read all about their unforgettable six-day crossing of Siberia, before traversing the infamous Gobi Desert in Mongolia —and how they are *kidnapped* and virtually held to ransom by a bogus taxi-driver and his compatriots in Beijing, China. Shaun also tells the unbelievable stories of how he and his daughter had a magnificent day in Xi'an, visiting the Terracotta Warriors, before meeting the farmer who had actually discovered them back in 1974 —and how the pair of them had their photo taken with the one and only living survivor of the notorious 'S21' Prison Camp from the 'Killing Fields' of Cambodia.

In Thailand all hell breaks loose as our intrepid explorers spend their days bottle-feeding leopard cubs, walking with tigers —and swimming with elephants, before climbing the 7-tiers of the Erawan Waterfall, crossing the Bridge over the River Kwai, riding the Thailand to Burma 'Death Railway' -and walking the length of 'Hellfire Pass'. To round-off their ten-week 'Voyage of Discovery' our *dynamic duo* pay a visit to the phenomenal PETRONAS Towers in Malaysia -and in Singapore they come face-to-face with the 'Lords of the Jungle', as they embark on their first ever 'Night-time Safari'.

To see an outline of Shaun and Hayley's incredible 20,000km overland journey from Cardiff to Singapore in full colour pictures, just go to Shaun's web-site: www.taffys-travels.com

On the web-site there are also links to all of Shaun's other books—which include lots more amazing and exciting journeys around the world.

and the Ngorongoro Crater, where they came face-to-face with the world-renowned Maasai Warriors, before rounding-off their incredible journey with a day-trip to the mystical island of Zanzibar.

To see an outline of Shaun and Hayley's journeys around India and Africa in full colour pictures, just go to Shaun's web-site: www.taffys-travels.com

On the web-site there are also links to all of Shaun's other books —which include lots more amazing and exciting journeys around the world.

A Note From The Author

The events listed in this book have been taken from personal diaries that I began writing after being made redundant on 23rd September 1991, and my subsequent idea to write a movie during this period of great uncertainty in my life. I have changed the names of almost all of the people who became involved in my adventure, apart from the immediate members of my family, simply as a humble courtesy to them, should they prefer to remain anonymous. My original reason for writing comprehensive diaries was to remind myself of my humble beginnings, should the film idea I started with ever reach fruition, and I was lucky enough to become wealthy, famous —or maybe even both? If all else failed then at least my diaries would represent a legacy for my children to read, in order that they might understand how determined I was to succeed, and comprehend my reasons for injecting one hundred and ten percent of myself into the project, in the hope of a better life for Caryl and myself, and subsequently a more prosperous future for our children.

Ted Simon created the inspiration for starting such an outlandish quest for fulfilment of life, and so no-matter what the future may hold I will always be indebted to him for writing Jupiter's Travel's. His amazing four-year journey around the world took him further than any mileage could reach, transporting his mind into the depths of his very soul, and now, after chasing my own dream, I, too, have reached the inner-sanctums of my own soul —through the journey of life itself. From the outset my mother cautioned me about the price that many people have paid for the pursuit of success or stardom, adding how she had been influenced by an old film called 'The Monkey's Claw', in which a man's wish is granted, only to turn out disastrously in the end, and so she prayed that a similar scenario would not happen to me, although she knew deep down inside that I would have to find out for myself.

Ironically, most of the great achievers in life whom I have stumbled upon over the years, have either had very sad beginnings to their lives, or suffered great traumas during their climb up the ladder of success –many have had both. What happened to my family and I during this period of uncertainty, be it good or bad, 'happened', and no-one can change that, for it is now in the past, but actions which can seem *wrong* at the time, can suddenly turn out to be *right* several years later, and as no-one can predict the future with ultimate accuracy, then chances have to be taken. If everyone followed the same attitude of leaving the unknown well alone, and no-one ever took any risks in life, then the world would be a much duller place to live in. As for those of us who are lucky-enough to have been born with an adventurous spirit, well it is our job to amuse and entertain both avid readers and *armchair travellers* alike with our magical tales and endless stories from the four corners of the globe –and beyond.

I rest my case.

This book is lovingly dedicated to my wonderful children, Liam, Carl and Hayley Donovan, the three most important people in my life.

THE LONG ARM OF THE LAW

A Honda Gold Wing 'Executive' has just glided past my window, the driver of which is relaxing back in his great *armchair* without a care in the world it seems -oh how I envy him. *(Right there and then I promised myself that one day I would do 'Route 66' on a Harley-Davidson motor-cycle).* Having listened to dozens of barmy advertisements, all bellowing-out at me from the speakers in the dashboard, I have decided that enough is enough, and so I have reverted once again to serenading myself with my *Love Songs* album on my Sony Walkman. Relaxing back in my chair, with my right hand holding onto the steering wheel and my left arm nonchalantly resting on the window ledge, I interlock my mind and my body as one, systematically entrancing them both into a world of tranquil bliss. Singing along with the music, but thankfully being unable to hear my own *chronic droning*, I give the usual sporadic glance in my rear-view mirror, expecting to see nothing more than hundreds of cars lining the freeway, but to my great surprise I am temporarily blinded by a myriad of flashing lights, that are coming up behind me at a rate of knots.

Immediately realising that it is a police car on my tail, as both the blue and the red lights are flashing intermittently, I naturally assume that the officer's are in hot-pursuit of some notorious gangster, and so I pull over from the fourth lane, into the third lane in order to let them pass. Suddenly I hear sirens wailing, and now I can see lots of other cars in my rear-view mirror, all of them pulling over to one side, in order to let the police car through, but still the officers in the said vehicle have failed to overtake me? A few seconds later and I can see that the police car is now driving directly behind me, its fusion of lights almost blinding my view in the central mirror. The penny has finally dropped; these guys are not after any bandits or bank robbers...they are after me! Not knowing what the hell is going on I cross two more lanes, before finally pulling over onto the hard shoulder up ahead. As I sit there waiting for the *cops* to close in on me, my mind is now in a somewhat deranged state, and I start to panic.

Any minute now they are going to open-fire on me, riddling the car with bullets and exploding all four tyres, in order to stop me from getting away, I think to myself, whilst biting my nails like there is no tomorrow…maybe there won't be? By now the whole freeway surrounding me is virtually at a standstill. Tentatively removing the headphones from around my ears, I lift the Walkman off my lap and place it gently on the passenger seat, before sitting absolutely still in my chair, waiting patiently for the police to come and tell me what it is that I have done wrong? The police car has pulled-over about a cars' length behind me, and now a figure has appeared in my side-view mirror. A fierce looking dude in an officer's uniform is now making *his* way (very slowly) towards my car, and a second officer is following closely behind *him*. Any minute now I am expecting to be dragged out of the car by two burly policemen, one of them brandishing a Magnum 44 at my head, as he throws me against the side of the mustang and tells me to keep my hands on the roof of the car, while the other guy kicks my legs apart, before frisking me from my armpits to my ankles. (I have definitely watched too many *cop* movies!)

With this *somewhat frightening* thought in mind I decide to release my seat-belt, just in case I end up getting strangled in the impending scuffle. The officers are now walking around my car (one on either side) and so I use the electric switch to lower the window on the passenger side. As a uniform appears in the passenger's mirror, I lean over towards it, making sure that both of my hands are still wrapped firmly around the steering wheel, and with a look of pure, unadulterated innocence I utter those immortal words: "What seems to be the problem officer?" My question is obviously treated with the contempt that it deserves, as it is left unanswered and I am met with another question instead. "Could I see your driver's licence sir?"

My cowardice is compounded by the fact that this great 'beast' (which I had brought to life in my mind) who was going to tear me limb-from-limb, before dragging me off to spend the rest of my days rotting-away in some grotty prison cell on the island of Alcatraz, only turned out to be a woman -and a very polite and attractive one at that. Unfortunately, my mind has been thrown completely off balance by this rather unexpected pleasure, and so confusion abounds in my brain, as I try to remember where I had put the blessed thing?

"Um, yes, certainly officer; now where did I… oh yes, it's in the boot –I mean the trunk -um, can you please tell me what I did wrong officer?" I burble.

Contents

PREFACE

It was late in the summer of '75 when I first climbed aboard a motor-cycle and rode off into the sunset, proud-as-punch at the thought of becoming a biker. How could I have even imagined as I covered those first few miles in and around Cardiff, that there was a fellow biker who had just reached the halfway stage of his 'Round-the-World' journey? One year later and the great heat-wave of '76 were upon us, as I celebrated passing my motor-cycle test on the 9th August -my 17th birthday, thus enabling me to ride any motorcycle I so desired. Within eight weeks I had sold my 50cc and replaced it with a 650, and so now the world was my oyster, and I intended to see it all. However, things rarely go as one has planned them to, and little did I know it at the time, but over the coming year I would learn some rather valuable and somewhat painful lessons about the wonderful world of 'super-biking'.

Apart from numerous mechanical problems which the bike suffered throughout the winter months, causing great heartache to myself, nothing would compare to the agonizing pain I suffered on Monday 17th June 1977, when I became involved in a near-fatal head-on collision with a car in the country lanes near to my home. The extent of my injuries, which included a snapped femur in my right leg, a broken coccyx bone in my back, and a crushed right foot, which was on the borderline of amputation, kept me in and out of hospitals, and certainly off my motorcycle (which was thankfully not the bike I was riding when I had the accident) for almost a year, before finally having the go-ahead from my father (I was still living under his roof, and therefore I respected his wishes) to start riding again. For the next three years I rode various motorcycles around the UK, which included popping over to the Isle of Wight on several occasions to visit my girlfriend, Jayne, who was training to become a nurse on the island.

Then in 1981, after borrowing an old school atlas from a fellow colleague in work, Jayne and I decided that we would take off on a circular tour of Western Europe. After taking an overnight ferry from Portsmouth to Cherbourg, where we made friends with two fellow (male) bikers, the four of us took a direct route along the western coast of France, stopping-off at Le Mont St Michel for a sightseeing visit along the way, before continuing southwards through Rennes, Nantes, La Rochelle and Bordeaux. After enjoying a hearty meal together Jayne and I said our fond-farewell's to the lads, as they took a south-easterly route towards France's southern coast, whereas we continued heading due south for the town of Pau, before traversing the incredible Pyrenees Mountains, and crossing borders into the deserts of north-eastern Spain. From here Jayne and I continued heading southwards, stopping-off for one night (and also crashing the bike the following day) in the town of Teruel, before passing through Zaragoza, and riding all the way down to the Mediterranean city of Valencia. From here we followed the coastal road to the seaside resort of Benidorm, where after four days of rest and recuperation the pair of us continued-on with our *Voyage of Discovery*, heading for the port of Alicante, where, at the stroke of midnight we boarded a ship which was bound for the paradise island of Ibiza.

After a short stay with some friends of mine, who owned a small bar on the island, Jayne and I sailed northwards to the magnificent city of Barcelona, where we took leave of our ferry, before moving on to a small seaside town called Rosas, where we had a great reunion with the bikers we had met on the ferry from *Pompey*. From here the pair of us travelled eastwards, crossing borders back into France, where we enjoyed many of the wonderful delights that St. Tropez, Cannes, Nice and Monte Carlo in Monaco have to offer, before dropping down into Italy and riding a further 250 miles to the magnificent city of Rome. After spending two nights in one of the most romantic cities in the world, Jayne and I began our 1,700 mile journey home, calling into Florence, Venice and Lake Garda along the way, before crossing over the Alps into Austria. From here it was northwards all the way, riding 500 miles through the heart of what was then West Germany, before crossing borders into Holland and catching our final ferry to the port of Harwich on the south-eastern coast of England.

Having traversed almost four thousand miles in just over three weeks, I immediately set-about planning two further journeys to broaden my horizons. Both of these new journeys would follow circular routes, similar to the first, only instead of covering such a vast area of Western Europe in one fell swoop, as I had done on my first jaunt, I decided to cover as many different countries and islands as possible on one half of Western Europe on the first trip, before likewise doing the same on the second trip. My ultimate destination on both of these journeys would be the northern shores of the mighty African continent. In May 1983 I set off once again on a new and exciting adventure, seeking out different cities, towns and villages in each new country that I visited, whilst enjoying spectacular landscapes along the way. I also took time out to study the locals in their own natural environments, chatting and mixing-in with these people at every opportunity.

Utilising the main *Auto-route* system through Central France this time, Caryl (my new girlfriend) and I headed for Sète on the Mediterranean coast, where, after a few days rest and recuperation, we boarded a ship bound for Tangiers in Morocco. With our first taste of Africa now firmly implanted in our memories, Caryl and I sailed across the Straits of Gibraltar, to the southern-most tip of Spain, before heading due west to the Portuguese border. To our great surprise we had to take a raft-like ferry across a small lake, before riding the final fifty miles to the coastal resort of Albufeira, where Caryl and I set up home in pre-pitched tents for four nights, before beginning our long journey home. Once again this was relatively straight-forward, as we headed for Portugal's capital city of Lisbon on the south-western coast, before riding northwards for a hundred miles or so, and then travelling in a north-easterly direction, before crossing borders back into Spain, as we headed for the town of Salamanca. Burgos and Vargas were next on the list, before finally coming to a halt in the city of Santander on Spain's northern shores, where, after two days of total relaxation, Caryl and I boarded a 24-hour ferry bound for the city of Plymouth. From here it was only a 150 mile ride back to Cardiff.

The *second* journey (described above) was actually less than 2,200 road miles, although half of that amount was covered again, only this time in nautical miles. The third and final journey would notch-up a total of 2,600 road miles, along with around a thousand nautical miles, although I only had eighteen days in which to complete this journey, as opposed to twenty on the last trip. This latest road trip began in Brussels, having already taken the overnight ferry from Felixstowe in the UK to the port of Zeebrugge in Belgium, where Caryl and I continued riding southwards through Luxembourg, France, Germany and Switzerland, before ending-up at the Italian seaport of Genoa. From here we took a 24-hour ferry across the Mediterranean Sea for our second visit to Africa, but our first to the country of Tunisia. After three days of relaxing and sight-seeing, we began our long journey home, catching an overnight ferry to Sardinia (via the island of Sicily) where we then rode the length of this barren island, before taking another (very short) ferry to the French island of Corsica.

Riding from south to north of this island proved to be the most difficult *mountainous* road journey I have ever encountered in my life, but eventually we made it. From here it was but a five-hour ferry crossing to the French mainland, followed by an eight-hundred mile ride to Calais, where we caught our final ferry back to the UK, before riding the last 250 miles from Folkestone to Cardiff. Having successfully completed all three of these journeys I felt very proud of my achievement.

CHAPTER 1

The Dream Begins

Although those journeys would be forever engrained in my memory, I decided to put pen to paper and write complete accounts of each trip, before coupling my memoirs together with the hundreds of photographs that we had taken along the way, just in case anyone else should want to do a similar journey and use my guides for reference. In 1984 I sold the bike, and Caryl and I bought our first home, before settling-in to married life together. However, there was still this niggling inside of me which would not go away, and I knew exactly what it was. The realisation that I was now living a *normal* existence hit me like a brick and I knew deep down inside that I badly needed an outlet. They say that a biker never throws his leathers away, he only hangs them up, and I must be a typical case, I thought to myself, but there was simply no way that I could afford to buy another bike, let alone the expense of touring on one, and so I wondered what would be the next best thing to actually doing a journey?

I suddenly remembered about an article which I had read in a bike magazine a few years ago, where the writer had given an outlying synopsis of a round-the-world journey, which had been successfully completed on a bike. The rider had apparently written an autobiographical account of the journey, and if I remembered correctly the title of his book was 'Gulliver's Travels'. "I stand corrected" I said to the bookshop assistant, as he informed me that the correct title was 'Jupiter's Travel's', written by a gentleman called Ted Simon, before adding that the paperback had been a bestseller for many years. Having purchased my copy, I found it almost impossible to put the book down for more than a few hours at a time, and as each page was turned my admiration for Mr. Simon grew to an almost unprecedented level. Unbeknown to me at the time, but 'The Dream' (of the future) had just been created.

How I was ever going to do anything about it though was an entirely different matter, and so onto the bookshelf it went for over five years, while I worked diligently on gaining employment promotions, whilst bringing up my two very young sons, Liam and Carl, with my beloved wife, Caryl. Unfortunately, on 21st September 1991, without any forewarning whatsoever, I was suddenly made redundant from the car sales industry. It was only five months earlier that I had given up an excellent job, after being literally head-hunted by the sales manager of the local Ford dealership to take up the appointment of 'Sales Representative' for his company. Little did I realise it at the time, but my world was just about to change... forever! Driving my car through the streets of Cardiff with Meat-Loaf's iconic 'Bat out of Hell' album blasting-out at me from the speakers, gave me constant reminders of my biking days, as thoughts of speed and blood-pumping adrenaline rushing through my veins, brought the memories flooding back to me. At that same time, and to my utter astonishment, I discovered that the current generation of young bikers, most of whom were at least ten years my junior, were also captivated by the very same album —and suddenly an idea struck me.

Reading Jupiter's Travel's for the second time I confirmed my suspicions about how several of the events which had taken place on Ted's *round the world* journey, could actually be portrayed by individual songs from Meat-loaf's 'Bat out of Hell' album, and so I decided to put my *theory* to the test. I began telling my friends and relatives —and also my work colleagues, all about Jupiter's Travel's, before *falsely* intimating that there could soon be a movie coming out which would not only tell the story about Ted's amazing journey around the world, but that it would also be using the songs from Meat-loaf's *Bat out of Hell* album as its primary soundtrack. The responses I received from my *little white lie* were phenomenal, and so I knew exactly what I had to do next. Buying individual maps of every continent on the planet, I firstly stuck them to huge sheets of cardboard, before carefully mapping-out Ted's journey from start to finish, by using various strands of multi-coloured wool and different types of locating pins to make the 'special' places stand-out.

Then, for the third time I read Jupiter's Travel's, only this time I logged every single event as it happened in turn on the journey. Once this had been completed I rang Penguin Books with a view to obtaining Mr. Simon's home address. However, they graciously informed me that Ted was now using a different publisher, before kindly giving me their telephone number. The gentleman I spoke to at the publishing company duly informed me that Ted had moved from the UK to California several years ago, before adding that he would be unable to give me Ted's full home address, as this was *confidential client information*. However he did offer to forward my letter on to Ted, if I sent it to him first. Now all I had to do was compose a letter. Although this sounds a very straight-forward way of communicating my idea, I knew that I dare not reveal my full intentions in writing, simply because I would be unable to express my true emotions to Ted unless I actually spoke to him in person.

However, I did manage to compose a five-page letter stating that I had a *workable idea* which I needed Ted's comments on, and so I sent this, along with a magazine article (which had been written some years earlier) about my three motorcycle journeys across Europe and North Africa, to the publishers in question. Included in the write-up was a colour photograph of me and Caryl, (taken when she was my fiancé), standing in front of a huge map of Europe, the pair of us clutching one of the three pictorial travel journals, which I had pieced together upon our return to the U.K. I could only hope that curiosity would get the better of Mr. Simon, as I patiently waited, day after day for a reply, not knowing whether my letter had even reached him in America, let alone interested him at all? Whatever the outcome of it all, at least I could say that I had tried. A week or so went by with no signs of any kind of reply in the post, and even though I had put my telephone number on the top of the letter, I did not expect Mr. Simon to spend the cost of an international phone-call, just to find-out what it was that I had in mind?

In desperation, I checked with the publishers, who confirmed that they had definitely sent the letter to Ted's home address in Northern California. It was during this period of uncertainty that a great friend of Caryl's, whom I have called 'Amber', introduced me to a very successful businessman called Jules Gilmore. Originally from London, Jules had recently moved to South Wales in order to set up a few partnership deals in various companies, and my name had been given to him by Amber, with a view to me possibly working for one of them.

Jules and I first met socially at a barbeque, and as neither of us wanted to *talk shop* whilst enjoying a relaxing evening, we arranged to meet later that week, to discuss future employment prospects. *(A job was something I so desperately needed, having spent the last month, or so languishing on the dole).* Having *broken the ice* at our first meeting, both Jules and I were able to chat freely and honestly about anything and everything, and even though there were no immediate vacancies in any of his newly acquired companies at this precise moment in time, we both knew that this would be the beginning of a long-lasting friendship. During our relaxing chat together, I happened to mention my little venture regarding Ted Simon, saying how I was hoping to turn his book into a movie, and to my great surprise, Jules was immediately interested, asking me to keep him fully informed with its progress. *(I must confess that I had half-expected this middle-aged, and very successful businessman, to laugh, or even scoff, at the mere thought of such an outlandish suggestion).*

On Wednesday morning I brought the milk in off the front doorstep, scooping the mail up as usual, before kicking the door shut behind me with the heel of my right foot. As I nonchalantly flicked through a selection of brown and white envelopes, I suddenly saw the one which I had been waiting for. Staring me right in the face was a personalised envelope, with our address (which had been handwritten) sitting in the centre of it, and a Californian postmark stamped clearly in the top right-hand corner. For a moment I stared at it in total disbelief, the dark side of my mind telling me that it couldn't possibly be my reply from Ted Simon, before the logical side of my brain finally kicked-in, and I began ripping open the envelope for all I was worth, my heart racing faster by the second. The letter inside had also been hand-written by Ted, thus giving it a friendly and more personal touch, and subsequently easing my fears of the dreaded printed letter, with photocopied signature, telling you straight away that the receiver is grateful for your interest, but unfortunately they are too busy to take matters any further, which basically means that they cannot be bothered, to reply to your request in person.

My letter had no-doubt intrigued Mr. Simon, as he enquired whether I intended making a board game about his journey, or possibly even starting a new religion. Whatever my plan was, he wanted to know, as it was obvious by the length of my *essay* to him, that I meant business, and that I would not be put-off easily. Ted had also included his home address in his letter, as well as his weekend address, along with his home telephone number, just in case I wanted to *spend more money* (as he put it) by calling him, to let him know what this 'great idea' of mine was all about? Rounding-off the letter, Ted said how encouraged he was by my "Good taste in wives!" (Perhaps that explains why Ted never returned the photograph to me with his reply!) Caryl could not share in my excitement, but then I knew what my next move was going to be —and she didn't! Knowing that Californian time was eight hours behind our own G.M.T., I waited until evening before phoning, only to be greeted by an answer-phone message. Having left my name and number, in the vain hope of receiving a return call, but hearing nothing within the next forty-eight hours, I tried again on Friday, Saturday and Sunday evening, but each time I met with that same recorded message from Ted, apologising for "Not being at home right now".

At 7am on the following Monday morning, Caryl called me to the telephone, saying that it was Jules Gilmore on the line, but she was mistaken; it was Ted Simon, returning my call, and asking me *in person* just what this mystery was all about? However, I was not about to *blow everything* by trying to pitch him over the telephone, and so I told him that I would only reveal my idea to him as and when I was talking to him 'face to face'.

"Well you had better come over here and see me then", was Ted's immediate reply. It was the very answer that I had been hoping to hear, and I knew right there and then what my next move would be. Asking Ted if I could call him again in a few days time, I Immediately phoned Jules, arranging to call around and see him at his place this evening. I knew that Jules was my only hope of getting to America, as not only had he shown great enthusiasm with the project, but also because he had a partnership in a local travel-agency. The $64,000 question was: 'How much faith did Jules have in me as a person —and also as a salesman?' Very soon I would have my answer.

As I stood in the dimly-lit porch-way of his ground-floor apartment on the outskirts of the city, too nervous to even ring the bell, I softly recited the speech that I had prepared earlier, half-wondering if I was about to make an absolute fool of myself, and maybe, worst of all, throw our wonderful friendship down the drain? Finally I plucked up the courage to press the doorbell, and as Jules answered the door, we immediately shook hands, as always, his ever-smiling face seeming to lighten my heavy load. I knew that I would have to put this to him as a business proposition otherwise I would lose all credibility as an entrepreneur, and no-way could I let that happen. As I ventured into the hallway, Jules led me into a side room, where the most elegant oval dining table that I had ever seen in my life took pride of place in the centre of the room. The table was surrounded by eight high-backed, wooden chairs, each of them graced with plush leather-seating and a pair of ornately decorated armrests that were fit for a king.

Jules and I sat at one end of the table, and while he said nothing, I began my *choice-close* sales pitch, offering him an equal partnership in my venture, so long as he was willing to be the full-time *sponsor* of course, or alternatively a percentage of the profits in exchange for a return flight ticket to San Francisco? Jules sat there for several minutes, just glaring at me with his whisker-lined eyes, before jumping to his feet, and as he gently combed his quiff of grey hair back into perfect position, he calmly suggested that we meet for lunch at a restaurant in the centre of Cardiff the following day. "At that point we will discuss this matter further", Jules added, letting me know that our time together for this evening was now at an end. I had assumed that Jules would consult a few business associates about the offer, before giving me my answer, and so I left him, feeling comfortable in the knowledge that he was at least giving this *'golden opportunity'*, as I saw it, some serious consideration.

That night I barely had a wink of sleep, as my mind turned over and over, continually worrying about the outcome of Jules' decision, knowing full-well that if he said 'no', then *The Dream* would be over long before it had even begun. Throughout our meal the following day, Jules said nothing regarding the proposition, which left me feeling somewhat nervous, and I knew that I dare not mention it first, otherwise it would seem like I was begging him to join me. After Jules had paid the bill, which he always insisted upon doing, we left the restaurant, before crossing the road together, and walking directly into the nearest travel agency, where a young lady was busy scouring the lists on her computer for late availability seats on airline passenger planes. Within a few minutes she came up with a *return* flight from Gatwick Airport to San Francisco, via Houston, Texas, which would be leaving in ten days time –it sounded perfect.

After confirming the date and flight times, Jules signed a cheque for £328, before handing me the confirmation booking form, and sighing "I only wish that I was going with you my friend". As we vacated the travel agency, Jules told me to call into his office on Friday afternoon, and he would arrange a car for me to drive to the airport, as apart from being a partner in the travel agency business, Jules also owned his own franchise with Budget Rent-a-Car International. That evening I rang Ted Simon to give him my E.T.A. (estimated time of arrival) in the U.S.A. Ted said that he was very much looking forward to meeting up with me, before adding that he was a very busy man, and therefore he would only be able to spare me a few hours of his precious time. However, I had already planned to stay in America for a fortnight, just in case he liked the idea and needed me to *stick-around* for a while?

If not, then my friend, Daniel, who lives in Orange County, just outside Los Angeles, (whom I had already rung and told about my proposed trip, just prior to ringing Ted), had insisted that I could stay with him and his lovely family for as long as I wanted to. 'California, here I come'.

CHAPTER 2

California, Here I Come

My beloved wife, Caryl, who is currently seven months pregnant, was not overly thrilled at the thought of me going over to America for two weeks, but she also knew that if this outlandish idea of mine worked out and the movie was made, then it would be the end of all of our financial problems –forever. Three days later I called in on Jules as arranged, and within minutes my transportation to the airport was sorted. Jules also said that there would be a car waiting for me upon my return to the UK, which, if I am brutally honest, I had half-expected him to say. However, Jules' next statement put me in a state of shock! Opening the top, right-hand draw of his desk, Jules produced a wallet that was packed with bank-notes which I instantly recognised. "You will obviously need spending money Shaun, and so here's twelve hundred dollars –that should see you through to journey's end", Jules said in a completely *matter-of-fact* way, as if he had just handed me a fiver.

What could one say to that? I had not asked for any cash, or a car, for that matter, and yet here was this guy, whom I had not even heard of a month ago, suddenly handing me over hundreds of pounds of his own money, along with paying for expensive flights, and arranging private cars to take me to and from Gatwick Airport. What seemed even more bizarre was the fact that there was no contract for me to put my signature on, and there was certainly no question of me ever having to repay the money. In fact, Jules had not even asked me to sign a receipt when he had handed me over the cash. Either he is a very generous person, or my question regarding his faith in me, had just been answered. Whatever the reason, I had no intentions of letting Jules down, and his latest gesture had made me even more determined to succeed –for both our sakes.

The following week I compiled a complete dossier of Ted's exploits around the world, which included extracts from his follow-up book, which Ted had originally entitled 'Riding home', but has since been re-titled 'Riding high'. Also, to give my sales pitch (to Ted) more impact I produced a *theoretical* advertising poster for the proposed movie, by cutting out a map of the world, and after sticking it to a huge sheet of card, I surrounded it with several pictures from various holiday brochures. To round-off my *masterpiece*, I used the record sleeve from Meat Loaf's new 'Bat out of Hell' album (the 're-vamped' version) as a centre-piece, with the heading **'FREEDOM RUN'** running along the top of the poster –and the sub-heading **'This is a true story'**, written in bold letters underneath. To complete my little *sales package* I took my 'Bat out of Hell' cassette and placed it snugly inside my Sony Walkman, before winding it on to the desired position, in readiness for the biggest sales pitch of my life.

On the day prior to my departure from the house, I took my briefcase from off the top of our wardrobe in the bedroom and gave it a good clean, both inside and out. It was only a plastic case, (not leather or aluminium), but it was a present from my parents when I had first ventured into the lucrative world of sales and marketing, and so it had always held great sentimental value to me. Placing the dossier at the bottom of the case, along with the walkman and a copy of Jupiter's Travel's, I then folded the poster very carefully, and placed it on top of everything, before closing the lid and locking it. Caryl had packed my suitcase to the brim with everything she could conceivably think of, and so I felt fully prepared for whatever lay ahead. My father-in-law had suggested that I stay with his nephew, Anthony, at his flat in Caterham the night before my flight, as Gatwick Airport is only a stones-throw away from his place, and therefore it would save me having to drive around 200 miles on the morning of my departure.

With approximately fourteen hours of flying time ahead of me, this seemed the logical course to take, and so after collecting the car from Jules in the afternoon, I loaded-up the boot with my bags, and filled-up the consul with a handful of cassettes, before throwing my jacket onto the back seat, in readiness to be retrieved at journey's end. Leaving my pregnant wife and my two beloved sons Liam, who was nearly four, and Carl, who was only sixteen months old, was undoubtedly the hardest part of this whole venture, but I so desperately wanted them to have a secure future. With South Wales being one of the worst hit areas during this current recession, and unemployment levels in the U.K. rising in their thousands every month, this seemed like the one and only opportunity I was ever going to get to try and pull us out of this financial crisis –before it was too late!

Jumping into the driver's seat and slamming the door shut, I took one last glance in the mirror at the family I was leaving behind, before giving them all a final wave. Using my right hand I then pinched the skin on my left wrist very tightly, just to make sure that this was all real, and not just a dream. Starting the engine and putting the car into gear, I released the handbrake first, whilst easing out the clutch very slowly, before taking off into the cold, dark, autumn night, somewhat excited, but more than a little apprehensive of what the coming days may bring. It was only a few miles to the main A48 dual carriageway, but no sooner had I joined it than I ran into a two-mile tail-back of traffic, which was virtually at a standstill. Slowly, but surely the jamb petered-out, and by the time I reached the 'Bryn Glas Tunnels', the roads were clear all the way to the Severn Bridge, as I crossed over from my home country of Wales into the lush green countryside of good old *Mother England*.

For the next hour the M4 seemed almost deserted, but the time drifted-by quite quickly as I sat and relaxed to the music. As I approached the outskirts of London I switched over to the M25, and began following the signs for Caterham. Following my directions as closely as possible I swerved off the road at the next junction, which only turned out to be the road to Croydon -and not Caterham? However, after traversing about five miles of country lanes, I managed to get back on the right road, and within no time at all I was pulling-up outside the 'Triple C' Service Station. The flat was located on the first floor, directly above the main entrance to the garage, and so after parking the car in the front forecourt, I made my way up the short staircase. It had just turned nine o'clock when Sebastian, Anthony's son, answered the door to me, and so after a very brief introduction, he duly invited me inside, where I was cordially introduced to his girlfriend, and also to his sister. (*Caryl's dad had made all of the arrangements over the telephone, and so I had never met, or even spoken to any of these people before today*). However, it wasn't very long before we were all getting on famously, and so Sebastian insisted that he take us all down to the local *watering-hole* for a couple of drinks.

My flight is due out at eleven, and so I am up by eight, and after drinking a steaming hot cup of coffee I ask Sebastian if I can make a quick phone call, to say a final farewell to my loved ones back in Cardiff. As soon as this was done I hastily packed-up the car again with my luggage, which I had decided to bring in from the boot last night, just in case someone decided to break in, or worse still, steal the car overnight, before saying my grateful thanks and final farewells to three people whom I will probably never meet again. Twenty minutes later, and I am pulling-up into this massive, bustling airport, with numerous freight terminals and dozens of departure gates everywhere I looked. However, the roads inside were well sign-posted (let's face it –they had to be) and so it wasn't long before I was handing over the car to its rightful owners. With my luggage trolley now fully loaded, I made my way over to my respective check-in desk.

This place sure was different to Cardiff Airport –I had never seen so many different nationalities of people or such a variance of costume in my life before. Even though the atmosphere was vibrant and the pace hectic, I could see from out of the corner of my eye, that to the Thomson representative who was sitting at a nearby counter, this was just another *normal* working day. Sitting in the departure lounge, I enjoyed watching all of the aeroplanes (especially the jumbo jets) defying the laws of gravity, as their immense carcases lifted high into the sky with such ease, that one could easily believe that *the hand of God* was underneath them, softly gliding them through the air, before gently resting them on the clouds above. An announcement has just been made over the loud speaker system saying that all passengers flying to Houston, Texas, must now begin boarding the plane.

Standing at the back of a very long queue, whilst waiting to receive my boarding card, I realise that for me, the DREAM has just begun. As I enter the aeroplane and begin walking down one of the main aisles in search of my allocated seat, I can see a huge screen to my left, and there are so many seats surrounding me that I truly feel as though I am walking into a small cinema. Sitting in an aisle seat in one of the many empty side rows, I begin making a few notes, when suddenly this little old lady starts tapping me on my right shoulder! "Do you mind if I sit in that window seat, next door, but one to you young man?" she asks politely. "Not at all", I answer with a huge smile, before undoing my safety belt and standing up, in order to let her pass. As much as I want to continue getting my notes up to speed, this lovely lady is hell-bent on striking-up a conversation, and so as I do not want to appear rude, I immediately put down my pencil, and we begin chatting to one-another.

She tells me that she is on her way to meet a friend of hers in Mexico City, who is a retired diplomat, and that she intends staying with her for the next three weeks. I reciprocate, of course, by telling her all about what I am doing, and before we know it, it is time for our breakfast-cum-brunch. For the next half an hour there is not a word exchanged between us, but just as I am about to start tidying-up my tray, in readiness to give it to our stewardess, the old lady suddenly places this relatively flat, A5 sized package on top of my pull-down table. "You wouldn't mind posting this book for me would you?" she asks in a kind of pitiful manner. Before I had a chance to answer her request, the lady is handing me over two, one-dollar bills, in order to cover the cost of postage. However, being a *soft touch (and a pure gentleman of course)* I refuse to take the money, but accept the menial task with pleasure, after all, if you can't do a small favour for an old lady, then who *can* you do one for?

Call me a 'sucker', and tell me that I am a gullible person if you like, but the thought that there may well be undesirable substances contained within this small, brown, unassuming, package has not even entered my head. Also, any ideas that this frail *old dear* could well be rendezvousing with a Columbian drugs baron in Mexico City, or that she might be meeting her diplomat friend to collect a dossier of secret documents from her, before selling them to the Russians, are so inconceivable, that they are also a million miles from my mind! *(If my mother was here right now she would have severely rapped my knuckles, because it was only a few days ago that she had told me not to accept any parcels from anyone, especially whilst I was on board a plane, or in an airport, or crossing a border, and I had simply replied to her "Come on mum — what do you think I am — stupid or something?")*

Within minutes of clearing-up our trays the place is thrown into darkness, and the screen in the centre of the aeroplane is now alive with several excerpts from various action-packed movies, but unfortunately there is no sound? Realising that the headphones which are plugged into our seats are not only for tuning into the various radio stations, but they are also for listening to the movies, I immediately place mine on my head and begin searching through the channels for the correct frequency. Unfortunately, no matter how hard I try I am still unable to get any sound whatsoever? Looking across to my left, I can see that several people in the middle row are encountering the same problem as me, and so, (after begging my leave from *Granny-Smith*), I move back a few rows to where everyone else's earphones seem to be working perfectly. I was surprised to see that the film, which incidentally, was excellent, was a new release, and even more surprised to discover that the second film (which I watched after lunch) was also just as recent, and even more entertaining than the first. *(At least that is what it says in my '27-year-old' manuscript, which I am now transposing into this book, although I cannot understand for the life of me why I failed to write down the names of both of the films concerned!)*

After the second film had finally finished I decided to strike up a conversation with the guy sitting next to me, whose name is Peter. Peter is a bricklayer from Liverpool, and he tells me that he is on his way to Arizona for six months, where he will be staying with relatives for the winter. Normally I would be very envious of Peter, but even he is happy to admit that my adventure sounds a lot more exciting that what he intends doing over the next few months. I am totally absorbed with the whole event of meeting people with their own tales to tell, along with visiting different places for the very first time, and so I think it is fair to say that variety is certainly the 'spice of my life'. This has undoubtedly been the longest flight that I have ever encountered, but ten hours soon passes by quickly when one is living in one's own *fantasy world*. As we begin our descent into Houston, Texas, Peter and I start filling-in our respective immigration forms for entry into the United States of America.

The city of Houston, which was named after General Sam Houston, who was the former president of the Republic of Texas, is the most populous city in the state of Texas, and the fourth most populous city in the whole of America. It is located in South-eastern Texas, near Galveston Bay and the Gulf of Mexico. At the turn of the 20th century, Houston's port trading flourished with the building of the Houston Ship Canal, along with its expanding railway industry, both of which were critical during the Texas oil boom. By the second half of the century Houston had diversified its economy, with the building of the 'Texas Medical Centre', which at the time had the world's largest concentration of healthcare and research institutions. Houston has also been nicknamed 'The Space City', as it also houses NASA's 'Johnson Space Centre', which encompasses NASA's Mission Control Centre.

At 3.55pm local time (9.55pm in Cardiff) our plane touches-down on American soil, and according to our captain the temperature outside is a healthy 73 degrees Fahrenheit, as opposed to somewhere near freezing-point in the UK right now. By the time we have both cleared customs Peter and I have barely enough time to take a few photographs, before making our way to our respective terminals, in order to catch our onward flights. (Peter is heading for Tucson, Arizona and I am going to San Francisco, in Northern California, of course). *The pair of us had already exchanged home telephone numbers and addresses on the aeroplane, although whether we will ever come into contact with one-another again lay purely within the lap of the gods?*

Apart from the fact that this is an internal flight, and therefore the aeroplane is a lot smaller than the last, I have just been informed by my *friendly* stewardess that all food and drinks must be paid for (whereas they were free on the last flight) and also if I wish to listen to the radio, or hear the words to the movie (which is now being shown on the big screen), then the cost of a pair of earphones will set me back four dollars! "Welcome to the United States of America", I whisper *sarcastically* under my breath. The lady I am sitting next to on this flight is from Miami. She tells me that she has two sons, who only turn-out to be the exact same ages as Liam and Carl. However, what is even more coincidental is the fact that her husband is from 'Mountain Ash' in South Wales! (*The world is getting smaller by the minute*). The two four-hour flights to the island of Tenerife in the Canary Islands, and back home again to Cardiff Airport, which Caryl and I took back in 1984, seemed never-ending at the time, and yet now, after having *survived* a ten-hour flight, my four hour flight to San Francisco seems like *a walk in the park*. Once again our aircraft has made good time, landing almost half-an-hour ahead of schedule; it is now 7.30pm local time.

San Francisco, which is Spanish for 'Saint Francis', was founded on June 29th 1776. Apart from being the second most densely populated large city in America, it is also the financial, cultural and commercial centre of Northern California. The California Gold Rush of 1849 brought rapid growth to the city, making San Francisco the largest city on the West Coast, and by the turn of the twentieth century, almost a quarter of the whole population of California were living within its boundaries. At twelve minutes past five on the morning of April 18th 1906, a devastating earthquake struck the city of San Francisco, bringing hundreds of buildings crashing to the ground, and rupturing numerous gas lines, which immediately ignited blazing fires that would rage on for several days. Although reports at the time claimed that the death toll was less than five hundred, it is believed that thousands of people actually perished on that fateful day, as three quarters of the city lay in total ruins.

During World War II, San Francisco was a major port for servicemen shipping out to join the Asia-Pacific War, and in 1945 the city became the birthplace for the United Nations. San Francisco is also renowned for several political factions, such as the 'Peace Movement', which focussed on the opposition of America's involvement in the Vietnam War, the 'Sexual Revolution' and the 'Hippy counterculture'. The 'Summer of Love' and the 'Gay Rights Movement' were also founded in San Francisco. Today, San Francisco is a major tourist destination, where people flock from around the world to visit its amazing Chinatown District, or to drive across the famous Golden Gate Bridge. Taking a boat trip out to the Infamous Alcatraz Federal Penitentiary is also a major tourist attraction. As I write this in December 2018, San Francisco is currently the highest rated American City on World Liveability rankings.

San Francisco is two hours behind Houston (where it is now 9.30pm, of course) and *Frisco's* time-zone is eight hours behind the UK –where it is now 3.30 in the morning –well, at least that is what my *body-clock* is telling me. With no customs bureaucracy to worry about this time, I am through the terminal within minutes. Wearily I drag my suitcase out into the street, before plonking my butt on one corner of it, whilst resting my briefcase on my lap –I am absolutely knackered! In most of the major airports around the world, *free-busses* are in abundance. This is a service where the drivers of these *mini-buses* pick up people who have just arrived on a flight, before giving them a complimentary ride to whichever hotel owns the bus –hence the name. Jules had already advised me to get on the

'Holiday-Inn' bus, adding that their hotels were really nice, and not too expensive, and so I really wanted to adhere to his advice, after all, it was *his* money that I was spending.

However, I was so tired that I had already decided to get on the first bus that came along, and as luck would have it, I did not have to wait too long. Not wanting to get ripped-off, I asked the driver how much it would be to stay at *his* hotel for the night. Unfortunately he said that he did not have a clue, as he was *only the driver*, but he then offered to radio through to the hotel, to find out what the (minimum) cost would be? While he was doing this, the Holiday-Inn Bus pulled-up right behind him (would you believe) and so I immediately shouted over to the guy to "forget-it", apologising profusely as I dragged my suitcase off the back of his bus, before handing it over to my *new* driver. As we drove through *the streets of San Francisco* (sounds like a good title for a series) I could not help feeling really guilty for wasting that first drivers' time, simply because I am *that* type of person.

However, justice prevailed, as the concierge at the Holiday Inn *kindly* informed me that the hotel which the other guy would have taken me to, was considerably cheaper than staying at the Holiday Inn "…but it is nowhere near as nice as it is here", the guy immediately added, as soon as he saw my face drop! It took every ounce of my reserved energy to carry my suitcase (I must get one with wheels next time) along with my briefcase of course, all the way across the rear courtyard to room 212, but in the end I made it. 8pm it may be in California, but to me it was four o'clock in the morning –and so, *like it or lump it*, I was going straight to bed for the night.

CHAPTER 3

Looking for Ted

It is still the early hours of the morning when I first awake, and so it is pitch black outside. However, by the time I have had the proverbial *shit, shower, shave and shampoo* the sun is just beginning to show its glorious face over the distant horizon. I am now ready to face the world. Last night was by no means cold, but this morning it is wonderfully warm. The air outside feels so fresh and invigorating and the aroma of scented flowers is so wonderful, that I feel as though I am *in seventh heaven*. Also, it wasn't long before my dreams of a utopian paradise were greatly enhanced -by the smell of cooking! Beyond the rear-entrance doors of the hotel sits the kitchen; the place where the odour of frying bacon had emanated from only seconds ago, before gently wafting its way across the courtyard, and straight up into my nostrils. The chefs timing could not have been better, for I am absolutely starving.

Following the pathway around the main garden area I pass the swimming pool, where the attendant is delicately removing the unwanted leaves that have fallen from the overhanging trees during the night. I say a friendly "Good morning" to the guy, but a gentle nod of his head is my only acknowledgement that he has heard my greeting. "Perhaps he doesn't speak any English", I say to myself, before gently pushing open the small pair of saloon doors that lead into the main restaurant. The décor is very plush inside, with oak beams and mahogany worktops everywhere. A good, solid breakfast is my first order of the day, and then I will arrange my transportation to Mr. Simon's place. According to Ted, his home

17

is in a place called 'Elk', which is apparently 'three hours north of San Francisco'. (People never talk in *mileages* in America, preferring to quote the anticipated amount of time it should take a person to drive from one place to another).

According to the map on the wall of the diner, *Elk* is just off the main highway '1', which basically runs the length of the Californian coast, and so I figure that my best bet will be to take a Greyhound Bus, as apparently they go 'everywhere' in America. Thankfully, the telephone number of their office *downtown* is in the phonebook, and so I called them as soon as I had finished my breakfast. "Sorry sir, but Greyhound Buses don't go to Elk; you will need to contact another company for that service. If you hold the line a moment I will give you their telephone number!" said the very polite lady on the other end of the line. I then called the number that the woman had kindly given me, but unfortunately all I got to talk to was an answer-phone? Feeling somewhat dejected, I decided to call Mr. Simon, to see if he could help me? Thankfully, Ted had already checked-out the transportation I would need to get to his place, and so he told me to *get pen and paper at the ready.*

"First you must catch the 7B bus to 5th Street and Mission, where the Falcon Bus Line at the Amtrak Train Station leaves at 3.30pm. The journey north will take around three hours, and you must ask the driver to drop you off after 'Philo', at the end of Greenwood Road, which is where I will pick you up at around 7pm, to take you to my home –have you got all that?" Thankfully, Ted had spoken slowly, and very precisely, and so I had managed to write down all of the main details in my notepad. However, after all that effort Ted then advised me to hire a car, saying how much easier it would be for me, before adding that the cost would not be *that* expensive either. Whilst contemplating my next move I decided to have a walk around the surrounding area. Dressed in a pair of white shorts, with matching tee-shirt, socks and training shoes, I strutted along the pavements, quietly *contemplating the universe*, whilst soaking-up the wonderful Californian sunshine.

Suddenly I spot a Budget Rent-a-car sign across the way. It is hanging high above this relatively small building, which has a mountainous backdrop behind it. Thinking about what Mr. Simon had said, I crossed over the road (which was about the same width as your average motorway in the UK) and ventured inside the building, with a view to finding out the cost of hiring a car for two weeks? Although part and parcel of the Budget Rent-a-car Company, the guy behind the counter duly informed me that this particular depot was for truck hire only. However, *my learned friend* was more than happy to contact the car-hire depot on my behalf, in order to find-out the cost of hiring the cheapest car available. (*$1.5 was approximately £1 Sterling back in 1991*). After scribbling lots of numbers onto a note pad on the counter, the guy informed me that the cost of hiring a *regular* car would set me back $35 (around £24) per day, or $138 (around £92) for one *whole* week, plus $35 for each day thereafter.

He then added that should I choose to take up any of these offers, then I would need to go over to their office at the airport, in order to confirm the booking, before collecting my car from the parking lot. After working out all of the *pro's and con's* of the situation, I decided to take my friend up on his offer. On my way back to the hotel I spotted this big, burly police officer, who was standing next to a magnificent-looking motor-cycle. As I drew closer I could see that the bike was a Kawasaki Z1000, exactly the same as the bikes used in the television series 'Chips'. The officer's uniform was also identical to those worn by our *heroes* in the programme, and from where I was now standing I could just-about make out the words 'California Highway Patrol', which had been stitched in an arc across both of the shoulder-blades of his grey coloured, short sleeved shirt.

I approached the officer very cautiously, as I could also see clearly the gun hanging from his right hip, which could well be a Magnum 44, the most powerful hand gun in the world; a weapon which could probably blow my head clean-off and so I had to ask myself this one question "Do I feel lucky?" (*By now you have probably guessed that one of my all-time favourite films is 'Magnum Force', starring the one-and-only 'Mr. Clint Eastwood'*). I ask the officer (very politely) if he can tell me where the nearest post office is, as I have a small parcel that I wish to post. The policeman hesitated for a moment, before taking his sunglasses out of his left-hand side breast pocket, and opening-up the two stems, one-by-one he then stared me straight in the face for a few seconds, before answering: "The post office is right behind you my friend". He then pointed across the road to a building which had the words **'Post Office'** standing-out in bold letters on the front of it. Now I felt really stupid! The officer then looked me *up and down* for a few more seconds, before tentatively placing his shades over his eyes. "Thank you officer, you see I am from *out-of-town*", is all I could think of to say, as the guy climbed aboard his motor-cycle.

"Before you go officer, I wonder if would you would be kind-enough to pose for a photograph for me please." I tentatively asked my new found friend, even though I was feeling a tad concerned that my humble request might have been pushing my luck just a bit too far?

"Sure thing my friend; say, you're from England; right?" He politely asked.

"Well, Wales actually," I replied, defending my home country to the last.

"Yeah, right...okay where do you want me to stand buddy?"

I took two photos of the officer, one standing next to his bike, and one with him straddling it. While he was doing his best poses for the camera, the officer asked me what I was doing in America, no doubt expecting me to simply say that I was "On vacation", but instead I decided to *come clean* and tell him all about my little adventure. Taking out a few business cards from his right-hand side breast pocket this time, the guy then handed me one of them, before saying, quite excitedly, I might add:

"Hey, if it works buddy, you be sure to send me two tickets for the movie".

After posting the *so-called* book, (as I never did find out what was actually inside the package) I called into the hotel gift shop, in order to purchase another roll of film for my camera, along with a packet of spare batteries for the Walkman, before taking a few more photos of the picturesque gardens which surrounded the hotel. However, when the reel reached '29', I suddenly realised that this was only a '24' frame reel –and so I changed the film immediately. (*Needless to say that the photos I thought that I had taken with the first film never came out!*) As I do not have to check-out of my room until 2pm, I decide to take the Holiday Inn free-bus to the airport, before taking a second free-bus to the Budget rent-a-car depot, as I figured that if I could pick up the car *before* I checked out, then this would save me having to drag my gear around with me. Unfortunately, things didn't go anywhere near as I had planned them to!

"Sorry sir, but if you do not have a credit card, then I will need to *Xerox* a copy of your flight ticket!" said my fastidious counter assistant.

(My credit cards had long-since 'ceased to function', as I had exceeded the credit limits on both of them many months ago, and as I had been unable to make payments on either of them for several months, I had duly left them in the top drawer of my bedside cabinet back in Cardiff). To make matters worse, my flight ticket was currently locked away in the safe in my hotel room, along with all of my other important documents, and the rest of my cash, of course, and so now I would have no choice but to revert back to plan 'A'. However, before I left the building I thought that I might as well *get the lowdown* on everything. As I would be in California for two weeks, the assistant recommended that I hire the car for the full fortnight, adding that I could always return the car earlier and get a refund if I didn't need it for the duration of my holiday.

As I had no idea of how long I would be staying with Ted, coupled with the fact that Daniel and his wife live in Orange County, just outside Los Angeles, which is over six hundred miles south of Elk, this sounded like a very reasonable proposition. The cost for the two weeks would be $297 including taxes (don't they just kill ya!) However, next to rear its ugly head was the deposit, which would be $800 "Because *you* don't have a credit card", we both say in unison, (actually I *sarcastically* said '*I*' and not '*you*', but does it really matter?) Working out that this would leave me with less than $50 I politely declined the guys offer. However, just as I was about to leave the place, the assistant called me back to the counter, saying that he would have a word with the manager, to see if he could get the deposit reduced. My *friend* then disappeared into a side room for a few minutes, before returning with the good news. "The manager will accept $425 deposit sir, if that is okay with you of course?"

Figuring that I could probably survive on $420 I told the assistant that I would accept the managers' 'more realistic' offer, before calmly leaving the building and returning post-haste to the Holiday Inn, in order to collect-up all of my belongings. It was just after one o'clock by the time I got back, which meant that I still had time for a spot of lunch, after-all I hadn't eaten for over four hours. After devouring a mountain of salad, I went back to my room, to pack-up my gear, before carrying my bags back across the courtyard and into the foyer, where I immediately returned my key to the receptionist, and duly collected my $20 deposit. By now the mini-bus driver and I were on first name terms, and so as a parting gesture, he kindly took me directly to the Budget rent-a-car depot. As a 'Thank you' for his kindness, I duly gave him a $10 tip. Upon signing the car-hire agreement I was pleasantly surprised to read that the $425 actually *included* the cost of the fortnight's rental, and was not *just* the deposit, as I had wrongly assumed earlier. This meant that I still had over $700 in my pocket, which should be ample for the two weeks ahead. The guy then offered to sell me the 'C.D.W.' (crash, damage, waiver,) policy.

This additional insurance would cost me another $100, and although the guy virtually insisted that I purchase it, in the end I decided to decline the offer, even though it meant that I could now be liable for any damage to the car, including writing it off, which would set me back around $10,000! As soon as all of the paperwork had been completed, I am escorted out into the parking lot by a fellow colleague of my *new-found-friend*, who happens to be from the Philippines. The guy speaks very good English, and he also has a great sense of humour, telling me dirty jokes as we walk across the yard to where my rental car is parked. Presenting me with a very sporty looking two door Mustang, in black, the guy begins going through the fundamentals of the car, which seemed pretty straight-forward, until I suddenly noticed that there were only *two* pedals, instead of the usual *three*, sticking up from the floor in front of the driver's seat? With everything that was going on I had simply forgotten the fact that the majority of American cars have *automatic* transmissions –i.e. they have no clutch or clutch-pedal.

Having only ever driven one automatic car before -and that was only for a couple of miles, I was a little nervous to say the least, but not wanting my friend to realise my apprehensiveness I climbed swiftly aboard and started the car without any hesitation. My friend then kindly pointed me in the right direction for getting onto *Highway 5 North*, and within minutes I am venturing out onto the main road –and *slap-bang* into the centre of an endless stream of traffic!

CALIFORNIA

FORT
BRAGG
MENDOCINO
ELK
BOONEVILLE
SANTA ROSA

COVELO
WILLITS

CLOVERDALE

SAN FRANCISCO

SAN JOSE
SAN ARDO
SAN LUIS OBISPO
SANTA BARBARA
MALIBU
BEVERLY HILLS
IRVINE

BAKERSFIELD

SANTA MONICA

LOS ANGELES

SAN DIEGO
TIJUANA

MEXICO

My primal aim now is to get into the correct lane, without upsetting too many commuters in the process. For nearly half-an-hour I go around in circles, trying desperately to read the signs above my head correctly, without veering across the white lines, or crashing into another car in the process. Apart from having no clutch and no gearstick to play with, I am also trying to come to terms with driving on the *right-hand* side of the road in a *left-hand* drive vehicle.

Feeling totally bewildered by it all, but as yet unscathed, I finally manage to make it onto the correct freeway…at last. Apart from my rental agreement, along with various other pieces of paper that are now safely tucked in my 'official documents' wallet, Budget has also given me a comprehensive road map of California, and so I have already planned-out my *scenic* route to Elk. At present I am following all of the signs for Cloverdale, which is approximately eighty miles north of San Francisco, and as soon as I reach there I will be turning off the main highway, heading in a north-westerly direction for about another twenty miles, until I come to a place called Boonville. Five miles past Boonville is Philo and just after Philo is Greenwood Road. This is where I will turn off for the final eighteen mile run into Elk. Suddenly a shocking thought occurs to me –I have forgotten to call Mr Simon back!

After he had given me all those directions, along with his sound advice about hiring a car, I had promised him faithfully that I would ring him back as soon as I had sussed it all out, and decided which method of transport I was going to use, but in all the confusion…..oh bollocks! My only hope now is that I am able find his house, or at least get to a phone before seven o'clock? I am surprised to see that there is no cassette player in the car (*8-track cartridges having long-since had their day -and CD's still yet to come into fashion*) and so I am forced to listen to the local radio station as I drive. If I was an ardent fan of The Beach Boys then I am sure that I would be in my element right now, because their songs are being played every other flipping record. Unfortunately, I have to confess that apart from their major hits in the UK, such as 'Cotton-fields' and 'California Girls', the group has never done an awful lot for me.

To be fair to the band, I think that it is listening to all of those blasted advertisements over and over again which is really driving me crazy, and not so much the music. Suddenly a simple answer springs to mind; I'll listen to my music tapes on the Walkman instead. After twenty-odd miles of fairly dense traffic, the road widened into three lanes, systematically spreading-out the cars on a *less-claustrophobic* scale, thus making me feel a lot more comfortable whilst driving on these unknown highways. The speed limit is 55mph –well at least according to the signposts which are dotted every few hundred yards or so along the freeway, it is. However, as I have been travelling at just under that speed since I got onto the highway, during which time several dozens of cars have gone whizzing past me in the outside lane like I am standing still, I figure it safe to assume that the American's enjoy breaking the speed limits just as much as we Brits do in the UK –and you know what they say "If you can't beat 'em –then join 'em".

For some inexplicable reason I had expected to see Kilometre markers, similar to the ones they have in Europe, as opposed to mileage signs like they use in the UK –and I was even more surprised when I discovered that the petrol Stations sell their fuel in quantities of pints and gallons, rather than litres?

By four o'clock I am at the end of the highway, and minutes later I am on the road to Boonville. At 4.30pm dusk is beginning to settle in, and as it is now late in the Autumn I can safely assume that darkness will be upon me before I know it. Another half-an-hour goes by, and I have just passed a turn-off, which is not sign-posted, and yet something in my mind is telling me to go back to the fork in the road and check it out...call it intuition if you like. It was a good job I did, because hidden under a shroud of bushes is a small (pointed) sign which simply reads 'Elk'. The roads are now very narrow and twisting, and I am down to less than 30mph.

The tarmac surface has also been replaced by a dirt track, and I am surrounded on both sides by dense forest, as I sink deeper and deeper into the abyss. There is no street lighting whatsoever, and only the occasional mailbox stands out at the side of the road, to assure me that there is *hopefully* life nearby? With the sky now as black as pitch, and no signs of any coastline in the distance, I begin to feel very uneasy. All it will take now is the sound of a howling wolf, or a loud hoot from an owl and I will be wasted. Suddenly I see a light shining in the distance, and as I draw nearer to it I can see that it is emanating from the window of a large cottage. Slowing my speed down to a virtual crawl, I can see the figure of a man through an open window –he is standing in the middle of a rather small kitchen. In front of the house there is a large driveway, with a huge pick-up truck covering at least half of the floor space. I am not sure if the guy has seen me, but I decide to pull over anyway, as I am totally lost, and in desperate need of assistance right now.

Traversing the few steps that lead up to the front doorway I curse my imagination, as thoughts of the *Texas Chainsaw Massacres* run amok in my mind, immediately sending one big icy shiver down my spine. Knocking firmly on the wooden door, I stand back, my body half twisted in readiness to flee from the prevailing attack, which my nerves have also conjured-up in the last few seconds. As the door creaks open I am confronted by this huge guy, who is either in his late teens, or his early twenties; it is difficult to tell in the dark? To my great relief the lad is softly spoken, as he politely asks me if he can be of any assistance. I quickly explain my current situation to him, and he immediately beckons me to follow him through a small hallway and into the main living quarters.

From the inside the place looks like a huge barn, with its great stone walls and high thatched ceiling, and as there is only one other door at the back of the room, which I naturally assume leads into to the one-and-only bathroom. I gather that these living quarters must be where everyone in the family eats, sleeps and does whatever else one does out here in the wilderness. In the far corner (from where I am standing) I can see another lad, slightly younger than the first, along with a middle-aged woman. They are sitting at a type of picnic bench, which has been built out of old logs and small tree stumps, and the pair of them is staring in complete silence at this *stranger from out of town*! The whole ambiance of this place (including the clothes that these guys are wearing) reminds me of the hit series, the *Beverley Hillbillies, simply* because these are obviously *simple country folk*, who live *simple* lives, in *simple* accommodation, and I cannot help but be completely enthralled by the delicate primitiveness of their lifestyles. Maybe one day these guys will also strike oil on their land, and become 'filthy rich' –like their television counterparts.

As soon as I mention Ted Simon's name the *lady of the house* suddenly comes alive, saying that she knows Ted's ex-wife very well, and that Ted only lives a short distance up the road. "In fact, you have probably driven right past his place young man", the woman adds, in a very broad Northern Californian accent. The lady then kindly handed over the house phone to me, telling me to give Ted a quick ring "Before he goes out looking for you in the dark", she added, as if it was unheard-of to go out into the forest after sundown. While I am doing this, I can see the eldest lad reversing that huge pick-up truck out of the driveway, his rear bumper narrowly missing the back end of my car by inches. "My brother and I are going out in a few minutes, and we will be driving past Ted's place, and so if you want to follow us in your car, we'll be happy to show you where to turn off", said the youngest son, adding his little bit of assistance to the situation. I am so grateful to the whole family for their help, and also for the kindness they have shown me, and I am now feeling very guilty about being such a coward earlier, and for even questioning their wonderful *northern hospitality.*

It has just turned six o'clock when the lads finally pull over at the side of the road. As I pull-up alongside their truck the eldest son leans out of the driver's window, and pointing over to the right he says "Follow that hill to the very top, and when you get over the ridge just ignore the house on the left –and Ted's DOME is on the right!" I really did not know what to expect after that last statement, and so I just tooted my horn, before giving the lads a final wave, and then I continued climbing the muddy embankment until I reached the very top. The house on the left I spot immediately, but I can only see a cluster of trees on my right and nothing else –so what the hell do I do now? As I stop to ponder my next move a light appears beyond the trees, and so I drive delicately forward, in order to get a closer look. I can now see that it is a small lantern, which is hanging on a long cable, and as my eyes follow the cable to its originating point I am able to make out the outline of a great wooden dome –it is the home of Mr. Simon.

CHAPTER 4

The Pitch

Driving through an opening in the trees, I spot a large dinghy resting against the side of this huge wooden igloo-shaped structure. To the right is the main dome of the building, and attached to this is a long hallway-cum-outhouse, which has a small veranda on the end of it. On the veranda there is a wooden bench, along with a small table and a stand-alone barbeque unit. Sitting in front of this uniquely-shaped building is a blue saloon car, and standing adjacent to it is a large BMW motorcycle. Switching off the engine, (but not my lights), I am able to see my *host,* along with a young lad, and the pair of them are now walking slowly towards me. I know that it is Ted because I recognize him from his picture on the cover of Jupiter's Travels. (*Even though Ted is now nearly 60 years old, he does not seem to have aged a great deal in the last 15 years*). Realising that the headlights are blinding my *welcome party,* I immediately switch them off, before getting out of the car.

Ted greets me with a warm handshake, before introducing me to his son, William, who is eleven years old. He then added that had I phoned him on the way over, then he and *Wills* (Ted's pet name for his son) would have met me in Boonville, where we could have treated ourselves to a juicy steak, in the main restaurant in town. However, food was the very last thing on my mind right now, as the realisation of where I was, who I was with, and what I was doing here, had all merged to form my Utopia. Inside the dome, the place was even more fascinating than the outside. Taking centre-stage was this monstrous-looking boiler, which would not have looked out of place in a *Terminator* movie. Sitting alongside the *Tin Man* was a large log rack, stacked to the brim with chopped logs of all shapes and sizes, each one of them waiting patiently to be cremated and consumed by the *furnace from hell*. (*Perhaps I have gone a little-bit over the top with my description here?*)

Located behind the boiler is a wooden, spiral staircase, leading up to the one and only upstairs bedroom, and standing against one of the walls is an upright piano with its lid wide open, and a set of music sheets resting above the keyboard, as if someone had recently been playing melodies on it. Numerous bookshelves stacked with all kinds of literature adorned another wall, and in front of this stood a large wooden desk, laden with paperwork and office stationary on the one side, and a fairly new computer on the other. Sitting beside the computer was a colour printer, waiting anxiously no doubt to spew-out more works of art, such as Ted's previous literary masterpieces. To the left of me there is a compact kitchen, complete with a square table and two chairs, and beyond that stands a solitary door, which I naturally assume leads into the one and only bathroom.

Whilst I am tentatively surveying my surroundings Ted walks over to a small armchair, which is situated underneath one of the domes' *triangular* windows, before easing himself into a comfortable position. Wills is sitting quietly at the kitchen table, no-doubt wondering what will happen next, and so I immediately place my *butt* on a two-seater sofa, sitting directly opposite Ted. Nervously fumbling in my briefcase for a few seconds, I then extracted the Sony Walkman, before replacing the *Bat-out-of-hell* tape, which I had taken out earlier to play in the car. Placing the headphones over my ears, I quickly reset the tape to the desired position, before switching the machine off and setting it down next to me on the sofa. While Ted continues staring at me in total silence, I take a deep breath, before beginning the *sales pitch* of my life.

"Firstly let me humbly apologise to you Ted, for keeping you in suspense for so long. However once again I will ask you to bear with me, but for only a few minutes this time, as I request your assistance in proving whether my theory is justifiable or not? I can assure you that I have not travelled over six thousand miles to get here in order to play games."

Ted just smiled at me, before extending his right arm out with the palm facing upwards, as if gesturing me to take to the stage and perform.

29

Standing to my feet now, I lean over towards Ted, handing him the Walkman, which he immediately places on his lap, before lifting and adjusting the headphones to the required position. By now I am pacing the floorboards, in readiness to continue-on with my speech.

"Before switching the tape on Ted, I want you to delve deep into your memory, primarily to two very special events which occurred during the course of your journey around the world. The first one I would like you to recall is when you were imprisoned in Fortaleza in Brazil, and you thought that you would never get out alive. This first piece of music I have related to that incident, and so I would like you to listen very carefully to the lyrics in the song, as I believe that they are so poignant as to what you must have been thinking, and going through at the time. We will deal with the second event and its corresponding song in a few minutes time".

Ted sits gently back in his armchair before pressing the *play* button as instructed. He then closes his eyes, as if in a trance. While Ted sits listening to the song *Heaven can wait,* I glance over at William, who is completely transfixed by the whole situation, and seemingly more anxious than his dad at finding out what my *dark secret* is. Having already timed the duration of the song, I know when it is coming to a close, and so after checking on my watch I ask Ted to return the Walkman to me, in order that I may wind the tape on to the second and final piece of music. This time I ask Ted to recall the time when he had to leave his new-found love, Carol, to continue-on with his voyage of destiny.

Replacing the headphones once more, Ted's mind delves *deep into the bowels of his past,* as he recalls that fateful day, way back in 1975, when his ship sailed away from the west coast of America, bound for the continent of Australia. I am hoping that Ted will be able to recapture his feelings of guilt and remorse -and utter despair, as he leaves Carol at the quayside, feeling all lost and alone, as she continues begging him not to leave her. *'For crying out loud'* is the song that I have chosen for this very emotional scene in the film, not only because it is a beautiful ballad, but also because the lyrics actually relate to the state of California, which is absolutely perfect for the storyline. The burning question is whether or not the scenes which I have conjured-up in my mind, along with the songs that I have chosen, will reflect Ted's factual memories? Unfortunately I will have to wait eight minutes and forty-five seconds (the length of the song) before finding out my answers on both accounts?

"I have to say that these songs are certainly not my kind of music, although you are correct in your assumption that they could easily portray those two particular events on my journey. However, there is one small flaw in your match-making of words and music, which you would never have known, had you not been sitting here with me right now Shaun! Carol actually came with me to Australia, although I never mentioned it in the book, of course, and so I am sorry to say that this very powerful -and rather sad love song could never have fitted in with my true thoughts at the time. By the way, what is the name of the artist, or the group that is singing these songs?"

I was astounded that Ted had never even heard of *Meat Loaf* before today, and also slightly saddened that the songs weren't a perfect match, especially as I am about to *spill the beans* as to why I wrote to him in the first instance. Pulling the poster from out of my briefcase, I proceeded to open it out in full, before holding it aloft in the air, so that my *captive audience* can see it quite clearly. "This is what it is all about", I announce, with great passion and a certain degree of excitement in my voice. "A movie", exclaims Ted, gazing at my artwork in sheer wonderment. "It all fits. Now I understand why you put so much effort into your letter...what a surprise".

I can see that Ted is enthralled with my idea, and as I begin packing away my gear, my whole body gives one massive sigh of relief. William is a little overawed by our joint enthusiasm, and also slightly bemused by it all. He has never read Jupiter's Travels, and apparently he is only now beginning to understand the scale of his fathers' achievement, and so I suppose the idea of turning *the old man* into some kind of cult figure, (or movie icon perhaps), does sound a wee-bit eccentric to an eleven-year-old boy. Ted is now in the kitchen, busily preparing a huge pizza for the three of us to share.

"Help yourself to a beer from the ice box Shaun...and get one for me too please", says Ted in a friendly manner. I begin looking around the kitchen for a large *cool-box* (the kind that one takes on a picnic) but I am unable to find one anywhere? Realising our minor misunderstanding, Ted soon puts me straight: "Sorry Shaun, I mean the fridge -see what ten years living in America has done to me!" *(Having originally been brought up in the UK by a German mother and a Romanian father, Ted had moved to the USA sometime after completing his epic journey).* Now that I am feeling totally at ease with my new friends, I open my heart out to the guys, talking freely about my life back home in Wales, along with telling them a handful of stories about my ten years riding motorcycles, including the three journeys I completed around Europe and North Africa back in the eighties. Opening-up my wallet I then took out various pictures of my two lovely sons, Liam and Carl, which I immediately showed to Ted and Wills. I also showed the guys a few pictures of my wonderful wife, Caryl, telling them that she is currently seven months pregnant, and that even though we do not know the sex of the child as yet, I have this gut feeling that I will be the father of a precious baby-girl before Christmas.

Ted does not possess a television set, and so he relies totally on his radio to find out what is going on in the world. Without a TV there is nothing to keep William up half the night, and so by nine o'clock he has gone to bed. Ted and I continue *chewing the fat* for an hour-or-so, before my host announces that he, too, must *hit the sack*, as he has to be up early in the morning, in order to get William off to school. Leading me into a small room, which is located underneath the staircase, Ted informs me that this is Wills' bedroom, adding that he had insisted on giving it up for me as soon as he knew that I was coming -how kind. Made entirely out of timber...like everything else around here, the room is oddly-shaped, with not much room to *swing a mouse* -let alone a cat, but it does have the one commodity that I so desperately need right now and that is a bed.

A very big day in my life had finally come to an end, and I was feeling somewhat optimistic about what the future may hold, but as for now, well it was time once again to rest my totally worn out, and seriously jet-lagged, body. In the morning I am the first to wake, or so I believe, and so as I do not want to disturb Ted or William, I remain silent in my bed until I hear the sound of footsteps gently pacing the floorboards above my head. As I pull back the quilt cover and begin fluffing-up my pillows, I can hear a lot of whispering going on between father and son, as they very delicately make their way down the stairs, so as not to disturb their sleeping guest... if only they knew.

After putting on my *shell suit* (*yes I humbly admit to being one of those 'sad people' who owned no less than two of those thin, multi-coloured tracksuits that were all the rage -for 'anorak's' like me anyway, back in the late eighties / early nineties*) I vacated Williams bedroom and went to join my *fellow inmates* in the lounge area of our humble abode. Ted is in the kitchen, once again, and William has just returned from the garden, where he has apparently been enjoying the delights of an outdoor shower. Wrapped in nothing but a large bath towel, with his hair still soaking wet, Will's teeth are now chattering at an immeasurable rate, as the temperature inside the dome is quite cool. Ted asks me if I would like the same refreshing start to the day, to whit I politely decline his offer, knowing full-well that a team of wild horses would be unable to drag me under any kind of waterfall in these *Arctic-like* conditions. (*Okay, once again I am exaggerating, but you can believe me when I tell you that at 6.30am in the month of November, Northern California is definitely 'not' the warmest place on the planet*).

The forest which surrounds the dome is very dense, and the trees are so tall that the sun has great difficulty in penetrating the many thousands of branches (and millions of leaves) which engulf the entire area, and so the temperature outside is really cold in the morning. However, as soon as the sun is directly overhead and its powerful rays are beaming-down directly over our little *escarpment*, it doesn't take too long for the place to warm up to a much more acceptable level. Ted's energies are focused entirely on getting William ready in time to catch the school bus, which picks his son up from the main road at 7.30am sharp every weekday morning. Should William miss the bus (which, according to Ted, has happened on more than one occasion) then Ted will have no alternative, but to drive him all the way to Mendocino, which is over an hours' drive from the dome.

Thankfully, both William and the bus driver are on time this morning, and so my host can now relax for five minutes, before concentrating his efforts on much more important issues –such as making us a hearty breakfast. (*I'm only joking Ted!*) Drinking my coffee out of a soup bowl, as Ted does not possess any cups, or mugs, was certainly a novelty for me, but as Ted's hospitality knows no bounds, and his culinary skills are second-to-none, suffice it to say that we both rolled-away from the kitchen table, the pair of us having been well watered and fed. Ted has kindly offered me a second night at his home, on the proviso that I leave him on his own today, as he has a lot of work to catch up on. He also told me that he will try and contact a few of his past acquaintances, in the hope that they may be able to help us with our cause. To my great surprise, Ted then suggested that I collect William from school this afternoon, saying what a lovely drive it is along the coast to Mendocino.

I am more than happy to accept Ted's offer of another *free* nights' accommodation, and I could not be more pleased that he wants to do whatever he can in order to help me with my proposed production, but the fact that he has now entrusted me as guardian (albeit temporarily of course) of his beloved son, well this immediately tells me that Ted's trust in me must be one hundred percent. Thankfully, the roads are considerably more inviting in the light, than they are in the dark, and within no time at all I have a clear view of the Pacific Ocean -the largest ocean in the world. Above the waterline its surface is seemingly calm, with only a few small waves to break-up the peacefulness and serenity of gently shifting seas, and in the distance I can see a perfect horizon, stretching-out endlessly in both directions. Turning right onto *Highway 1*, California's main coastal road, which is about five hundred miles in length (well, according to my map, it is) I immediately pull over into the gutter, before getting out of the car and walking across to the other side of the road.

Taking my camera out of its case, and lifting it up to eye level, I am just about to take my first photograph of this magnificent coastline, when suddenly I hear this great thunderous roar behind me, causing me to jump out of my skin! Swiftly turning around, I come face-to-face with this monstrous-sized log-hauling truck, which is literally *thundering* down the highway, its sheer bulk causing my poor mustang to vibrate like a *jack-hammer on heat*. Running back to my car, my heart now firmly lodged in my mouth, I begin to wonder whether I should have taken out that *crash, damage, waiver* policy after-all. Luckily no physical contact has been made with the vehicle, and so my precious little mustang will live on to fight another day. As for me, well I daresay that I will get over this rather *traumatic* incident as soon as I have had a long, cool beer to help me calm-down my nerves.

Ted was certainly right about the road to Mendocino -it was very picturesque indeed, and by eleven-thirty I am pulling-up in the town centre, having enjoyed a very relaxing, and totally carefree drive. My first objective is to find out where William's school is located, as I dare not be late when it is time to collect him, otherwise Wills may have already left on the school bus when I get there -and then what a fool I would look when he turns up at his dads place without me! The directions Ted has given me are perfect, and so not long after entering the town of Mendocino I find myself driving past the schools' main entrance, before parking-up the car a few blocks down the road. Mendocino is not what I would call a large town, but it certainly is a very pretty one. Most of the houses have two storey's and the majority of them are made up of rectangular wooden panels, all varying in size, along with the usual array of windows and doors, all of which are supported by huge beams, running both vertically and horizontally.

However, there are a few brick buildings dotted around the town, which help to break up the monotony of timber framed homes in every direction, even if each house does have its own kind of individuality. Nearly all of the gardens are surrounded by picket fences, the majority of which have been painted in white, of course, and each fence averages-out at around two to three feet in height. In almost every garden there is an abundance of multi-coloured flowers, some of which are rather short -and others that are very tall indeed, but every one of them is in full bloom, and so they are all looking quite radiant, as they continue soaking-up this wonderful Californian sunshine. Although Mendocino is relatively small, especially in comparison to any of the major cities in America, it is still very spacious, and I love the way that the houses have all been neatly segregated around the town, rather than being piled on top of one-another, like so many of the homes in coastal towns and seaside villages that I have visited throughout Europe.

William finishes school at 3.15pm, which means that I have plenty of time to treat myself to a spot of lunch, and so I decide to go in search of the local bars. Should I fail to find any that serve-up pub-lunches, then I might have to resort to eating in a café -or I might even go the whole hog and splash-out on a meal in a restaurant...who knows? If truth be known, any establishment that serves food will do me right now, so long as it fills that small gap in my tummy. It is now one o'clock, and I have just discovered the *Seagull Inn,* standing proud on one corner of the main street which runs through the centre of town. Immediately, I order a roast chicken dinner -which, for some inexplicable reason, turned-up fifteen minutes later as a *fish* meal? However, I am so hungry by now, that I decide to say nothing, before devouring the food like there is no tomorrow.

The owner of the pub has recognised my *English* accent, and so now he is asking me to tell him all about Europe, and what the people from each country are like -as if we *Brits* are supposed to know everything there is to know about our continental counterparts. The reason he is being so inquisitive (he says) is because he and his wife are planning on taking a six-month vacation in Europe next summer, during which time they hope to explore several countries, and so he wants to know which ones have the friendliest people, and also where are the most interesting places to visit. Thanks to the three trans-European journeys I completed back in the eighties, I am able to give him a decent insight into where to go, and what to look for in almost a dozen countries. However, considering that this guy is around fifty years old (and therefore assuming that his wife is probably about the same age) I did not bother recommending any of the various campsites which I had stayed at along the way.

Anyway, the man had already boasted that him and his wife intended spending around $40,000 on the trip, and so with that amount of money to play with the pair of them could easily afford to stay in very nice hotels for the duration of their holiday. As soon as I had finished my lunch I decided that it was time for me to take my first dip in the Pacific Ocean. Following the pathway which leads down to the beach, I am soon at the water's edge, and having already kicked off my training shoes, I am now playing the proverbial balancing act, as I try ripping each one of my socks off with one pull, for fear of losing my balance and keeling over into the water. Two small boys, not a lot older than my lads, are scrambling over a pile of rocks just to the left of me, while their parents sit peacefully relaxing in their deckchairs nearby, the pair of them basking in the warm, autumn sunshine, while at the same time keeping a keen eye on both of their son's adventurous activities.

Standing up to my knees in the water for several minutes, I begin to feel a numbing sensation in my feet, and so I decide that it is time to return to warmer and somewhat *drier* ground. I had only ventured into the water in the first place, so that I could say that I had been *in* the Pacific Ocean, whereas in reality the outcome of my efforts was no different than if I had been standing in the Atlantic Ocean, because the water was bloody freezing. Whichever way I wanted to look at it was immaterial right now anyway, as the hands on my watch are telling me that it is nearly three o'clock, and therefore it is time to go and retrieve my car, before collecting William from his school. Driving barefoot back to the building, I arrive just as several school buses are pulling-up in the main courtyard. A few minutes later the front doors of the main school building suddenly burst open, and before the teacher who was *supposedly* overseeing their safe exit had time to say "See you tomorrow kids", dozens of them had already poured-out into the playground area.

Several of the children are being met by various friends and relatives, but the majority of them are making their way over to their respective buses for the long (or maybe not-so-long) journey home. My eyes have been peeled for any signs of William for the last few minutes, although it seems that the younger children have been let out first, because the older students are only now beginning to emerge. A few minutes pass, and by now most of the children have boarded their buses, and so I begin to fear that William might be one of them? Suddenly I spot him emerging through a set of double doors at the rear of the playground area. He is walking in-between two other lads, and all three of them are happily chatting-away to each-other. I begin waving frantically to William, in the hope of attracting his attention, but he hasn't seen me. However, one of his friends *has* spotted me in the distance, and so after a swift tap on the shoulder and a pointed finger, Wills is soon waving back at me.

Halfway across the courtyard William stops to have a few last words with his mates, and to say his farewells to them, before turning and making his way over to where I am standing. Greeting each other with big, beaming smiles, I admit to William how nervous I was that I might have missed him, to whit he nonchalantly replied "No chance of that; I knew that you were coming to collect me; dad had rung the school first thing this morning and told the administrator, who in turn told my teacher, and he then passed the message on to me at lunch-time!" As William was now my *official navigator*, I asked him to suggest a place that he particularly liked, which he thought that I might also like to see. After a short deliberation, William recommended that I drive further on up the coast, to a place called Fort Bragg, and so I did. Within minutes of reaching the town we pulled over at an eatery called *Jack's Muffler Shop,* where I treated the pair of us to a glass of soda (that's *pop* to all you British kids) and William to a hamburger, because he said that he was feeling quite hungry.

On our way back to the dome William insisted on showing me a short-cut through the forest, where the road was nothing more than half-baked mud for much of the way. William said that his dad often uses this route, but only when it is dry, and not when they have had persistent downpours of rain. "Sometimes when we have heavy thunderstorms and torrential rain for a few days the roads flood-over, and the whole area becomes completely impassable", William added, in a kind of adventurous tone. All-in-all I found William to be a very polite and intelligent young man, who could make conversation on almost any subject, and so I enjoyed being in his company immensely. Ted must be very proud of his son. William and I eventually pulled up at the dome just as dusk was settling-in, the pair of us having watched the sun set on *the* most glorious golden horizon only a few minutes earlier. After warmly greeting the pair of us, Ted told me that he had tried to contact an ex-colleague today, whom he knew would be interested in our little venture.

Unfortunately, the guy was currently off work through illness, and so Ted had left a message with his secretary for him to please contact Ted as soon as he was well again, and had subsequently returned to his office. Sitting in his usual armchair, Ted then beckoned William and me to sit on the sofa opposite him, while he continued-on with his story:

"The man's name is Harvey Edwards, and he used to be the editor of the Sunday Times. He is the guy who gave the go-ahead for me to begin my jaunt around the world back in 1973, which the newspaper duly sponsored of course. Harvey left the Sunday Times and moved to America several years ago, where he took-up the position as the editor of *Traveller's Magazine,* but since then he has moved up in the world, and he is now a film producer. Harvey used to ride a motorcycle back in the *good-old-days,* and we have been great friends for many years, although I have to admit that I haven't spoken to him for a very long time".

Things were beginning to move forward; I could feel it in my *water*! Earlier on this morning Ted had told me that he would be *wining and dining* William and I at a local restaurant this evening, which I had been looking-forward to all day. Unfortunately, William has just confessed to his dad that he has forgotten to finish-off a project for school, which must be handed in to his teacher by tomorrow morning. William estimates that the outstanding work will take him approximately four, or possibly even five hours to complete, and so Ted is left with no alternative, but to cancel our *big-night* out. As luck would have it, Ted still had a few steaks left in the freezer, along with his last bottle of Californian red wine...which was looking rather lonely, sitting there on its own in the wine rack, but thankfully not for long...and so all was not lost.

As soon as we had eaten, Ted and William spent the rest of their evening battling-it-out on the word-processor, while I continued working on more scenes for the movie. At midnight I rang my lovely family (with Ted's permission of course) as I knew that they would now be having their breakfast. It was so wonderful to be chatting with Caryl and the boys again -especially Liam, who just wanted to know how soon his dad would be coming home to him. The following morning William went off to school with a beaming smile on his face, and a completed project in his satchel. I had said my goodbyes to him before he left, not knowing whether we would ever meet again, as Ted had already told me that William's mother would be collecting him from school this afternoon, to stay with her for the weekend. I was also quite sure that by the time William returns to his dads place on Monday, I will probably have long-since said my goodbyes to Ted.

CHAPTER 5

The Round Valley Indian Reservation

Last night Ted had told me that the place was too small for two adults, let alone *three people* to share, and so I was expecting him to *give me my marching orders* today. However, and to my most pleasant surprise, instead of sending me on my way, Ted cordially invited me to spend the weekend with him at his farm in Covelo. (*According to my map, Covelo is about twenty miles inland, and roughly seventy miles north-east of Elk on the outskirts of the Round Valley Indian Reservation*). Ted tells me to pack-up all of the gear that I have brought with me, and then put everything into the trunk of my car, as he wants me to follow his car in the mustang all the way to the farm.

"Then when I return to Elk on Monday morning, you can continue heading southwards to your friends place in Orange County, rather than having to detour all the way back to Elk, just to collect your stuff" Ted adds.

I respect the fact that by this time Ted will have spent five nights living with someone whom he had never met a week ago, and so I am sure that he will want nothing more than to get his life back in order again. (*Actually, I am very grateful that Ted is willing to spend another three days with me, having already put up with me for two!*) For the next hour or so Ted and I get stuck into cleaning the place from top to bottom. While Ted worked diligently in the kitchen, washing-up all of the pots and pans, before sorting out any items of crockery and cutlery which we might need for the weekend, I swept the floor of the dome from stem to stern, before stacking-up the log rack for the boiler from the mountain of chopped wood outside. When everything was finally *done and dusted* Ted put a huge padlock on the door, before kindly posing for a few photographs for my *ever-increasing* photo album.

Reversing the mustang around, in order to face in the direction of the main road, I took one last look in the mirror at the place which I had called 'home' for the last two nights, and as I drove away, I could have sworn that a very small lump had momentarily welled-up in my throat. However, sentimentality would have to be put to one side for another day, as I now have to concern myself with keeping up with Ted, who is tearing through the country lanes in his car like a man possessed. Rounding a sharp bend, I spot a relatively small creature lying dead at the side of the road, and from this distance it looks like a large rat. I would love to get a closer look at this hairy rodent, but I am going too fast to pull over, and I daren't slow down (even a touch) for fear of losing Ted, who is still driving at an erratic pace. Suddenly a deer appears from nowhere, running across the road right in front of me, and I am forced to brake fiercely, in order to avoid a collision.

"This certainly is 'animal territory'", I said to myself, as soon as I had regained my composure. Resuming a normal pace, and naturally assuming that Ted would be long-gone by now, I rounded the next corner, only to see Ted's car parked at the entrance to someone's driveway. Ted is standing alongside his car, and he is waving his hands in the air, trying to attract my attention, as he obviously wants me to pull over as well. "I have to drop some clothes off for Wills for the weekend; can you wait here in your car for me Shaun -I won't be five minutes" Ted says, ever so politely, before disappearing up the long pathway which leads to the house. Having brought the Mustang to a grinding halt only seconds earlier, I am happy to let the disc brakes cool down for a while...along with my shattered nerves. I realise that this must be where Ted's ex-wife lives and I also wonder how hard it must be for Wills, having to commute between his parent's homes, rather than all three of them living together as *one big happy family*.

If Caryl and I ever got divorced (heaven forbid) then I would be nothing less than devastated -and if it happened while the children were still very young, well, being separated from my kids would simply tear me to pieces. In fact, I have only been away from my family for a week as I write these notes, and I am already missing them all terribly. After a few minutes Ted returns to his car, and so together we *hit the road* once more. (*I told Ted about the critter I saw earlier at the side of the road, and he said that it was probably a possum. Like hedgehogs in the UK, thousands of these harmless marsupials are killed every year on the roads in the US*). Our last port of call before joining the main highway is the local outdoor fruit market, where Ted has just finished putting his order in for next week. I am not sure whether the company runs some kind of delivery service, or if Ted intends picking-up the fruit and vegetables on his way back to Elk on Monday, but either way, I simply forgot to ask him?

Willits is the first major town we arrive at, which is roughly halfway between Elk and Covelo. Ted has indicated to pull over directly outside this small ice-cream parlour-cum-mini-cafeteria at the side of the road, and so I likewise follow his lead. Getting out of his car, Ted says that it is quite *hot and stuffy* driving in the car, and so we should treat ourselves to a cool, refreshing drink, before continuing-on with the second half of our journey. I could not agree more. Sitting on a picnic-bench inside the shop, each of us guzzling a can of Coca-Cola, Ted tells me more about his place in Covelo. He says that he is currently rebuilding his barn, which had accidentally burned-down last winter, adding that he can only work on it at the weekends, because he has to stay at the dome in Elk during the week, in order that Wills can attend his school.

"One of my farm-hands rung me last week, to say that there is a break in the main cable, leading to the water pump, and so I will also have to take a look at that when we get there -the work is never-ending Shaun," Ted adds with a sigh. For once in my life I am unable to say anything in return, which is definitely a 'first' for me. Ted paused for a few moments, and then looking me straight in the eyes, he said "You're really flying high here, trying to take all of this in, aren't you Shaun?" I was more than happy to admit that I felt as if I was walking on air right now. "Well now you know exactly how I felt when I was doing my journey around the world", Ted answered reminiscently. By two-thirty Ted and I are high in the mountains surrounding Covelo, and once again he is waving me down, with a view to pulling over at the side of the road. We stop next to this huge boulder, which has a large plaque attached to it with the following inscription:

ROUND VALLEY

The first inhabitants of Round Valley were the Yuki who resided here for thousands of years in harmony with their natural surroundings. In 1854 European settlers entered the valley. In 1856, conflicts between settlers and Yuki escalated and to protect local tribes the entire watershed was designated a reservation. Additional tribes were subsequently forced on the property; Nomlacki, Wylaki, Lassik, Sinkyone, Pomo (including Cahto, Kabeyo, Shodakai, Yokayo, Shokawa, Shanel, Kashaya, and Habenapo, among others), Wappo, Concow, Maidu, Kalusa, Colusa and Achumawi. In 1864 the government reduced the reservation by four-fifths, to its current size.

CALIFORNIA REGISTERED HISTORICAL LANDMARK No. 674

Plaque first placed May 30th 1959. This plaque placed by the state department of parks and recreation in cooperation with the people of Mendocino County March 21st 2002.

It was only after I had finished reading the words on the plaque that I gazed above the rock for the first time -and what my eyes beheld made my chin drop to the floor. The beauty and sheer enormity of the landscape stretching-out in front of me was awesome, and for once in my life I was totally speechless. I had never seen such a vast open-space before. (*Ted obviously knew that I would be well-impressed with what I saw*). One more stop-off for Ted at an ATM machine, and then one for me at a gas station, as I was now running low on fuel, and by 3 o'clock the pair of us are pulling-up at the farm. 'Den', one of Ted's farmhands, has come to greet us, and he seems to be a really nice chap. All of the other workers have gone home for the winter, and the majority of them will return again next year for the summer season, but Den has built himself a *shack* on Ted's farm, and so he stays here permanently, primarily to look after the place in Ted's absence.

Den and Ted begin talking about digging-out the pump this weekend, and I have this strange feeling in my *water* that my services are going to be required for this major operation, which has just been scheduled for first thing tomorrow morning. As we make our way over to the farmhouse, Ted points over to the barn that he is currently rebuilding, which is no more than a skeleton structure with a roof on it at present, and he tells me that all of the hay which is normally stored underneath this massive canopy, has recently been sold by Den, and so it certainly looks as though Ted's farmhand is earning his keep. Opposite the main farmhouse is a smaller building, which Ted originally had built for his mother to live-out her final years in. Unfortunately, after a couple of years his mum decided to go and live with Ted's sister, and so now it is used to house some of the summer workforce. In the wintertime it is usually left vacant.

As we enter the single storey building I am introduced to Ronaldo, who is from Mexico, his wife, Shauna who is from the US, and their two-year-old son, Aaron. Ronaldo is only about five and a half feet in height, but he is powerfully built, and his tanned skin colour is the envy of all sun-worshippers. In complete contrast to Ronaldo, Shauna is as thin as a rake, and as white as a ghost. Both Ronaldo and Shauna have wonderful personalities, and I could tell straight-away that they were very-much in love. As for Aaron, well he is like most boys of his age -full of life and utterly adorable. The couple rent the house off Ted all year round, on the understanding that one bedroom is kept free for Ted whenever he visits, which is mostly on weekends and during the school holidays. There is also a spare room at the back of the house, and Ted has just informed me that this is where I will be sleeping for the next three nights.

A quaint little cottage suite sits in one corner of the room, and an old armchair in another. On the right as we enter the room there is the heating boiler, which reminds me of the tin man from the Wizard of Oz (a kind of 'Minnie-me' of the boiler in the dome) and alongside our *over-sized bean-can* there is a pile of empty cardboard boxes. Behind these boxes lies my bed -or should I say my *mattress*. Ted disappears for a few minutes, and when he returns he is carrying in his arms a thin, white cotton sheet, along with two thick, multi-coloured blankets. As Ted set-about preparing a comfortable sleeping area for me, I began the laborious task of unloading everything which I had unceremoniously bundled into the car this morning. (*Unbeknown to me at the time, but this ritual would be par-for-the-course in the days that followed*). Ted and I have both been thinking a lot about the movie over the last couple of days, and we have a slight conflict in ideas.

I want the opening scene to show Ted (preferably), or if not, an actor who is portraying Ted, standing on a rostrum in the middle of this huge auditorium, which is packed to bursting with hundreds of people who have all come from far and wide to listen to Ted telling the true story of his epic journey around the world. I also want Ted to write the opening scene, of course, as no-one could write the story better than the man who actually did the trip. However, Ted believes that the film would be better received if the storyline revolved around someone who wanted to do the journey (like me perhaps) and then we could use extracts from Jupiter's Travels to compliment the movie. Whichever way we choose to go, Ted says that he would like to get in touch with Meat Loaf and Jim Steinman, with a view to them writing (and also singing) more material, especially for the film. Ted then admitted to me that he now likes Meat Loaf's music, and that he listens all the time to the *Bat out of Hell* tape, which I had given him yesterday as a memento of my visit.

While Ted was in a somewhat *submissive* mood, he also conceded to the fact that he has neither the time, nor the skills required, to write a script for the opening scene from the portfolio which I have *painstakingly* prepared for him. "However, if you can get hold of a professional scriptwriter, then I would be willing to give him all the assistance he requires" Ted adds positively. *(Apparently, some whizz-kid from Hollywood had suddenly appeared on Ted's doorstep a few years ago with a similar idea to mine, but after talking to the guy only a couple of times, Ted ended-up sending him away with a flea in his ear).* Of course I appreciated the fact that we needed to have people who were professional's in the film industry involved in the project if we were hoping to have any chance of the movie reaching fruition, but where the hell do I begin looking for those kind of people? As a 'thank-you' for all his help and support, along with the wonderful hospitality he has shown me so far, I have just told Ted that I am treating him to a steak supper in town this evening —providing that he knows where there is a half-decent restaurant, of course, as the place looked somewhat barren when we entered it this afternoon?

Ted graciously accepted my offer, saying that seven-thirty would be an ideal time to eat. Den and Ted then went off to check-out all of the other work that needed doing on the farm, and so I decided to go for a spin into town, to acquaint myself with the layout of the place. Driving down the dusty road I truly believe that a horse would have been a much more appropriate mode of transport to come galloping into town on, as the whole place looks like something out of a *spaghetti western*. Pulling-up outside the 'Buckhorn Saloon', I sit in the car for a moment, staring ominously at the building, my outlandish imagination half-expecting Billy The Kid to come flying-out through the saloon doors at any given moment, as the sound of several gunshots fill the air, and the County Sherriff comes running across from the local jailhouse, followed by a posse of mean-looking deputies, all of them armed to the back teeth with high calibre rifles and double-barrel shotguns, in readiness to blast this band of no-good gunslingers into oblivion! *(Sometimes my overactive imagination really worries me).*

Walking through the dimly-lit porch-way and into the main bar, I feel like a total outcast, as a dozen pairs of beady eyes are immediately focussed on this *stranger from out-of-town*. A guy who is sitting at the bar, right next to where I am now standing, gently lifts his Stetson from off the top of his head, before tentatively mopping his brow with a napkin... and the pool game, which was in full swing only a few seconds ago, is now at a standstill, as the deathly silence has caused the guy who was just about to play his shot, to stop and glance over his shoulder at the *unwanted intruder.* I am feeling very uneasy about the hostile atmosphere which I seem to have unwittingly created just by being here, the hero inside of me telling me that discretion might just be the better part of valour here, and so I decide to leave while I am still in one piece?

"Howdy", says a female voice from behind the bar, stopping me dead in my tracks, having already made my way to the door, in readiness for a sharp exit. "Can I get you something to drink my friend?" *(Young lady, you will never know how glad I am right now to hear that magical word 'friend').* "Beer please" is all that I can think of to say to her, as my eyes try to focus on anything in the room which does not resemble a human outline. Casually placing my posterior on one of the bar stools, I stare directly at the pump handle in front of me, completely oblivious of my surrounding audience –he lied! Thirty seconds later, this rather lovely looking barmaid hands me a glass of Budweiser, before asking me for

$1.75, and so I hand her a five dollar note. Upon returning my change the young lady introduces herself to me as 'Sienna' (a lovely name) and so I reciprocate by giving her my name. With the eerie silence now broken, the tense atmosphere begins to dissipate, and I can feel an air of normality slowly, but surely, being resumed.

The barmaid and I begin chatting-away together like we are old classmates at a school reunion party, and pretty soon she is asking me all about Wales and the UK in general, and so after answering her numerous questions, I duly return the favour, by asking her all about California and the United States of America –a country which I am finding more fascinating with every day that passes. Without posing the obvious question, Sienna simply removes my (by now empty) glass from the counter, and refills it with another half-litre of Budweiser, before pouring herself one as well. "Cheers", she says with a huge heart-warming smile, before chinking my glass with hers. "God bless America" I reply, with a view to keeping our personal Anglo-American friendship as wonderful as it is right now. Pulling another five dollar note out of my pocket, I hand it to Sienna, to pay for the drinks, but my offer is politely refused by the young lady.

"No siree, this one's on me -but you can buy the next round… if you really want to Shaun," she says with a beaming smile. Thanking Sienna for her very kind gesture, I cannot help but wonder what a pub landlord would say in the UK if he caught one of his barmaids buying a complete stranger a bottle of beer? Sienna tells me that she has always found European men to be a lot more courteous than American guys, "Which is probably the reason why I am engaged to a Dane", she adds, in a kind of *matter-of-fact* way. One of the regulars has just invited himself into our private conversation, having had enough, no doubt, of listening to the *nosey foreigner* asking his fellow-American all kinds of inquisitive questions. Dressed in a pair of scruffy jeans, complete with a black leather belt and a red chequered shirt -and donning a black baseball cap on top of his head, and a huge pair of mountain boots on his size fifteen feet, this giant *Neanderthal* is your archetypal American (or should I say 'Canadian') lumberjack –the kind of man who eats *three shredded wheat* for his breakfast every morning…twice!

Being in this guys' company was actually a pleasure, because he turned-out to be a real gentleman, and I also found him very interesting to talk to. (*At least that is what he made me write in my notebook, otherwise*

*he was going to tear me limb from limb...*only joking!) After listening to
countless country and western singers, all of whom had been *wailing-away*
on the television set above the bar, the majority of them droning-on about
their heart-rending break-ups, or pouring their hearts out with tear-jerking
stories, the ones that make you want to go outside and slit your wrists,
I decided to head for home. This afternoon had been an 'enlightening
experience', shall we say, but now I was more than ready for a *good-old
nosh-up*. It was nearly seven o'clock by the time I got back to the farm, Ted
having already washed and changed for the evening, and so I had better get
my skates on. Ted is quite happy to take his car out this evening, as he does
not intend on having more than a couple of drinks, and so by seven-thirty
the pair of us is sitting at a table in the one and only restaurant in town,
which incidentally, only opens on weekends...and closes at ten o'clock.

The food is excellent (I had pork-chops, by the way...watch this space)
if a little pricey at $45 for two meals, but then as this is in essence the only
time that I have had to put my hand in my pocket since first meeting Ted
three days ago, I am more than happy to foot the bill. Walking back to
the car Ted and I pass the local community centre, and as Ted recognises
the lady who runs the place, he decides to call in for a quick chat. The
games room is about the size of your average lounge area, and sitting right
in the middle of it is this full-size pool table, which actually leaves very
little room for any other kind of activity to take place. However, this is
never a problem, because, just like the restaurant, the place only opens
on weekends, and I have just been told that tonight has been a busy
night —because a total of 'five' children have turned-up for their evening
of fun and games.

By ten o'clock Ted and I are back at the farm, and I am feeling all
excited because Ted has offered to show me photographs and film footage
from his journey, which he has rarely shown to anyone else. As soon as my
private viewing had finished Ted brought out a video from his cupboard,
which he says, is a documentary relating to his journey, which was made
back in 1986. As the credits began appearing on the screen Ted gave me a
brief introduction as to what the documentary is all about:

"The purpose of the documentary was to give something back to all
of the people who had written to me over the years after they had finished
reading Jupiter's Travels. The producer and I decided to take my bike out

of the British Transport Museum in Coventry, and then I rode it around various parts of the UK, visiting several people who had sent me letters. I then chatted with my readers about what kind of impression, or influence the book had made on their lives. Anyway, you will see what I mean in a moment".

Ted had obviously made this documentary for a company in the UK, which surprised me at first, because with it being only five years old, I had naturally assumed that it would have been made by a television company in the United States. Anyway, enough had been said, and so now it was a case of 'on with the show'. As the title 'Riding Home' becomes visible on the screen, the soundtrack from the cult film 'Easy Rider' plays gently in the background, and as the music and the title fade out simultaneously, Ted appears on the screen. He is riding his famous Triumph Tiger, the bike which carried him a total of 63,000 miles, through 54 counties around the world. After a few seconds Ted begins his narration, outlining the statistics of his achievement first, before going into detail about how the journey had affected his soul. As soon as the introduction is over Ted is seen chatting to various people in their own homes around the UK. Ted is explaining to them how going on this journey completely changed his life, and the people then reciprocate, by telling Ted how his book, Jupiter's Travels, had influenced their lives in so many different ways.

The programme lasted for nearly an hour, and the documentary finishes with Ted giving a final narration about travelling in general, and what life holds for him in the future. I must confess that I had mixed thoughts about the documentary, although I did not share these thoughts with Ted at the time. On the one hand I fully appreciated the way that it had been set out, and I could easily understand the various comments, which were mostly compliments that had been bestowed upon Ted by the people involved in the show...after-all, I am probably one of Ted's biggest fans. However, I also felt that just talking about Ted's incredible achievement could never do justice to the journey itself. This is why I aspired to making this movie in the first place; because I wanted to recreate all of the amazing adventures and unbelievable encounters which Ted had experienced throughout the journey, and then put them all on the big screen for everyone else to enjoy as a kind of *factual reality*, so that they could truly feel as though they had actually been a part of the journey themselves.

Also I wanted to give this *epic* production a great soundtrack; music which not only coincided with each scene individually, but by using songs which bikers throughout the world have been playing for over twenty years. (*I rest my case*). After removing the tape from the video-player, Ted fumbled-around for a few seconds, before producing a piece of paper with a name and address written clearly on one side. "Here it is Shaun, the name of the television company that produced the documentary; 'BBC Wales in Cardiff'...do you know them?" Well, to coin a certain phrase '*you could have knocked me down with a feather*'.

"Do I know them Ted? My brother, Gary, has only been working for the company for the last twenty years!" I replied, practically stuttering my words in total disbelief.

(The BBC Wales Television Studios are situated at the southern end of the LLantrisant Road, in a place called Llandaff, which is less than two miles from where I was born). Ted tells me that the man who produced the documentary is called 'James Gerwyn' adding that he had flown over to see Ted about five years ago, in order to proposition him in a similar way to what I had done this week. During his visit to California, James had stayed in a hotel in Mendocino -unlike me, of course who had graciously been given the honour of being Ted's guest for nearly a week, and so I was feeling well chuffed about that side of things. "When I went back to the UK to do the filming, I stayed at James' house, which is in a place called *Canton*, I believe, or was it called *Riverside*? Anyway his house was very nice", Ted added.

The house where I was born (in the front room), and that I lived in for over twenty-five years, was originally classed as being in the constituency of 'Canton', but then the city council changed the boundary lines some years later, and so the house is now located in the 'Riverside' area of Cardiff. The world is getting smaller by the minute.

Apparently, James had told Ted that he had read Jupiter's Travels several times, and that he was a great fan of his, and so Ted humbly believes that he will be more than happy to help us with our mission.

"When I get back to Elk on Monday I will give James a call, to see if I can arrange a meeting between the two of you upon your return to Cardiff", Ted added in a somewhat confident tone. Things are definitely starting to move, and I truly believe that this could be the breakthrough that I have been waiting for all along. It is nearly midnight by the time I finally get to bed, but it seems that the boiler is hell-bent on keeping me awake, as it continues clicking, clacking and clunking (with not a minute of respite) all frigging night. What I wouldn't have given for an industrial can-opener right now!

CHAPTER 6

Down On The Farm

The following morning I am forced to *eat my words,* as everyone is now complaining about how cold it was last night and yet my room was still as warm as toast. However, outside my window it was a completely different matter. Last night the temperature had plummeted below freezing, leaving a heavy frost on the ground this morning, and an extremely cold nip in the air. Ronaldo has just assured me that it will soon be warm again "As soon as the mist has lifted and the sun has broken through for the day", he adds in a very confident tone, as if he, himself, was controlling the weather. Shauna has kindly offered to cook us all a special 'Mexican' dish tomorrow night, as a kind of farewell meal for Ted and me, even though Ted will probably be back there the following weekend. However, before she can do this, Shauna will need several items, including lots of herbs and spices, from the local grocery store "And that is where you two come into the equation", Shauna adds, before handing me a lengthy shopping list with a tender-sweet smile that is impossible to refuse.

Going shopping to buy all kinds of tasty morsels has never been a chore for me, especially when I am going to reap the benefits of my hard work, and as Ted knows where the shop is located, we have decided to do the shopping together. While I had been busy sampling the local *firewater* yesterday Ted had dug four small holes, at various intervals along the route of the damaged cable, which is connected to the pump at one end...and to lord-knows what at the other end? As soon as we have returned from the shops, Ted informs me that I will have the pleasure of finding-out if any of

Ted's previous patchwork repairs have come unstuck, or indeed if there are any new breaks in the cable? As this will entail digging-out a pile of earth from underneath the (now visible) lengths of cable in these mini-trenches (which Ted has already excavated) before giving each one a massive tug, to test its solidity, I am not overly thrilled about the idea, even though Ted has given me a pair of thick rubber gloves for my protection.

The good news is that it took me less than an hour to dig out and check all of the joints, but the bad news is that I now have to continue digging-out a trench along the rest of the cable, in order to find out exactly where the new breakage is? By two o'clock I have excavated approximately fifty feet of trench, and my back is in half, and so while Ted continues working on his barn, I retire back to the house to indulge myself in a huge mug of steaming-hot coffee. After a short rest I then spent yet another two hours continually digging-away at the earth, and tugging-away at the cable, but still I cannot find that blessed break? Ted has made it quite clear to me (several times, in fact) that he did not bring me here to work on his farm, and that at any time I wanted to quit, I am at liberty to do so.

Den has also assured me that he will take over the job from wherever I leave off, adding that there is no real urgency to sort out the problem before Ted and I leave the farm on Monday. However, I have never been a quitter, and so I continue-on, *digging my life away*, in spite of the fact that it has now started to rain! (*I have also never taken anything off anyone in my life, without being able to give something back in return, be it physically, spiritually or financially*). Within an hour those bands of white, fluffy clouds which were delicately hovering above our heads earlier-on, have now been replaced by a dark grey, ominous-looking sky, and those first few droplets of rain that had fallen from the heavens earlier, have now amassed into a torrential downpour. As the trenches begin to fill with water and my feet start sinking deeper into the mud, I can see that I am going to make very little progress here, and so I decide to exit my *soggy* trench, and go and join Ted in his *dry* sanctuary...in other words under the shelter of his semi-built barn.

Ted is currently in the middle of building a workbench out of a pile of burnt timber, which has been left by the side of the new barn -as a pertinent reminder of the old one perhaps, and so I spend the next hour or so assisting him in separating the good wood from the bad. Unfortunately the light -or should I say the lack of it, finally puts an end to our *grafting*, and so, somewhat reluctantly, we retire back to the farmhouse, the pair of us feeling *sort-of* satisfied with our afternoons' work. As we enter the lounge area, Ronaldo and Shauna begin giggling to each other like a pair of cheeky school kids. Ted and I are both bemused by their chuckling, but it is only after looking directly at one-another, that we immediately realize what they are laughing about. Our faces are as black as soot, the pair of them having been plastered in charcoal fibres from the burnt-out embers which we have been working with for the last hour or so. Ted is first to use the bathroom, to try and wash out the charred remains of a thousand dust particles, which have somehow managed to find their way into every orifice in his body...and then I am next. What a state to be in!

In recompense for laughing at us, Ronaldo has given Ted and me a bottle of ice cold Mexican beer each, which we immediately devoured without question. "If one is wet on the outside, then one deserves to be wet on the inside -that is my motto Ted," I say chinking his now seriously empty bottle. "I couldn't agree with you more my friend" replies Ted, before disappearing into the kitchen, in order to snaffle two more of Ronaldo's beers from the fridge. (*Well we did buy a case of them from the store for him when we went shopping for Shauna*). Ted is fully aware that I intend heading south, to my friend, Daniel's place in Irvine, Orange County, on Monday morning, and so he has invited me to use the house phone to call him, in order that I can make Daniel aware of my departure from Ted's farm, and also to give him my approximate E.T.A. (estimated time of arrival) in Irvine, Orange County.

Knowing that I have approximately a 600 mile journey in front of me, I have already accepted the fact that it is going to take me the best part of two days to get to Irvine, and so I told Daniel to expect me either late in the afternoon, or early in the evening on Tuesday. Daniel immediately gave me his works' telephone number, telling me to ring him if I reach Los Angeles before five o'clock, in which case he would come and meet me wherever I was, in order that I could follow him back to his home in Rockwood, Irvine. If I arrived in L.A. after 5pm, then my orders were to ring his wife, Zara, at home, and she would do likewise. Ted had told me that the roads would be very good once I was on the main freeway heading south, and so I suppose I could have considered driving non-stop, all the way to Los

Angeles, thus completing the journey in only one day, but I was in no rush whatsoever, and besides, I wanted to relax and enjoy the landscapes, and to make the most of the second week of my California '*fly-drive*' holiday.

In my first week I had more than accomplished what I came here to do, and so I knew that my sponsor, Jules Gilmore, would agree with me entirely... simply because I deserved it. As soon as I am back the UK the real work will begin in earnest, and both Ted and I have agreed to continue pursuing the project from both sides of the Atlantic. Returning to that same restaurant for supper this evening, I am left on my own for a few minutes, as Ted disappears across the room to have a chat with some other friends of his, whom he has just recognised. Within minutes Ted is beckoning me to come on over and say hello to this middle-aged couple, and so I immediately wander across to their table. The people are very nice to talk to, and I am so pleased that Ted has introduced me to them as his "Good-friend", rather than describing me as a *work colleague*, a *business associate*, or simply an *acquaintance*. Last night I had talked myself into having the pork chops, instead of a steak, but tonight I am going for the duck sausages -or at least I think I am?

The waitress, whom Ted and I have given our order to, is a little batty, to say the very least, and so it is no real surprise when I am presented (for the second night running) with a plate of delicious looking pork chops! "Déjà-vu" I say to Ted, having already told him about the chicken dinner that I ordered in Mendocino, which turned-up as fish! Ted assures me that it is not a Californian custom to serve foreigners with the wrong meals "Otherwise they would soon be out of business", he quips, jokingly. However, when I tell the waitress about the error, she asks me if I will have the chops instead, adding that the restaurant only cooks to order due to the lack of customers!

"The chops are more expensive than the duck sausages, but seeing as it was my fault, I am willing to let you have the chops for the same price" the waitress then added, in a vain attempt at making me feel as though *I* am the one who should be grateful for her kind offer!

I am not a happy bunny, but as I have no intention of causing any kind of scene, and I certainly do not want to get the poor woman into any trouble, I accept her offer with a placid smile...or was it a grimace? With that small matter having been *amicably* resolved, Ted and I now have another far more important issue to deal with! You see, Ted wants to drink red wine with his steak, whereas I want to drink white wine with my chops.

"To save any arguments, we'll simply order one bottle of each" I say jovially.

"It's a good job that we walked into town this evening" Ted answers, before shifting the decorative pot of flowers and the condiments rack over to one side of the table, in order to make room for our onslaught of *vino*. When Ted first arrived on his motorcycle in California he said (*in Jupiter's Travels*) that he had gone to a party with, *quote* 'Real stars' and so I asked him if anyone he had met at that function was still in show business, and if so then did he still have their contact numbers? Ted said that he could not think of anyone off-hand, but that he would check in his address book when he returns to Elk on Monday.

I am a great believer in the old adage that '*It is not WHAT you know, but WHO you know*', that really counts in life, and therefore one simple acquaintance that Ted may have stumbled-upon in the past, could be the whole key to our success. (*After all, I had been introduced to Jules Gilmore by a mutual friend at a barbeque, which had started this whole thing off in the first place -and now, with a little help from Ted, I may well get to meet the top brass in the BBC as well*). Before I become too excited about everything that is happening in my life -most of which is going on inside my head, Ted decides to bring me back down to the '*world of reality*'. "You must be patient Shaun, and you cannot expect things to happen overnight -especially when you are dealing with people in the television or film industry. Also, you have to understand that I am very committed to various other projects, apart from this movie idea, and so you are going to have to do most of the groundwork yourself"

Ted then went on to explain how he has invented an agricultural labour-saving device for planting intricate and awkward seeds into the ground. At present Ted is experiencing difficulty in getting his idea patented because of a similar proto-type drawing which had been submitted to the patenting office by another inventor some years earlier, which they claim is identical to Ted's idea. Ted vehemently refutes their claims, insisting that not only is his invention "Nothing like" its counterpart, but that it also performs an entirely different function. In order to justify his statement, and hopefully resolve the issue, Ted will have to go to Washington as soon as possible, because he already has a Scandinavian company interested in manufacturing his product, and so they are pushing Ted hard to get all of the legalities sorted.

Apart from battling with bureaucracy, Ted is also trying to finish his latest book, which he says that he has been working on for the last six years, along with running a farm, keeping a home, and also looking after an eleven-year-old son. "So lord knows when -or even 'if' I will ever have the time to do another bike journey?" he adds, somewhat despondently. Returning to the farmhouse after our meal, I lie awake in my room for hours, wondering whether all of the pieces of this enormous jig-saw puzzle will eventually fit into place, or am I destined to make a laughing-stock of myself, with only photographs and memories to remind me of my foolhardiness. The rain continued beating-down mercilessly upon the farmhouse all night, and by the morning the torrential downpour was still showing no signs of letting up. At least this meant that the ground which I had worked so hard on yesterday would be a lot easier to work with today -if and when it ever stops raining, of course? While Ted is enjoying the delights of a Sunday morning lie-in, I am busy preparing my southward route, in readiness for tomorrow mornings take-off.

Studying my map very carefully, I work out my proposed route to Irvine. There are only two roads that I can travel south on, one of which goes directly through the desert, or the other one which primarily follows the coastline virtually all the way to Los Angeles ('City of Angels'). I decided to choose the latter. Ted has now emerged from his *pit*, and so, after seeing me drawing up my route-plan, he has one or two suggestions which could assist me in my decision-making:

"When you get to San Francisco it will probably be better for you to take the 19th Avenue turnoff onto the '280' South. This way you can avoid all of the repair work which is still being carried out on the freeway, after it was devastated by last years' earthquake. Alternatively, you might want to consider driving the car all the way to Los Angeles tomorrow, where you could drop the car off at Budget's office's in downtown L.A., and then claim back the $138 for your second weeks' rent. You could then use this money to pay for your flight back to Frisco in a weeks' time. Not only will this give you an extra days' leisure-time, but it could also save you some money as well."

Assuming that I will not need any transportation once I have reached Daniel's place, I decide to take Ted up on his idea, and so I ring the Budget office to see if this is possible? Unfortunately, the guy in the office tells me that because my hire car is not in the 'higher grade' category, an alternative drop-off point is not an option for me. My other alternative of course, is to drop off the car at the Budget offices in San Francisco tomorrow, and then fly down to Los Angeles for the week, before flying back to San Francisco the following Sunday. However, as I have it in my heart to drive the length of California at least *one way*, if not both ways, I am not enamoured with that idea at all.

Curiosity had gotten the better of Ted, and so he decided to ring the local travel agents anyway, just to find out the prices of the flights both ways. Thankfully they turned out to be a lot more expensive than Ted had anticipated, and so now I would not have a guilty conscience about keeping the car for another week, which will mean spending even more of Jules' hard-earned money on fuel. (*Mind you, when you consider that the cost of a gallon of petrol in America is less than a quarter of what it would cost in the UK, my total expenditure on gas was quite minimal*). After breakfast Ted went back out to the barn to continue his work, while Ronaldo concentrated all of his energies on chopping more wood for the boiler, just in case we had another cold snap overnight. Shauna is playing the role of the dutiful wife, by gathering-up all of the laundry and putting it into the washing machine, before cleaning-up the house and preparing lunch for everyone, while Aaron is peacefully sitting on the window ledge, happily watching his big, strong daddy at work.

Thoughts of my own two sons waiting for me back home have now clouded my mind, and suddenly this *euphoric plateau* that I have been living on for the past week seems non-existent. Also, as this is only the second time that Caryl and I have been away from each-other since becoming man and wife, I am now missing her company terribly, and I feel nothing but emptiness inside of me, along with a loneliness which I have never experienced before. Maybe it is the great distance between us this time, which is making it so much harder than the last (when we were only 100 miles apart) or perhaps it is because she is seven months pregnant, and therefore I should be there for her and my two boys, and not out here chasing some outlandish dream? My selfishness knew no bounds, having

58

blinkered me from what was real in life, and all I wanted right now was to get out of this dream world, and return home to *the land of the living.*

Having a personal belief in something which no-one else really shares is very dangerous, and it can also be quite costly...in more ways than one. What drives people like me to follow their beliefs I will never know -perhaps it is some kind of external force, or maybe even an extra-terrestrial, or super-natural being, telepathically hypnotising us on a wavelength which no-one else can hear or intercept —or am I being hypothetical here? In reality I can only assume that it is no more than creative thought-patterns, which everyone has at one time or another —some more than others, but the only difference between *people 'A'* and *people 'B'* is what they actually do about it, and nothing to do with how 'brilliant' their ideas were in the first place. Anyway, enough moralising for one day -I have a movie to make, and so I had better snap out of this negative frame of mind, before I talk myself into getting on the next plane home. At that moment Den came into the house. He was carrying a pile of papers, and he asked me if I had seen Ted?

When I told him that Ted was out working on the barn, Den said that he did not want to disturb him, and that the paperwork could wait until later. (*That is what I like about Den -nothing in life is ever urgent*). Looking a little lost for things to do, Den then asked me if I would like to go on a tour of the farm with him, to whit I immediately accepted his kind offer. The rain had eased-up considerably by now, and so I was pleased that we would not be forced to put our *wellies* and overcoats on, in order to save us from getting soaked through. Den firstly took me over to his *shack*, which was located right at the far end of the main field, and as he showed me around the place, he proudly announced that he had built it completely on his own. Considering that he had made the whole structure purely out of a pile of old pieces of disused timber, along with several sheets of corrugated iron, I have to say that I was somewhat impressed.

Next on the list was 'the pen', where Den keeps his two goats, several waddling ducks, and a couple of very noisy geese. Behind the pen Den had also (single-handedly, apparently) built a storage shed-cum-workshop, which was nearly twice the size of his beloved shack. Made from a stack of wooden fork-lift pallets, along with dozens of sawn-off logs, and a few panes of glass, this huge hut had a unique decorative feature which I had never seen before. Sticking out of one of the side walls were hundreds of different sized jars, along with various shaped wine and beer bottles, which Den had somehow cemented into the wood "To give the place a touch of class", Den boasted. With my grand tour of the place now at an end, Den and I wandered over to the barn, where Ted was busy working away -and getting even blacker than he was yesterday. Before we left the dome in Elk, Ted had given me an autographed copy of Jupiter's Travels, and so I asked Den to take a photo of Ted handing me over the book, as a kind of confirmation of our joint venture together.

Tomorrow morning, before I leave, I will ask Ted to write a short message inside the front cover, not only as proof of my visit to his farm in Covelo, but also to confirm his approval of my proposed production. By now the rains had returned with a vengeance, and so I decided to go back to the farmhouse, and work on the script for the rest of the afternoon, leaving my friend, Den, to sort out the cable problems next week, by which time the weather will hopefully be a little more conducive. Inside the house Shauna was busying herself in the kitchen, taking it in turns to feed Aaron his dinner, and to prepare our meal for this evening. By five o'clock Ted was back in the house with the rest of us, saying that he had had more than enough of the barn "At least for this weekend" he added, before heading straight for the bathroom. Ronaldo had probably taken down half of the rainforest in North America by now (*okay, maybe I slightly exaggerated there, but looking at the huge woodpile outside the window, I was sure that they would have enough logs for the entire winter period*) and so he was ready to quit working as well.

By six o'clock the five of us had settled-in comfortably in the living room, and now Ted was giving us all a slide-show of even more pictures from his epic journey. This included several shots which he had taken of The Palace of the Maharajah of Baroda, (where Ted had actually spent the night), The Taj Mahal in India, a selection of snaps from remote African villages -and so much more. It had undoubtedly been the film show of a lifetime, and now we were all ready to indulge ourselves in "The meal to end-all meals" —well, according to Shauna anyway. The spicy chicken wrapped in tortilla and garnished with avocado and a hot-peppered dressing, was absolutely magnificent, and I know that it delighted everyone's palate -especially mine. Our communal jug of water was constantly being refilled by Ronaldo, as each mouthful of food warranted at least half a glass of this *precious coolant*, not only to help wash the grub down, but also to stop one's tonsils from catching on fire -or someone's wind-pipe from burning-out!

It was great being with my *new-found-friends*, celebrating, laughing and joking with the guys, as if I had known them all for years, and yet it was also quite sad for me, knowing that this was in essence a 'farewell send-off', and not knowing whether I would ever see any of them again, was really tugging at my heart-strings. Ted says that when he has completed the rebuilding of his barn, he is going to invite all of his friends around for one massive *barn-dance*, before using the place to store his annual hay crop once again. 'Boy, would I love to attend that bash', I thought to myself, hopefully assuming that Ted was including me in that last statement of course? As the night wore on Aaron became increasingly impatient with the *unruly mob* that surrounded him, and so Ronaldo and Shauna took their leave of us, hoping to get Aaron to sleep as soon as possible.

As soon as they had gone to bed Ted and I cleaned-up the mess that we had all made between us, before drifting back into my room for a nightcap, and our final drink together. For the first time in a week we both spoke frankly and openly to each-other about families, work, and life in general, as opposed to waffling-on about bikes, journeys and movies all the time. And then it was time for us to take to our beds. During the night I was forced to use the bathroom, which is situated at the end of a short corridor, but upon my return to my room I can see through the small window above Ted's bedroom door that the light is on in his room. Walking up close to the door, I gently whisper "Are you awake Ted?" to whit I am told to "Get some sleep" by my host, who was obviously *not amused* at being disturbed. Even though it is still pitch black outside, I decide to check the time on my watch -it is 5.30am. As I intend making tracks the minute it gets light, I decided to pack up all of my belongings and pile them into the boot of the car, in readiness for the off. With that job done I then went into the kitchen to make myself a cup of coffee.

I then gently tapped on Ted's door, asking him if he would like a cuppa, but he graciously refuses my offer, asking me instead to bring him the copy of Jupiter's Travels, which he had given me earlier on in the week, which I am happy to do, of course. By the time I have finished my drink,

Shauna and Ronaldo are up for the day, and so I say my farewells to the two of them, before asking Shauna to give Aaron a goodbye kiss for me when he awakes. As the sky begins to brighten, Ted wanders into the living room, and as he hands me his book (for the third time in as many days) I say an emotional farewell to an extraordinary man, a caring father —and most of all, someone who I am proud to call 'friend'. Giving one last look at the place which I have called 'home' for the last three nights, I take my leave of the farm, and as soon as I am out of sight from everyone, I pull over to the side of the road, switch off the engine, and proceed to read Ted's inscription in the book. It says as follows:

Dear Shaun, your idea of matching my story with the songs of Meat Loaf seems promising, as the basis for a movie project. I hope you can pursue it, and I will give you what help I can in bringing it to fruition, provided only that I retain control of the use of my own copyrighted material. It's been a pleasure having your company and I look forward to hearing more news.

Good luck and best wishes.
Ted Simon. (Ted actually signed his name, of course).

The clock in the dash reads 8.06am, and yet the 'correct' time on my watch says 7.06am, which rather confused me, as I made one final check on the map for my directions to Los Angeles. Starting the car up once again, I take a quick glance at the fuel gauge, before pressing 'play' on the radio, and then I am off. I am already one hour ahead of schedule (according to my watch anyway) as I had intended taking off at around eight o'clock this morning, but I already know from previous experience that something will crop up along the way to lose me that hour. The fog is very dense as I pass through the mountains, making my way out of Covelo, but my day has just been enhanced, as the newscaster on the 7.30am news announces that Terry Waite has just been freed. "What a great way to start the day" I said with a smile —and just as I did so, the sun showed its face for the first time —as if announcing its approval too.

CHAPTER 7

The Long Journey South

By eight o'clock the clouds have all disappeared, and it looks like I am in for clear blue skies for the rest of my journey south. Within no time at all I have reached the town of Cloverdale, and so I decide to call into one of the roadside cafeterias for a spot of breakfast. (*There goes that hour I was talking about earlier. However, as I only intend stopping once more for fuel on the way, I am quite happy with the way things are going so far*). By ten o'clock I am heading south once more and it is still morning-time, as I cross over the Golden Gate Bridge back into San Francisco, only now I have to pay a $3.00 toll. Adhering to Ted's instructions, I find myself travelling along this massive five-lane freeway, and every lane is as packed with as much traffic as your average three lane motorway would be back in the UK. However, no matter how many vehicles are surrounding me, they still cannot hinder the beautiful scenery and fantastic landscapes which I can see in every direction, and every five minutes I just want to stop the car and take a few more photographs for the album.

At San Jose I fill the car to the brim for the second time. If I keep to a steady 55mph, a full tank should be good for around 400 miles, or maybe even a little more if I am lucky? (*With fuel costing around one dollar a gallon, it works out that each fill-up is costing me less than £10, which is almost unbelievable, compared to what a tank of gas would cost me back in the UK*). The freeways in America are so vast, and the lanes are so wide, that driving on them is an absolute pleasure, rather than being a chore, and because the majority of the road surfaces are made up of solid concrete (instead of the ubiquitous Tarmac, which we see on nearly all of the motorways back in the UK) it also means that there is considerably less maintenance required, and therefore far less roadwork's are going on all of the time. A Honda Gold Wing 'Executive' has just glided past my window, the driver of which is relaxing back in his great *armchair* without a care in the world it seems -oh how I envy him. *(Right there and then I promised myself that one day I would do 'Route 66' on a Harley-Davidson motor-cycle).*

Having listened to dozens of barmy advertisements, all bellowing-out at me from the speakers in the dashboard, I have decided that enough is enough, and so I have reverted once again to serenading myself with my *Love Songs* album on the Walkman. Relaxing back in my chair, with my right hand holding onto the steering wheel and my left arm resting on the window ledge, I interlock my mind and my body as one, systematically entrancing them both into a world of tranquil bliss. Singing along with the music, but thankfully being unable to hear my own *chronic droning*, I give the usual sporadic glance in my rear-view mirror, expecting to see nothing more than hundreds of cars lining the freeway, but to my great surprise I am temporarily blinded by a myriad of flashing lights, which seem to be heading towards me at a rate of knots. Immediately realising that it is a police car on my tail, as both the blue and the red lights are flashing intermittently, I naturally assume that the officer's are in hot-pursuit of some notorious gangster, and so I pull over from the fourth lane, into the third lane in order to let them pass.

Suddenly I hear sirens wailing, and now I can see lots of other cars in my rear-view mirror pulling over to one side, in order to let the police car through, but still the officers in the said vehicle have failed to overtake me? A few seconds later and I can see that the police car is now driving directly behind me, its fusion of lights almost blinding my view in the central mirror. The penny has finally dropped; these guys are not after any bandits or bank robbers...they are after me! Not knowing what the hell is going on I cross two more lanes, before finally pulling over onto the hard shoulder up ahead of me. As I sit there waiting for the *cops* to close in on me, my mind is now in a somewhat deranged state, as I start to panic. Any minute now they are going to open-fire on me, riddling the car with bullets and exploding all four tyres, in order to stop me from getting away, I think to myself, whilst biting my nails like there is no tomorrow… maybe there won't be?

By now the whole freeway surrounding me is virtually at a standstill. Tentatively removing the headphones from around my ears, I lift the Walkman off my lap and place it gently on the passenger seat, before sitting absolutely still in my chair, waiting patiently for the police to come and tell me what it is that I have done wrong? The police car has pulled over about a cars' length behind me, and now a figure has appeared in my side-view mirror. A fierce looking dude in an officer's uniform is now making *his* way (very slowly) towards my car, and a second officer is following closely behind *him*. Any minute now I am expecting to be dragged out of the car by two burly policemen, one of them brandishing a Magnum 44 at my head, as he throws me against the side of the mustang and tells me to keep my hands on the roof of the car, while the other guy kicks my legs apart, before frisking me from my armpits to my ankles. (I have definitely watched too many cop movies!)

With this *somewhat frightening* thought in mind I decide to release my seat-belt, just in case I end up getting strangled in the impending scuffle. The officers are now walking around my car (one on either side) and so I use the electric switch to lower the window on the passenger side. As a uniform appears in the passenger's mirror, I lean over towards it, making sure that both of my hands are still wrapped firmly around the steering wheel, and with a look of pure, unadulterated innocence I utter those immortal words: "What seems to be the problem officer?" My question is obviously treated with the contempt that it deserves, as it is left unanswered and I am met with another question instead. "Could I see your driver's licence sir?"

My cowardice is compounded by the fact that this great 'beast' (which I had brought to life in my mind) who was going to tear me limb-from-limb, before dragging me off to spend the rest of my days rotting-away in some grotty prison cell on the island of Alcatraz, only turned out to be a woman -and a very polite and attractive one at that. Unfortunately, my mind has been thrown completely off balance by this rather unexpected pleasure, and so confusion abounds in my brain, as I try to remember where I had put the blessed thing?

"Um, yes, certainly officer; now where did I... oh yes, it's in the boot –I mean the trunk -um, can you please tell me what I did wrong officer?" I burble.

"You cannot wear headphones whilst you are driving in this state sir", said the young lady, pointing to the Walkman on the passenger seat.

"I am so sorry officer", I grovelled quite sheepishly "You see I am from England; I mean Wales -and I didn't know that it was against the law to wear headphones whilst driving...honestly!"

To my utter surprise, my feeble excuse is accepted by the woman, as she folds up her citation booklet, before slipping it back into her breast pocket. "You have a good day sir, and remember 'no using headphones whilst you are driving in this state'", she commanded, before walking around the car to her male colleague, who has been standing by the boot of my car throughout this *minor ordeal.* In my rear-view mirror I can see the pair of them exchanging pleasantries, and I can just imagine her saying to him "Just another dumb-ass tourist!" (*I know now that my fears were unfounded, and that I may have exaggerated a little on how scared I really was at the time, but believe me when I say that until something like this actually happens to an individual, it is impossible to say how one would react, and what kind of mixed-up emotions they will go through until it is all over –please trust me on this*). Obviously the Walkman remained on the passenger seat for the remainder of today's journey.

Unfortunately, this means that I am now back listening to the radio, once again, and because I am covering so much ground, I have to keep adjusting the frequency in order to get it onto the right station...and what a *pain-in-the-ass* that is. By two o'clock I have reached San Ardo, and so with three hundred miles safely tucked under my belt I decide to pull over and give my legs a good stretch. Had it been the middle of summer right now, then I could easily have made it to Irvine before nightfall, but as I only have until 5pm (instead of 10pm) to make the most of the light, I have set my sights on reaching the city of Santa Barbara instead. My next major destination is a place called San Luis Obispo, which I should reach within the next hour or so, and by then I will have crossed the halfway mark to Daniel's place –and by the time I get to Santa Barbara, I will have covered around five hundred miles. The landscape has not changed that much since Frisco, with giant hillsides and huge mountain ranges surrounding me on both sides for best part of the journey, my original idea of following the coastal route all the way to Santa Barbara having taken a nose-dive due to time constraints

Another hundred miles has now been covered, and I am literally chasing the sun before it dips below the mountains in front of me. Half-an-hour later and it has disappeared from the sky, forcing me to put my lights on, as dusk settled into nightfall. Santa Barbara is over forty miles away, and I have been on the road for over ten hours, and so I have decided to take the next exit off the freeway, with a view to finding a hotel, and graciously admitting defeat. Set back into the hillside I can see a huge building on the left hand side of the road, just a few hundred yards up ahead, and a junction road leading straight to it. From a distance it looked quite inviting, but the closer I get to it the more *sinister* it appeared! Following the signs for 'Visitors Entrance' I decide to take a photograph of this extensively lit, and rather oddly shaped structure, and so pulling up at the side of the road I start searching in the boot of the car for my camera, when I suddenly hear a car pulling up behind me, and within seconds a male voice is enquiring "Can I help you sir?"

As I turn to face my inquisitive motorist, my stomach gives another *sickening* turn, as my eyes beheld a truly unwanted sight. Another police car is now yards away from me, and leaning out of the drivers' window, is another police officer. As cool as a cucumber I reply "Yes please, I wonder if you could tell me where I can park my car, as I would like to book a room here for the night". *No answer was his stern reply*, but his look was enough to un-nerve me. The police officer then got out of his car, before slowly walking up to where I was standing, and placing his left arm around my shoulder, he gently swivelled me around to face the building in question, before saying (in a rather cynical tone) "Really sir, and why would you want to spend the night in power-station?" Feeling very stupid, and totally humiliated, I asked him (rather sheepishly) if he could possibly tell me where the nearest hotel was. He said that it was twelve miles back up the road to the nearest town, or thirty miles if I continued heading in the same direction. No-way was I going to backtrack, and so it looked as though I would be making it to Santa Barbara after-all.

Within a couple of minutes I had turned onto the main coastal highway, where ample lighting lined my route for miles, thus making the night driving both easy and pleasurable. By 6.30pm it was all over. I had covered a total of 528 miles today, and so now I desperately needed to relax and unwind in a steaming-hot bath, in order to loosen-up my body, as my muscles felt as though they had set rigid in the driving position. My eyes were also very tired, and my throat felt as dry an *abbo's arm-pit,* probably from all of the dust particles that I had swallowed during my journey southwards, but once again I had made it.

Santa Barbara, which is Spanish for 'Saint Barbara', is the county seat of Santa Barbara County. The city is located in the South of California, in between the Santa Ynez Mountains and the Pacific Ocean. Graced with a Mediterranean climate, Santa Barbara is often referred to as the 'American Riviera'. At the end of the 19th century, oil was discovered at the Summerland Oil Field, and in a very short period of time, numerous derricks for oil drilling were constructed in the Atlantic Ocean, a few miles east of Santa Barbara. This was the very first offshore oil development in the world. Between 1910 and 1922, when silent movies were all the rage, Santa Barbara boasted having the world's largest movie studio, producing well over a thousand films, of which less than ten percent have survived to this day. At 6.44 am on June 29th, 1925 an earthquake measuring 6.3 on the Richter scale struck Santa Barbara, destroying most of the downtown area and causing the collapse of the Sheffield Dam. Due to the early morning timing of the earthquake, only thirteen people died in the disaster, which was the first major earthquake since the 1906 San Francisco quake.

Many of the city's famous buildings were completely rebuilt after the quake, including the 'Santa Barbara County Courthouse', which has often been described as 'The most beautiful public building in the whole of the United States'. During World War II Santa Barbara housed the 'Naval Reserve Centre', as well as the 'Marine Corps Air Station', but neither of these military bases were able to stop a Japanese submarine surfacing offshore on February 23rd, 1942, before firing a volley of shells at the Elmwood Oil Field. During the second half of the 20th century several fires raged across the area of Santa Barbara, including the Coyote Fire of 1964, which burned 67,000 acres of backcountry, and destroyed over one hundred homes, the Sycamore Fire in 1977, which burned-down around 200 homes, and the devastating Painted Cave Fire in 1990, which completely incinerated over five hundred homes. On January 28th, 1969, due to an undersea break, around 100,000 barrels of oil leaked out into the Atlantic, fouling hundreds of miles of the ocean, and blackening the entire coastline from Goleta to Ventura.

On each of my bike journeys I had covered over five hundred miles in one day, but this was my first time in a car. Santa Barbara turned out to be a very clean city, and the place was alive with activity. Neon lights lit up nearly every building in the town centre, as music cascaded all around me, and local commentaries of today's big (American) football game rang out from every bar room. I can also hear the sound of pool balls cracking into each other, followed by jubilant cheers from the soda-swigging spectators, who are obviously too young to join their dads in a little *alcohol indulgence*. This is undoubtedly a holiday place, and I am really looking forward to seeing what else it has to offer. Jules' money is lasting really well, but that doesn't mean to say that I am going to abuse my spending this week.

As I continue driving around the many blocks, in search of any hotel that is quaint, but clean, I notice a sign advertising rooms for only $29.95, and so I pull over to check it out. Following the desk clerk up the wide staircase and through a set of double doors into a very long corridor, with doors every six feet or so apart on either side, I am duly informed that the toilet is at the very end of the hallway…well, aren't they always! However, the room is fine, and so everything that I had neatly loaded into the boot of car this morning is now unceremoniously dragged back out again, before being hauled up the stairway, and then dumped in one corner of the room. Thankfully there is a customer's car-park adjacent to the hotel, and so at least the car should be safe for the night.

A shower was adequate enough to dilute the stiffness from my limbs, and so armed with a fresh set of clean-smelling clothes, I ventured-out into the streets of *Southern California*. Even though I am still in the same American state, the climate is very different in the southern half, compared to what it is up north. Santa Barbara is certainly warmer than San Francisco, and it is very much warmer than Covelo and Elk. There is absolutely no chance of seeing a ground frost here in the morning, and the clear blue skies are also telling me that seeing any overnight rain is also out of the question. One bar is all I will see tonight, simply because of the individual beer prices. Half a pint of lager will set me back two dollars, whereas a three-pint 'pitcher' is only six bucks, and so my choice is obvious. That journey must have taken more out of me than I thought (well either that, or the lager was a lot stronger than it tasted) because for the first time I have managed to sleep until it is light. I am also not feeling jet-lagged or over-tired any more —in fact I reckon that I would have slept for a couple more hours, had I not been woken by the sound of ringing bells!

Being able to see properly out of my window in the *cool light of day* I soon discover why the room was so cheap. Running adjacent to the hotel is the main railway line, which cuts through the heart of the city, and sitting smack-bang outside my room is the main level crossing, enabling drivers to cross the line when the coast is clear. However, whenever a train is approaching the junction, it is suddenly a case of "Wake-up all inhabitants in the immediate vicinity", as half-a-dozen red lights begin flashing like mad, and those pesky bells start ringing for all they are worth. Wearing a full tracksuit to walk about two hundred yards to the beach was a big mistake, as the temperature is already hovering around the seventy degree mark, and it is only nine o'clock in the morning. Realising my error of judgement, I returned post-haste to my room, only to reappear ten minutes later suitably attired for the weather conditions. Thankfully, I have until noon to check out of my room, and so I intend making full use of the glorious sunshine until the very stroke of midday.

The huge sandy beach is broken-up by a wide concrete pathway, which runs the length of it. The pathway is also divided by a yellow line, which runs through the centre of the concrete, and on one side of the line are dozens of white arrows pointing to the opposite half of the walkway, which I have to admit confused me somewhat? However, it wasn't long before I found-out the reason for the arrows —the hard way! Ambling-along with not a care in the world, graciously admiring the young and the not-so-young, as they come jogging past me in their shorts and vests, I smile and say a polite "Good morning" to each person in turn. Suddenly I am forced onto the sand by a hefty shoulder-barge which I haven't seen coming, and for a moment I am in shock. A guy on a bicycle is now wobbling down the pathway in front of me and he is shouting at me to "Read the arrows buddy!" and only then did I realise that I had been walking in the 'cycle-lane'—oops!

I want to apologise to the guy, but he has long-since gone by now, and so I decide to treat this little incident as one of life's great learning-curves. Adjacent to the beach is a beautiful harbour which is full of luxurious yachts of all shapes and sizes, their decorative masts and colourful sails dominating the skyline for the entire world to see. This place is truly a paradise on earth, and I am humbled by its sheer magnificence. Palm trees line the promenade, and a circular fountain containing two dolphin statues sits as a centrepiece at the main entrance to the beach. Surprisingly, there are no deckchairs or umbrellas in sight; no body-builders or surfers *doing their funky-stuff*—and no volley-ball games being played on the sand, but the answer to all of my questions is a very simple one. It is now November, and so the summer season has long-since passed —well, in a Californian's eyes it has anyway. However, to a *stone-cold Brit'* it is the middle of July, and so if I want to strip-off and have a sun-bathe then I darn-well will.

After two hours of rest and recuperation, firstly on the golden sand, and then on the finely manicured lawn which encompassed the opposite end of the beach, leading to the harbour, I returned to the hotel, to recheck my route to Irvine. According to my calculations it is only about one hundred miles away, and so two hours, more or less, is all I should need to get there. Daniel has told me to ring him when I get to the outskirts of the city, and he will come and meet me, but as both he and his wife work full time, I do not want to drag him out of his workplace in the middle of the afternoon, and so there is no point in me arriving before tea-time, as no-one will be at home to answer the telephone. (*We didn't possess mobile phones back in the late eighties / early nineties*). I can then follow Daniel's or Zara's car, in my car, back to their home. I have never met Daniel's wife, Zara, or their daughter, Darcy, beforehand, and so I am really looking forward to seeing them for the very first time.

Filling the car to the brim with fuel once more, I begin to make my way through the traffic, when suddenly I spot a 'one-hour' photo processing laboratory on one corner of the street, at the set of traffic lights which I am currently approaching. I am more than anxious to know whether or not the first film I had used had worked okay, and I also wanted to see if the second film, which has now been fully used-up, and which is currently sitting in the camera, has also worked correctly, as both films had been purchased by me several years ago. Turning right at the lights I pull up at the side of the road, only to find an empty police car parked-up on the opposite side of the street, with three burly police officers standing adjacent to it. Not wishing to be arrested for *kerb-crawling* or *jay-walking* (whatever that is)

I decide to confront *them* before they confront *me*! Walking boldly over to the guys, I politely ask them if it is okay for me to leave the car where I have currently parked it for an hour or so, before telling them openly that I simply wish to have some photos developed.

"Sure thing my friend, so long as the warden doesn't catch ya –hell, even us three are afraid of her! If I was you I'd park on the next block buddy; your car will be fine there for an hour", said one of the officers, in a kind of patronising, but friendly tone.

Thanking him kindly for his advice, I immediately turned around and began walking back to my car, when one of the officers suddenly asks me where I am from? Turning back to face them I nonchalantly replied "Wales –but then I don't suppose that any of you gentlemen have ever heard of the place?"

"Hell I know Wales –that's where good-old Tom Jones is from –right?" replied the officer who was standing in the middle of the other two.

"That's correct, my friend, and so I daresay you will also have heard of Princess Diana –she is the Princess of Wales, even though she is actually English, of course" I replied.

"Hell, everyone knows Princess Diana –she's a wonderful lady", said a second officer...with true sincerity, I might add.

Although these guys are generally friendly, and openly joke around in front of me, I am given the impression that once they are out of uniform, all three of them would have more fun causing trouble, rather than *upholding the law!*

"Thanks for the advice gentlemen", I say, before getting back into my car and driving off down the road...about one-hundred yards in total.

It is now twelve thirty, and as the photos will not be ready until one-thirty, it means that I now have one hour in which to enjoy a spot of lunch. Directly opposite the photo laboratory is a small cafeteria, its front terrace area encompassing half-a-dozen sets of tables and chairs, and so, after stripping-down to my waist, in order to make the most of the afternoon sunshine, I *moseyed-on over* to see what delights they had on their menu? For the first time on my travels in California I am confronted by an unfriendly waitress, who has obviously gone AWOL (absent without leave) from the local cattle-drive.

"Please put your shirt back on sir", she pouts, before *yanking it* (no pun intended) off the back of my chair and unceremoniously dumping it in my lap. "And can you please move your camera off the table sir, in order that I can lay out the cutlery properly", she added, before grabbing my camera case and handing it over to me...without so much as a smile!

"That *time of the month* is it?" I say under my breath, as she storms-off, my food and drinks order having been *reluctantly scribbled* on her little pad.

Thankfully, the excellent food easily made up for my new friends' attitude problem, and I even ended-up having the last laugh on her –quite by accident actually. As my watch struck one-thirty, I decided to pop across to the shop, in order to collect my photos, before returning to the cafe to enjoy a second coffee, whilst having a good look through the pictures. However, when I came out of the shop I could see this poor waitress, who was now running up and down the street looking for the culprit who had *done a runner* from her cafe without paying his bill -oops! Unfortunately I was right about the first film; it had not been correctly inserted into the camera, and therefore the negatives came out completely blank...shit! Thankfully the second film gave me thirty-seven photos, and a thousand memories of the most incredible week of my life to date. It was now two o'clock, and so I really must be on my way.

Within an hour I am passing-through Malibu, and the sign at the side of the road says "Malibu -27 miles of scenic beauty". Malibu, which is renowned for being home to the rich and famous, particularly those in the entertainment industry, such as movie stars, television celebrities and musical icons, also boasts having a wonderful Mediterranean climate. Add to this a twenty one mile coastline, which incorporates a plethora of beaches, including Malibu Beach itself, along with Zuma Beach, Point Dume Beach, Country Line, Topanga Beach, Surf-rider Beach (nicknamed 'The Bu' by locals and surfers alike) and Dan Blocker Beach, which was named after the famous actor, who played *Hoss Cartwright* in the iconic series 'Bonanza', and it is no wonder that thousands of tourists fly in from all over the world to visit this haven of sub-tropical beauty. Apart from its glorious beaches and endless sunshine, Malibu is also famous for its potteries, Malibu tiles having graced many of the mansions in nearby Beverly Hills hence they are a much sought-after commodity.

Beverly Hills, which is often referred to as '90210' -one of its major zip codes, was originally a Spanish ranch that harvested Lima beans, and it was actually named after 'Beverly Farms', in Beverly, Massachusetts, the word 'Hills' being added simply because it was located in the *Hollywood* hills. Ironically, it was a group of investors that had initially come looking for oil, but found water instead, that decided to develop the area. At the time of its inception, Beverly Hills was built as an 'all-white' community, with the sale or rental of its properties to black or even Jewish people being strictly forbidden. However, by the early 1940's hoards of black actors and businessmen began moving into in the area. In the decades that followed hundreds of famous people would buy properties in Beverly Hills, including Tony Curtis, Ray Charles, Frank Sinatra, Dean Martin –and even the king himself, Elvis Presley. 'Rodeo Drive' is the most famous shopping strip in Beverly Hills.

Oh how I would love to spend a few days just wandering around some of their fabulous mansions, looking at all of their magnificent rooms and sumptuous pools, although it is highly unlikely that I ever will? However, I have no time for day-dreaming, as it is now three o'clock, and I still have one or two places left to see before heading into Irvine.

Santa Monica Beach is my next stop for a meander on the sand, in order to give my *little* legs yet-another *big* stretch. This amazing beach, which is three and a half miles long, incorporates a selection of parks, picnic areas and playgrounds along its golden sandy shores, along with volley ball courts, the original 'Muscle Beach' area and separate walking and cycling paths. Lifeguards are almost always in attendance to watch over the many swimmers, surfers and paddle-boarding enthusiasts who make good use of the incredible waves that the Pacific Ocean regularly delivers along this glorious coastline. The beach has always been very popular with the locals, along with tourists from all over America, but it gained worldwide recognition when the American television series 'Baywatch', starring David Hasselhoff and Pamela Anderson, was filmed on this very beach between 1989 and 1999.

Just along the way is the famous Santa Monica Pier, which was erected in 1909 and boasts a National Historic Landmark –the 1922 Looff Hippodrome Carousel. Just a few steps down from the pier, is the International Chess Park, which incorporates a human-scale chessboard along its sidewalk. Around the turn of the twentieth century, Santa Monica Beach was one of the few places where blacks were allowed to take part in seaside activities, albeit that they were confined to a 200 feet roped-off area, which was commonly known as the 'Ink Well' or 'Negro Beach'.

Gazing out into the Pacific Ocean it seems quite surreal to me that I am now walking on the same beach that Caryl had walked on over a decade ago, long before we had ever met –and several years before I had even considered visiting America. Now I was here, but without Caryl –and so I swore right there and then that one day we would return here together. Los Angeles is soon within my sights, although I will not be heading into the heart of the city, only veering around it, in order to join the '405' highway to Irvine. As the exit lane for L.A. draws near I am able to see the infamous *smog* looming over the whole of the city, like a giant shadow, engulfing the skyscrapers and blocking-out the daylight from the people below. I know that rush-hour will soon be here, and so I want to be well clear of Los Angeles by then. The news announcer is churning-out warnings of accidents everywhere on the freeways, and so I can only sit back and pray that none of them have occurred along my chosen route.

Three car pile-ups, motorcycle crashes, and even an upturned trash-cart are causing mayhem on the roads, but as yet it has not affected the flow of traffic that I am driving in right now, thank goodness. Turning off at the next junction and following the signs for the '405' turned out to be my downfall –and boy was I going to pay for leaving things until the last minute. The whole of the southbound carriageway, as far as the eye can see, is nose-to-tail with traffic, of which I am but a minute part of it. However, the traffic in the outside lane is moving at a decent pace, but sadly I am not allowed to slip into it, because it is a 'car-pool' lane. To make use of this *special* lane there must be at least two people in the car, otherwise a fine of $250 will be imposed by the police on any solo drivers, who have no passengers in their car. (*Trying to cheat the system is no good, because there are video cameras placed at strategic points along the freeway, and so even if there are no cop-cars around the police will eventually catch up with the perpetrators*).

The 'pool-car' system was primarily designed to encourage commuters to share vehicles, thus cutting down the output of carbon monoxide fumes that are released from the exhaust pipes into the atmosphere, which effectively causes the smog in the first place. Signs are dotted all along the freeway, giving telephone numbers to ring if one wishes to take up the opportunity of saving on fuel bills, along with making new friends, and also cutting-out the traffic queues, of course. I imagine that this idea has boosted trade for the taxi drivers, because anyone who is in a desperate hurry to get home for some reason, and who can afford to leave their car in work for a day, will simply jump in a taxi and fly down the pool lane. Judging by the amount of taxis that flew past me during the hour I was stuck in this blessed traffic jam, there must be plenty of 'city-slickers' out there. (*However, I have it on good authority that the system doesn't work all that well, and so I suppose it is a case of 'back to the proverbial drawing board' for L.A.'s environmental and transport ministers*).

The waiting is tedious, as I go from dead slow to stop, then back up to 25mph, before having a thirty second burst at 55mph –and then it's all back on stop again. Suddenly I see a sign for 'Irvine City Limits', and so as the traffic starts moving away again I begin edging my way across three lanes of traffic, whilst simultaneously dodging the suicidal maniacs, who are obviously hell-bent on not letting me through. The area of Irvine, which was officially named in 1914, after the Irvine family, who were agricultural pioneers and prominent landowners back in the 19th century, is located within the boundaries of Orange County. The indigenous people originally lived in the area over two thousand years ago, until a Spanish explorer arrived in 1769, subsequently claiming the land for Spain. After Mexico's independence from Spain in 1821, the Mexican government took control of the land, before distributing chunks of it to Mexican citizens who applied for grants. The land was then ceded to the United States in 1848, after the Mexican-American War.

Development of the sixty-six square mile city did not begin until the 1960's, and on 28th December 1972 the city was formally incorporated into the metropolitan area of Los Angeles, California. Irvine's layout comprises of townships, or 'villages', as they are more affectionately known, which are segregated into long streets, with wide roads that encompass three lanes of traffic in either direction. The majority of the houses are of similar design, and each community includes a selection of schools, commercial centres and religious institutions. The landscaping of the areas includes bicycle corridors, parks and greenbelt areas, the greenery being irrigated with reclaimed water. The more affluent villages also offer additional amenities, such as 'members only' swimming pools and tennis courts. After the fall of Saigon (Ho Chi Minh City) in 1975, and during the course of the next decade, large numbers of Vietnamese refugees began settling in nearby 'Fountain Village', thus vastly increasing the amount of American Asians now living in the city.

The main problem with American freeways is that there is no such thing as *fast lanes, slow lanes,* or *over-taking lanes*; one simply drives up behind the car in front of them, before passing it in the first lane that becomes available...period. If a driver in America flashed his (or her) lights to the car in front of them (as drivers do in the UK when they want the car to move over into a slower lane) it wouldn't mean anything to the driver in front, and so he (or she) would probably just ignore it altogether?

Well I certainly did not expect to be driving in the dark again today, but at least I was near to my destination, and I also had a good chance that either Daniel or Zara would be at home, as it was now five o'clock. Pulling into the first petrol station I see, I ask for directions to the Rockwood Estate, but to my shock, horror, neither of the women who are working behind the counter has ever heard of the place –great! My only recourse now is to use the garage payphone to ring Daniel's home, and hope that someone is at the house? Remembering to omit the area code (as instructed by Daniel) I dialled the number, to whit I am greeted with a ringing tone almost immediately –fingers crossed. After three rings the phone is answered –but my heart then drops as I am met with an answer-phone message, apologising for nobody being at home –bollocks! Suddenly the message is interrupted, and a second later I can hear a female voice asking me "Is that you Shaun, it's Zara; can you hear me okay –if you're in Irvine tell me exactly where you are and I will come and collect you".

The attendant has told me that the name of the area which I am now standing in is called 'Michelson' and so I pass this on to Zara, adding that I am currently parked in a Chevron Petrol Station, which is adjacent to a large shopping mall. Thankfully, Zara knows exactly where I am, and so she tells me to stay put, adding that she will be there within the next ten minutes. At last my long journey south was at an end...well for now anyway.

CHAPTER 8

Rockwood, Irvine, Orange County

Ten minutes later, just as promised, Zara turned-up, and accompanying her was her lovely daughter, Darcy. Donning a mane of very long black hair, I can see through the shadows that Zara is a very attractive woman, and as for her daughter, well she is simply *edible*. Zara is currently five months pregnant with another little girl on the way. As Zara's occupation involves working with toxic chemicals, she had to undergo stringent tests to confirm the safety of her baby, and one of those tests revealed that she was carrying a little girl, which she was happy to be told in advance of the birth. (*I have to confess that as much as I love my two little boys to bits I am secretly praying that Caryl and I become the proud parents of a little girl when our new baby is born at the end of next month*). Upon reaching the house, Zara informs me that my brother, Gary, had rung them earlier-on in the week, just to see how I was getting on, and to find out what had happened with Ted Simon, as they thought that I would have left his place about a week ago....but little did they know?

Daniel has just arrived at the house, and I have to say that he looks a little different to how I remember him. (*Mind you, it has been three years since I last saw Daniel, when he came to Cardiff on business, and so he is a little older, of course...and my memory for faces has never been that good anyway*). "You don't recognise me, do you Shaun", Daniel suddenly blurted out, as if he had been reading my mind. "That's because the last time we met I was wearing glasses, whereas I only use contact lenses nowadays", he added. His outburst somewhat alleviated the small amount of guilt that

78

I was currently feeling inside. Thankfully, Daniel's personality had not changed one iota, and I knew that underneath that immaculately dressed, clean-shaven exterior, lurked an outspoken, hilariously funny, and totally wild human being, whom my friends and family had come to know and love in the short space of time that he had spent with us in *God's country* (Wales). Daniel's immediate task is to stoke up the barbeque, while Zara began preparing a sumptuous salad–and Darcy spent her time playing host to *yours truly*.

After filling-in each other with all that's been going on, on both sides of the Atlantic, which included swopping several family photographs, of course, I then told Daniel and Zara all about my project, along with how successful my meeting with Ted Simon had been. Both of them were enthralled with what I was doing, wishing me all the very best of luck for the future –'luck' being the operative word, as it was something that I was going to need a lot of –along with plenty of tenacity and an endless determination if I was going to make this thing succeed. Daniel then said that he and Zara were planning to visit Scotland in the New Year, and that they were also hoping to call in on me at some stage during their holiday, as Daniel really wanted Zara to see the beautiful city of Cardiff (my home town) and for her to meet the rest of my family, which he felt that he was already a part of...how nice is that?

As the clock struck 8pm, both Zara and Darcy headed off to bed, as the pair of them was shattered, and they would both be having an early start in the morning. At this point I brought out two cans of Budweiser which I had brought with me from Covelo, and after adding a couple of ice cubes from out of Daniel's freezer-box, the pair of us sat watching 'The outlaw, Josie Wales', starring Clint Eastwood, whilst *chewing the fat* over life in general, along with what the pair of us would be getting up to tomorrow evening. The sofa we had been sitting on all evening turned out to be a sofa-bed, and so that was where I would be spending the night. However, my thoughts of having a decent nights' sleep were about to be shattered, because standing adjacent to the television set is a very large cage, and in that *very large* cage sits a fully-grown, multi-coloured parrot, whose name is 'Baron'. Baron had an absolute *blue-fit* when Daniel pulled out the cantilever bed, as the clanging and clattering really frightened him, and so he obviously decided to wreak his revenge on me by squawking mercilessly for half of the night, before giving me an early morning call. By this time I truly understood the meaning of the term 'Early-bird'.

Both Daniel and Zara live very full lives, and so they have a set pattern every day which involves perfect organising and acute timing. Starting the day at 6.30am Zara is first up, showering and preparing herself for the day. At 7am she wakes Daniel, so that he can follow suit, while she is preparing breakfast, along with Darcy's clothes for the day. Darcy is woken by her mum at 7.30am, and that is when everyone sits down to breakfast, which usually consists of a bowl of cereals each, followed by pancakes, which have been smothered in syrup. I passed on the pancakes, as they were simply too heavy for my stomach at this time of the morning, preferring to settle for a bowl of good-old *crunchy* corn flakes, along with a glass of orange-juice. (*Daniel and Zara do not drink tea or coffee, which was a major bind for me, as I really need a good caffeine fix if I am going to start my day on the right note*).

Zara is first to leave the house, along with Darcy, who will be taken to Zara's mums for the day, as their daughter won't be starting school for another year, and so *grandma* has to play babysitter until that time arrives. Zara will be taking a few months off work after the new baby is born, but she plans on returning to work once everything is back to normal, and a new routine for all parties concerned has been fathomed-out. As for today, well, after dropping-off her darling daughter, Zara will then continue-on to her place of work, where, after an eight hour shift, she will do everything in reverse. Daniel is the lucky one, as his workplace is only a few miles from his home, although for the next few days he will be working from a different office, which is situated over on the far side of Orange County, and so he will be leaving somewhat earlier than his normal time. By 8.15 there is only Baron and I left in the house, or so I believe, until I accidentally discovered two small birds in another, much smaller cage, by the window, which must have been hidden by the curtains last night.

By nine o'clock I am literally gagging for a cup of coffee and a round of toast, but there are no caffeine-based beverages to be found anywhere in the cupboards —and as for an electric toaster, well lord-knows where that might be hiding, if they even have one at all? After twenty minutes of searching I finally *give up the ghost*, realising that it will be much easier to find a cafe somewhere, where I can obtain the required items. (*It will also save on the washing-up, which is a bonus, as I have much better ways of spending my very valuable time in this glorious autumn climate*). Directly across the road from Daniel and Zara's house is the local community leisure park, which not only boasts the usual array of swings, slides and climbing frames for the children, but it also has a huge swimming-pool, complete with a separate Jacuzzi...paradise found! Daniel has already given me his personal key (all of the local people have identical keys) which not only enables me to access the park, but it can also be used to turn on the outdoor hot-tub, thus warming the water up to a temperature of 104 degrees Fahrenheit –whoopee-doo!

Remembering to bring the house key with me, I begin searching around the area for any signs of a cafeteria, or even a food van, where I can get some sustenance, but there is absolutely nothing within walking distance. (*It was only after doing this that I appreciated the enormity of American cities, and why the car is such an important commodity to the people –indeed, it is a necessity*). With this thought in mind I set off in the car, feeling quite confident that I will become completely conversant with the surrounding area after a few miles of driving, and I will also be able to find my way back to the starting point without any difficulty whatsoever. However, I was about to find out that it is a lot easier finding ones way *into* a city than getting to know ones way around it! As soon as I had left the Rockwood Estate and joined the main road running through Irvine, my bearings deteriorated rapidly, and so it wasn't long before I became hopelessly lost. Every turning I took seemed to lead into another identical estate, and there were still no signs of any restaurants, newsagents, or even a blessed petrol station, where I could buy some refreshments, and, more-importantly, ask for directions back to the Rockwood Estate?

After half-an hour I end-up in what appears to be a massive business park. Tower blocks, warehouses and plush, glass-fronted buildings are in abundance, but still there is not a soul in sight, who could come to my aid? My only course of action now is to keep driving straight, and not to turn off, for fear of going around in complete circles. I cannot even see the end of the road, but I carry on regardless. Passing an airport on my right hand side, I suddenly notice a large Beefeater-type restaurant, set in its own grounds just up ahead, and so I quickly swerve into the left-hand lane, thus enabling me to turn into its car park at the next junction. Unfortunately the restaurant doesn't open until eleven o'clock, and it is now only 9.30am, but luckily one of the waitresses has come in early to prepare the tables, and so she happily gives me directions to 'Co-co's Diner, which, she says, will be open by now, and is only about a mile up the road. Apparently it is mainly used by people who frequent the airport.

It has taken me the best part of an hour to get to this place, and so I wonder how long it is going to take me to get back home again? Putting this thought right to the back of my mind, I immediately devoured a very enjoyable breakfast of ham and eggs on toast, washed down with two cups of percolated coffee. One of the bartenders at Co-Co's Diner has advised me to follow the main road around the coast, before spending the day at Newport Beach. However, I have set my heart on a dip in that wonderful swimming pool, and once I have made up my mind to do something, there is no going back. All I have to do now is find my way back to the blessed park? Every single street is miles long, and the roads are the width of an average British dual-carriageway. They also look identical to one-another and each road has dozens of turn-offs on either side, and so where the hell does one start? Road signs showing place names are virtually non-existent around this *neck-of-the woods*, and it also took me ages before I realised that I had been reading the *major road* and *minor road* signs the wrong way around, and so even if I had possessed a street-map of this place (which I didn't) I probably would have ended up *disappearing up my own arse?* I might add that my sense of direction also leaves a lot to be desired.

Again and again I get honked at, as I try to figure out which way to turn at junctions, but neither way looks familiar to me, and so I try to trust to luck, but it is all bad. Another hour in the car soon passes, with little success, and now sheer frustration has gotten the better of me, as I begin to boil in anger. I wasn't born with a great deal of patience for situations such as these, and my tolerance levels had long-since been surpassed. A poor, unsuspecting young lady is walking along the pavement just up-ahead of me, unaware that she is about to burden the brunt of my anguish, as I pull up alongside her. "Excuse me", I yell out of my driver's window, causing her to jump out of her skin with fright. "Can you help me please, you see I am looking for the Rockwood Estate, which I know lies off Main Street, which I have driven along several times in the last hour, or so, and so I know that I am close to it...aren't I --after-all, there cannot be another

Rockwood Estate, or more than one 'Main Street' in Irvine, surely to God"
I pleaded somewhat aggressively.

My *questionable* tone had obviously been treated with the contempt
that it deserved, as the young lady retorted her answer in an equally abrupt
fashion:

"My good man, this state has twenty-nine cities altogether, of which
every one of them has their own *Main Street*. Secondly, you are now in
Orange County –Irvine is about five miles north of here hun', and so I
humbly suggest that you turn your car around and get into the correct
area, before you even contemplate doing anything else!"

As the woman turned her back on me, before storming off in a huff,
I knew that I had been well and truly *shot-down in flames*, and deservedly
so. Unfortunately, this was *the straw that had broken the camels' back,* and
I now knew that if I intended getting back to the house in daylight hours
then I would have no option but to ring Daniel on his works' telephone
number, and ask him to give me directions back to the Rockwood Estate.
Finding a phone box took me another ten minutes, but finally I spotted
one outside the entrance to a food-store, and so I immediately pulled over,
jumping out of the car and running across the pavement to the said booth,
before anyone else had the chance to use it. After checking the location
address of the phone box, I rang through to the main switchboard of
Daniel's works, before asking the operator to put me through to his office.

"I'm sorry sir, but Daniel went to lunch about five minutes ago, saying
that he would be out of the office for a couple of hours...can I take a
message for him?"

There was obviously no point in leaving a message, and so I simply
asked the lady if she could help me with my dilemma, but unfortunately
she had also never heard of the Rockwood Estate? Profanities in the form
of a dozen curses are said under my breath, as I slump down to the floor,
letting the midday sun beam through the glass panel and roast me alive.
I am now feeling so desperate it is unreal, and I really don't know what to
do next? I have lost all of my confidence in finding the house again under
my own steam, and for the umpteenth time I curse myself for ever leaving
the estate in the first place. Suddenly the sun's rays are no longer shining
down upon me, and a shadow has now been cast over the top half of my

body, and so I look up, only to see a man, who is probably in his late fifties, early sixties, standing directly over me.

"Excuse me sir, but I happened to overhear your conversation over the telephone, and so I would like to help you get back to Irvine my friend, although I haven't a clue where the Rockwood Estate is, I must confess?"

Any help at this moment in time was more than welcome, and so I began writing down the instructions on my note pad, which luckily I had brought with me this morning, just in case I wanted to take down any notes. *Take the 55 freeway south until I reach Interstate Highway 5, and then follow the road for about three miles...and so on.* Unfortunately there are several temporary detours along the route, and road closures are in abundance, thanks to various maintenance works that are currently in progress, but after another half-an-hour of going around in proverbial circles I finally find my way back to Irvine...at last. My first port-of-call is the nearest petrol station, where I buy a detailed road map of the area, which thankfully includes the most-recently built 'Rockwood Estate' -which is probably why no-one has ever heard of the place?

By one-fifteen my troubles are over, and the longest four hours of my Californian jaunt have paled into nothing-more than a historic (and also quite humorous, now that I am able to look back on it) story for me to recall in my memoirs. With the sun still blazing, and the swimming pool looking more inviting than ever, I immediately strip down to my swimming trunks, before diving head first into the cool (but certainly not cold) crystal-clear water. Next in line was to try out the springboard, which was so springy that it catapulted me high into the air, my body almost somersaulting with the momentum of the final *bounce*. The Jacuzzi was next in line, as I baked my body in boiling bubbles for a good half-an-hour, before prostrating my wrinkled carcase on one of the many available sun-beds. (*Actually I was the only person in the park, and so I guess you could say that I was spoilt for choice*). "Ah –this is the life" I shouted, not caring for a moment about who might be listening to me, for I had found my utopian paradise –in a community park on the Rockwood Estate. By four o'clock dusk was beginning to settle in, and the air temperature had dropped to around the sixty degree mark, and so after doing a spot of reading on one of the park benches I returned to the house.

I then spent the next hour working on the script for the film, before the family returned home to lay claim to their precious dining-table, which was currently cluttered with all of my *movie paraphernalia*. However, I needn't have bothered clearing up my mess, because Daniel has decided to take us all out for supper this evening at one of his favourite restaurants. I am so glad that Daniel has offered to drive tonight, because the roads had confused me enough during the day-time, let-alone having to traverse them in the dark. (*By the way, everyone, including my good self, found this mornings' little episode absolutely hilarious*). Once again the food is of the Mexican variety, and it is very tasty indeed. It is a lot more expensive than I had expected it to be, but what does that matter when the company you are sharing is nothing less than perfect, and the whole ambience of the place is simply *out-of-this-world*. By eight o'clock we are back at the house, and so after dropping Zara and Darcy off for their early evening ritual, Daniel and I popped over to the local store to buy a few groceries, along-with a crateful of duty-free alcohol.

Stacking my shopping basket with great lumps of ham and cheese, along with a few tins of frankfurter sausages, several packets of biscuits, two loaves of bread, and a tub of full-fat butter, Daniel joked that it was no wonder that I had put on so much weight since the last time he saw me, as everything I had chosen was decidedly bad for ones constitution. I knew that he was right, of course, but as I had no idea of how to use any of his cooking facilities –and I certainly had no intentions of venturing-out into the wilderness, to go in search of a cafe or diner again, I was more than happy to settle for a load of *junk food* over the next few days. (*Living with Daniel and Zara I soon realised how health-conscious the pair of them are, which is how we all should be, of course –and it is certainly a good way to bring up their young daughter. As for me, well Daniel's jovial little quip in the store had made me swear an oath to myself that I would go straight on a strict diet...the minute I returned home*).

Sharing a bottle of Mateus Rose between us soon saw the rest of the evening disappear into oblivion, and so once again it was time for bed. Daniel had already let Baron out of his cage, so that he would not get frightened when we opened-up the sofa-bed again tonight, and so he had flown upstairs to perch on the tailboard of Darcy's bed, which, according to Daniel, is where he always heads for whenever he is let out of his cage at night, and so hopefully I should get a decent nights' sleep. The weatherman on the *morning news* programme has just announced that today's 'high' will be normal for this time of year, averaging around the seventy degree mark on the coast, rising up to eighty degrees inland –now that is what I call 'autumn temperatures'. The sky is grey at present, but according to our friend on the TV, this will have all cleared-up by mid morning, and then everyone can look forward to clear blue skies and warm sunshine for the rest of the day –*whoopee-doo* again. Zara has just left for work, and this evening she will be staying over her mother's house with Darcy –and so Daniel has just informed me that as soon as he gets home from work tonight he and I will be "Painting the town red".

Trying to ignore the amount of calories I have just consumed at the breakfast table, whilst forcing-down a very weak cup of percolated coffee (*a last resort, which I had made myself*) I continue to study 'Jupiter's Travels', along with Ted's follow-up book, 'Riding Home', which also consists of many chapters relating to his amazing journey around the world. My aim is to utilise the most profound and exciting events in both books to produce a fast-moving adventure storyline. However, I still wish to retain Ted's individuality as the main character, by using narration sequences throughout the movie, in order to express each pinnacle of mixed thoughts and emotions as they occurred in reality. The work is never boring, but it can become very frustrating, as I try my uttermost to personally visualise what Ted is actually describing on certain parts of the journey, along with remembering individual photographs from Ted's personal collection, most of which I have only ever had a fleeting glimpse of? Attempting to recapture his feelings, just as if I had been there and experienced the whole journey with Ted would certainly not be an easy thing to do either, but I am sure that Ted will be on hand if I need any assistance from him.

After a couple of hours of day-dreaming my life away, and digging-deep into the inner-sanctums of my own imagination, I inevitably end up with a blistering headache for my efforts, but it is worth it. It took a little longer than expected for that great fireball in the sky to break through the hazy skies this morning, but by eleven o'clock my *sun-worshipping* was once again well underway. Having the whole park to myself again today I took full advantage of all the facilities, which even included having a ride on the kiddies swing for a few minutes, while I *contemplated the universe*. I then proceeded to fall fast asleep on one of the sun-loungers for nearly two hours. Luckily I am naturally dark-skinned, and so those ultraviolet rays did nothing but enhance my already bronzed (from yesterday's little blast) body. Returning back to the house for a mid-afternoon snack, I slid open the patio doors, which lead onto the small terrace area, before placing one of the kitchen chairs in the one and only corner of the garden that was still catching the last of the suns' rays, before it slipped beyond the horizon.

Plonking my backside down on the chair, the top half of my torso having been stripped clean once again, I immediately plugged in the earphones to my Sony Walkman, before popping in my *ABBA's Greatest Hits* album. I then let my body and soul slip away into the land of make-believe...well, something like that anyway. In the few seconds of silence between one track stopping and the next one starting I suddenly hear Daniel's home phone ringing in the background. The phone is hanging on the wall just inside the kitchen, which is virtually within arms' reach...and so I duly answer it. It is just my good friend, Daniel, who is checking-in on me to make sure that everything has gone okay today, and that there has been no *repeat performance* of yesterday's farce to stress me out again? I assure Daniel that I am fine, and so he rounds off our conversation by saying that he will be home in about an hour "And that is when we will begin to party, my friend", he adds in a rather ominous tone.

As soon as I had replaced the handset the blessed phone rang again, only this time the call is coming from a lot further afield...it is Caryl, ringing me to see how the meeting went with Ted Simon –and she also wants to know what I've been doing with myself for the last ten days? I tell her how smooth everything has gone so far, before adding (thoughtlessly) how lovely and warm it is over here –and enlightening her on *what a swell time* I've had with Daniel, Zara and Darcy over the last couple of days.

"Well its midnight over here –and it's bloody freezing", Caryl replied in her best *venomous* voice, before adding that I had better ring Liam in the morning, because he has not been very happy at all since I've been gone. Once again I have managed to open my big mouth –and put my proverbial foot right in it... dumb-ass, or what? My guilty conscience forced me back to work until Daniel finally arrived home, which turned-out to be nearer the six o'clock mark, rather than around five, as he had intimated, but hey-ho.

As soon as Daniel walked through the door *all hell broke loose*, as we prepared ourselves for the evening ahead? As soon as Daniel exited the shower I dived into it, and the second that he stopped spraying his body with various deodorants and antiperspirants, I started doing likewise. While he had a shave in the mirror I got dressed into my cleanest clothes, and while he suitably attired himself, I splashed my face with a handful of my hosts' favourite after-shave. Forty-five minutes later and these pair of *cool-dudes* were ready to hit the town and 'knock-'em dead.' (*If truth be known we were two happily-married men in our thirties, going out for a quiet drink together, but there is nothing wrong with fantasising that we were still young, free and single, even if we weren't teenagers any more —coupled with the fact that neither of us ever wanted to be single again, as we both loved our wives and children more than life-itself*).

Our first port-of-call is an amazing shopping-mall, as I want to buy a few presents for Caryl and the children. However, after walking around the place, Daniel assures me that he can get much better deals at one of the local shopping warehouses. Our next stop is Newport Beach, where I am about to embark on one of Daniel's favourite cocktails, which apparently consists of no less than 'six' different liquors. Unfortunately, Daniel will not be joining me on this occasion, as once-again he is in the driving seat. After ordering an 'Adios mother-f*****' (the name of the cocktail in question) for me, and a bottle of Budweiser for himself, Daniel and I stood and watched the barman smiling all over his face as he prepared this lethal concoction for his latest *victim*! Thankfully, there are several fruit-juices included in the recipe, and so the drink actually tastes very nice, even though it is turquoise in colour. The drink lasted me nearly an hour, partly because of its potency, but mostly because of its price tag. (*Thank goodness Daniel was only drinking Budweiser's, otherwise it could have been a very expensive evening*). Our second bar is packed-to-bursting with youngsters, and so I, for one, was feeling somewhat out of place.

As these *teenage bucks* potted each pool ball in turn, their cheer-leader-type girlfriends clapped and cheered, with the occasional hip-hop dance thrown-in for good measure. Daniel has ordered me a large beer this time, along with a Bud' for himself, and I cannot tell you how glad I am to be drinking normal fluids, with *average alcoholic volumes,* once again. However, when my drink is handed to me by the barman I have to use both of my hands to take it off him, because the blessed glass is not only the size of your average *goldfish bowl,* but it also the shape of one too! (I.E. completely rounded, but with a flat base). Daniel thinks this is hilarious, of course, and once-again I have been caught, hook, line and sinker by my American friend...the swine! However, the good news is that Daniel could see how enthralled I was with the shape and size of this *somewhat unique* glass, and so he purchased it off the barman (for an undisclosed sum) as a memento of our evening out together, though lord-only knows how I am going to get the blessed thing back to the UK?

Our final venue for the evening is a complete contrast to the other two bars. Music is playing very loudly in the background –so loud in fact that it is drowning-out most of our conversation, even though our faces are only inches apart, as the place is heaving with party revellers of all ages. In the main lounge area there is one long bar which runs the length of the room, behind which seven bartenders are frantically pouring drinks from over forty different taps, in their vain attempts at keeping pace with the enormous demand from their multitude of customers. By eleven-thirty the *all-night party-animals (i.e. Daniel and I)* have thrown-in our individual towels, each one of us having our own personal excuse for not being able to continue-on partying into the early hours of the morning. Daniel has to be up early for work tomorrow, and I am feeling slightly worse-for-wear, thanks to the mixture of drinks which I have consumed over the last four hours.

At the stroke of midnight I ring my big son, Liam (courtesy of Daniel's home telephone), and he tells me to hurry up and come home, before asking me if I have bought him lots of presents? Oh how I love that little man... and his little brother, Carl, too, for they are the apples of my eyes –and now that I have spoken with Liam I am missing them both terribly, along with Caryl, of course (even if she does *nag-me-to-death* on occasions) because without her and the children at my side, my whole world would simply fall apart. Both Daniel and I have overslept this morning, but he is the only one who will have to make excuses to people, even though he is the man in charge –'lead by example' -and all that stuff. Thankfully, the Southern Californian weather has not changed one iota since I first arrived in this wonderful place, and so my itinerary for the day should be very similar to yesterdays. However, this evening will be vastly different to last night, because all four of us will be driving down to San Diego to stay with Daniel's father, Dexter, and his wife (Daniel's step-mum), Sophie, who both live in a mobile home, which, according to Zara, is situated next to a beautiful lake, with the most unbelievable views that anyone could ever wish to wake up to in the morning.

It sounded so perfect that for the first time since I had arrived in the Southern half of California, I simply couldn't wait for the day to end –and for the evening to begin. Last night Daniel had showed me an estate of magnificent houses near Newport Beach, which he said were worth over half-a-million dollars each primarily because of their situation near to the beach. Considering that our lovely three bedroom semi-detached house in Cardiff (which was less than six years old) was only worth around fifty grand, I knew that I could only ever dream of owning such a luxurious abode –but then-again, if the movie takes off then who knows what my family and I could end up living in -and also where in the world it might be? It is so nice having company at the poolside today, even though the gentleman can only stop to talk to me for five minutes, as he is currently in his lunch hour from work, and he insists on swimming twenty lengths of the pool before getting dressed again and returning back to his place of work.

He tells me that his name is 'Louis', and that he works in one of the main office blocks in the city centre, adding that lunchtime is the only time he gets to do any exercise, as he is a married man with two children "Who are very demanding", he adds with a smile. "Tell me about it" I jokingly replied. While Louis' burnt-off a few hundred calories in the pool, I continued working on the opening scene, which will show Ted (or an actor portraying him) getting dressed into a set of racing leathers and motorcycle boots, before walking to the end of his garden and opening the door to a large, but very old and somewhat weather-beaten workshop-cum-outhouse. In one corner of the workshop a large motorcycle will be standing upright on its centre-stand, a grubby sheet of green tarpaulin covering the majority of the bike, barring the front wheel and the bottom half of the rear wheel. After gently removing the dust-cover Ted will then throw it under the workbench on the opposite side of the workshop, before dropping the bike down off the centre stand, and manoeuvring it out through the open door, into the garden.

Ted will then sit astride the bike, before taking a set of keys out of his top pocket, and placing them into the ignition. After turning the ignition on Ted will pull out the kick start and turn the bike over on the second kick, before flicking down the side-stand and dismounting the bike, (with the engine still running) while he uses a separate key to lock the workshop door. Ted will then walk back into the house, where he will deposit the key on a hook inside the kitchen, before walking into the living-room. At this point he will pick up a crash helmet, a set of leather gloves and a zipper-type briefcase, which Ted will then carry back out to the garden with him. Flicking open the catches to a top box, which is attached to a rack at the rear end of the bike, Ted will then place the briefcase inside the box and lock it, before sitting astride the bike once more, only this time he will put his helmet and gloves on. After turning off the choke, Ted will ride the bike through an open side gate, before locking the gate behind him, and then blasting off down the road.

At this point the song "All revved-up with no place to go" by Meat Loaf, will play in the background, and after following the bike through the city and out into the countryside for a few minutes, while the credits are coming onto the screen, Ted will disappear over the horizon. The scene will then change to an outdoor parking area, where there will be about a dozen of the top touring motorcycles in the world, lined neatly along one side of the lot. As the screen broadens, a country pub will come into view, and exiting the main entrance will be a group of around twenty people, most of whom are wearing identical tee-shirts under their leather jackets, because they are all members of the same motorcycle club. Climbing aboard their respective motorcycles, the majority of them having pillion passengers, they too will start their bikes, before riding off into the distance, as the music resumes playing from where it left off with Ted.

The third part of the opening scene will show a young couple resting on a patch of grass next to a stream, enjoying the warmth of a summers' day together. After glancing at his watch the lad will immediately jump to his feet saying "Gosh –look at the time...come-on, hurry-up or we'll be late for the show". The pair of them will immediately grab their leather jackets and helmets from off the ground, before running to the top of a small embankment, where they will both mount a trails bike, before likewise riding-off into the distance with the same music playing in the background. The final part of the opening scene will show all three sets of riders, along with dozens of other bikers, converging in an underground car-park. As the music dies away, the bikers will all dismount and lock their bikes, before making their way through a set of corridors leading to the main terminal building. The next scene will be in a large auditorium, where Ted will begin his talk on the journey, with a view to promoting his newly released book 'Jupiter's Travel's', of course.

After a brief introduction to Ted Simon, the person, Ted will begin to tell his amazing story, as the camera begins focussing on the crowd, in particular the youngster on the trails bike, who will become a prime character as the movie progresses. Instead of simply re-enacting the actual events that took place on Ted's journey as one continuous feature, I have decided to split the movie into four individual sections, returning back to the lecture hall after each *quadrant* of the journey has been completed. Splitting the film into four sections will enable the director to enhance this secondary characters' persona, and each time the film reverts back to the classroom scene, where Ted asks his audience if anyone has any questions which they would like to put to him, our budding *round-the-world* traveller can pose a question which will immediately lead us onto the next scene.

I know that this latest idea of mine will also please Ted, as it veers towards his idea of someone wanting to do the journey, rather than the film focussing purely on Ted's accomplishment. For two complete amateurs in the film world I think that Ted and I might have *hit the nail right on the head*. As Daniel has given me *carte-blanche* to use his home-phone, I immediately return to the house and call Ted, to ask him what he thinks about my adaptation of his original idea, along with saying my final farewells to the man —well, for now anyway. Ted is enthralled with the whole idea, and he is also thrilled to tell me that he managed to contact James Gerwyn at his home in Cardiff last night. Apparently James told Ted that he would be absolutely delighted to have a meeting with me, and that I must contact him at the BBC studios in Cardiff as soon as I return home...the dream continues.

CHAPTER 9

San Diego And Mexico

Knowing that we will be staying overnight (or maybe even until Sunday) in San Diego, I decide to make a pile of sandwiches out of the food I have left, as it will save us having to mess around cooking meals this evening, and it will also give us a packed lunch for tomorrow, should we decide to go out for the day. Daniel has finished work early, in order to get his family's bags packed for the trip, and just as we are loading the car, Darcy and Zara turn up at the house. I had considered taking my car to San Diego, just in case we stayed until Sunday, and then I could leave early in the morning, for the long journey back to San Francisco, but Zara has assured me that we will be coming back to Irvine tomorrow evening, and so it would only be a waste of precious fuel if we took two cars. According to Daniel, the trip to his dad's place is around eighty miles from door-to-door, and so the journey should take approximately one-and-a-half hours. At 6.15pm I parked my car in Daniel's parking space under the car port at the back of their house, and then the four of us set-off (*on my next great adventure*) to the city of San Diego.

San Diego, which is Spanish for 'Saint Didacus', is the eighth largest city in the US. It is located on the Pacific coast of America, at the base of Southern California, bordering with the country of Mexico and the city of Tijuana. San Diego has often been declared as the 'Birthplace of California', as it was the first area to be discovered by the Europeans, after Spanish explorer, Juan Rodriguez Cabrillo, docked his ship in the harbour back in 1542, before setting foot on the shore and immediately claiming the land for Spain. Almost three hundred years later, in 1821, San Diego became a part of the newly independent country of Mexico, before being reformed a couple of years later as the First Mexican Republic. Following the Mexican-American War (25th April, 1846 -2nd February, 1848) California became part of the United States of America, and two years later, it was officially admitted as a state of the Union. During World War II, San Diego was a major military base, so much so in fact that it was the first of a number of American cities that the Japanese planned to attack with biological weapons on September 22nd 1945.

The plan was named "Operation Cherry Blossoms at night" and consisted of a squadron of Kamikaze pilots who would fly a handful of planes that were filled with plague-infected fleas into heavily populated areas, with the intention of killing tens of thousands of innocent people, thus causing chaos to the cities in question. Thankfully, the Japanese surrendered a month or so before this heinous operation could be perpetrated. Today, tourism plays a huge role in San Diego's economy, San Diego Zoo, along with its Safari Park, and Sea-world, of course, being high on the list of their major attractions. San Diego has also been acclaimed as 'America's Craft Beer Capital' with its famous 'Beer Tours', and its infamous 'Beer Week' which takes place every November. Cruises are also a major part of San Diego's tourist industry, including whale-watching excursions, which primarily observe the migration of Gray Whales, mostly in January of each year, although humpback and blue whales have also been spotted off the coast at various other times of the year. San Diego is also home to the largest 'Sport Fishing' fleet in the whole of California.

The journey went quite quickly (probably because Daniel was driving, as he does not like to *hang-around* on the open freeways) and so in just over an hour we are entering the mobile-home site. Mind you, I use the term 'mobile' very lightly, as some of these homes looked bigger from the outside than half of the houses I have stayed in on my travels. Driving around the huge communal swimming pool, Daniel then turned to the right, before pulling-up at this truly magnificent –and utterly huge, *mini-mansion*. When Daniel had first mentioned the fact that his father lived in a 'quote' "Mobile home" 'unquote', I had envisioned a caravan similar to the one that James Garner used in *The Rockford Files* television series, or one like Mel Gibson lived in, in the *Lethal Weapon* movies, but this place was really something else.

Dexter, Daniel's dad, greeted us all very enthusiastically at the door, before introducing his good lady wife, Sophie, to me, as we stepped inside their *humble abode* –yeah, right. Inside the place was completely *open-plan*, making the lounge-cum-dining room-cum-kitchen area, look absolutely massive. Half-a-dozen doors led the way to four huge double bedrooms and two spacious bathrooms, and the fixtures and fittings in every room were very lavish indeed. Dexter thoroughly enjoyed giving me a full tour of the place, and as we meandered around each room in turn, he explained to me how the whole house is pieced together.

"In theory the entire place could be transported to another location –with the help of three huge articulated trucks, of course, but in reality it is highly unlikely that it would ever happen", he said, before adding that because the place is surrounded by concrete patio's, brick walls and perimeter fencing, it would be far easier to just move themselves and their belongings to another mobile home site, if ever they decided to leave the place, which, by the sound of how much they both loved their home, was highly unlikely.

Escorting me out onto their main veranda, which boasts panoramic views of the entire bay, Dexter pointed up to the sky, saying that the whole place will be awash with a sparkling array of multi-coloured lights in about an hours' time. "This is due to the firework displays that go on almost every night in Sea World, which is just around the coast from us. Just be grateful that you can only see the fireworks, and that you can't hear the racket they make", Dexter added jovially. By now Daniel is calling us from his dad's personal bar in the living room, asking us what our *poison* is for the evening –and once again Dexter takes charge of the situation. "You cannot come to the border of Mexico without becoming a 'true Mexican' Shaun, and so watch what I do first, and then simply follow my lead –okay my friend?"

It seems like I had no say in the matter, and so I simply did as I was told -without question. After pouring out two shots of tequila into two small glasses, Dexter then cut two slices off a lime (which had already been cut into quarters) before placing them both on a small dish in front of us.

Picking-up a small salt-cellar, he then sprinkled a smattering of salt over the back of each of our left wrists. After licking the salt off his wrist, Dexter then took a bite out of one of the pieces of lime, before leaning his head back and throwing the tequila down the back of his throat. Dexter then finished his little party-trick by slamming the shot glass back down on the bar, before releasing the last breath of air that was left in his lungs and gasping: "Now you do it Shaun –if you want to become a true Mexican, of course".

I was quite happy to follow his lead, as it was something I had never done before, although I was a little apprehensive about how strong the tequila would be? However, and to my pleasant surprise, it wasn't too bad at all –in fact it was quite nice. Dexter doesn't keep any beer in his fridge, and so I am duly given a huge tumbler full of vodka and Coca-Cola, whereas Daniel treats himself to a large shot of rum with a mixture of various juices –and Dexter settles for his usual tipple of a large 'G. & T'. (That's 'Gin and Tonic' to all you non-alcoholics out there).

Sophie doesn't drink at all, and Zara is *off the sauce* because she is pregnant, and so the pair of them is quite content drawing pictures and playing games with Darcy on the living room carpet, while the lads get down to some *serious drinking*. For over an hour the '3 Amigos' continued putting the world to rights, and while we were on the subject of foreign countries, Daniel suggested taking a trip into Tijuana tomorrow morning and spending the day in Mexico, which was an opportunity not to be missed for me, of course. However, the bad news is that my passport is safely tucked-away in my briefcase, which is in Daniel's house back in Irvine, and without it I may well get into Mexico (as there is no passport control when entering the country) but without that all-important document, which proves that I am a *bonafide British citizen,* there is no way that I will be able to get back into the United States, that is for sure! Daniel has no intentions of driving me all the way back to Irvine, in order to retrieve my passport, and I am not insured to drive his car, and so our only hope is for Daniel to ring the border control office in the morning and ask them if he can use his I.D. card to guarantee my *safe* return back into the United States?

As the night rolled-on I fell into one of my *negative* trains of thought (which, believe-it-or-not, I do have now and again) about what might happen if the message wasn't passed down the line, and the guy at the border refused to let me back into the US –what on Earth would happen then? At the very least I would probably miss my flight back home, and then I would have to purchase another flight ticket –and that could be just the start of my troubles? After a few more vodkas I decided to express my deep concerns to the lads, and that is when Dexter intervened. "Tell you what Shaun, you can drive my car back to Irvine tomorrow and retrieve your passport, as my insurance covers any driver, and so you won't have any worries with the police". Dexter then proceeded to write out the directions back to the mobile home park from the main freeway, which seemed simple enough to follow, even though by now all three of us were *three-sheets-to-the-wind*, so to speak. Dexter then walked me out into the parking lot, before pointing out the route that I needed to take once I was back inside the caravan park, which also seemed pretty straight-forward.

As we walked back into the house (I use the term 'house', because effectively that is what it is –a 'house') Dexter put his arm around my shoulder, before asking me what he thought was a very straight-forward question:

"Now Shaun, you do know how to drive a 'stick-shift', don't ya?"

The look of total bewilderment on my face must have made Dexter wonder what the hell he had let himself in for, but sighs of relief soon abounded (on both sides, I might add) as he explained to me that he meant a car with a manual gearbox and a *gear-stick* (I.E. a 'stick-shift'). Even though Daniel had also written precise directions on how to get back to his house, without a road or street map in my possession (I had left the one that I purchased from the garage in my car, of course–*sod's law*) and and considering the fact that I had managed to get hopelessly lost within a five mile radius of Rockwood, it was fair to say that Daniel was less than confident that I would make it to his home, let-alone find my way back here again? However, after one final drink, all three of us agreed that retrieving my passport from *up-north* would be my safest option.

As the evening drew to a close, Dexter escorted me to my room for the night. It was certainly the most luxurious bedroom that I had ever slept in, and the queen-sized bed had the most-comfortable mattress that I had ever had the pleasure of prostrating my carcase on. Wrapping myself in the huge burgundy quilt, which was as silky as satin (maybe it was satin –I have absolutely no idea) I immediately fell into the deepest of sleeps. It is a darn good job that I borrowed an alarm clock from Dexter last night otherwise I would never have woken up at 6.30 am this morning, as planned. The sun is just showing its face over the brow of the horizon as I wash the remaining *sleep* out of the corner of my eyes, before vigorously brushing my teeth in an attempt to get rid of that *morning-after* taste. Having already dressed myself in the bedroom, I gently slip out through the front door, leaving everyone else still deep in the *land-of-nod*. Gazing out over the bay for the first time in the light is even more breathtaking than it was last night, when the sky was alive with a fusion of multi-coloured fireworks, and I cannot think of a more beautiful view to start my day...and also to begin the journey ahead.

Reversing Dexter's open-top sports car off the hard stand, I follow the contours of the bay (as instructed) until reaching the main entrance to the mobile home site. From here it is practically a straight line journey, all the way back to Irvine —well, according to the lads, it is. Having allowed myself three-and-a-half hours to complete the round trip means that I have until ten-thirty to traverse the one-hundred-and-sixty-odd miles, most of which will be on the open freeways, I am pleased to say. The roads are excellent, and also very quiet, probably because it is Saturday morning, and so, being the weekend, there is far less people commuting to their places of work than there would be during weekdays. It is also nice to be using my left leg on the pedals again, after driving an automatic car for the last twelve days, although very little gear changing is required on the freeways, of course. The first hour whizzes-by, and I am adoring the feel of the cool breeze on my face, blowing my hair in the wind, as I tear-up the tarmac —at 55mph! By eight fifteen I am back in Irvine, after a very pleasant and trouble-free drive.

Dexter said that there was enough fuel in the car to get me to Irvine, but once there, I would have to put a few more gallons in tank, in order to get me safely back to San Diego. Apart from putting gas in the tank, I was also desperately in need of finding a fuel station for an entirely different reason...I was in dire need of a bathroom! After driving around the town for a few minutes, I suddenly spotted the same garage that I stopped at when I first arrived in Irvine, and so after filling-up the tank -and emptying my bladder, I followed the same route back to Daniel's house that Zara had shown me, when I had followed her car, four days ago. It was now 8.35am. Apart from grabbing my passport from out of my briefcase, I also grabbed the pack of sandwiches from out of the fridge, which, after all my hard work, I had also forgotten to pack yesterday, along with my camera –another very important item that I had omitted to take with me to San Diego. (*Sometimes my brain just doesn't work as it should do, but that is probably because my mind is almost always elsewhere, rather than being on the job in question?*)

The return journey to San Diego was even easier than my journey northwards, and it was just as enjoyable too. I finally pulled-up outside Dexter's mobile home at 10.34am –less than five minutes behind schedule, and so I gave myself an honorary *pat-on-the-back*. By now Daniel and his family had all finished their breakfast, but Sophie kindly offered to make me a pile of pancakes with maple syrup. However, having already devoured the pack of sandwiches that I had retrieved from the house, I politely declined her offer. Dexter was really pleased to see that his car was still in one piece, and shortly after my arrival he and Sophie drove off to his pharmacy, as they were both working today. (Actually, they should have opened the pharmacy at 9.30am this morning, but they decided to wait for me to return with the car –bless them). Daniel tells me that there is a major football game being shown on TV at 2.30pm this afternoon. "It's my dad's old college team, against a bunch of no-hopers", he adds confidently, as if Dexter's team are unbeatable.

"So do we need to be back here for two-thirty then?" I ask in all innocence.

"Not at all Shaun...dad will make sure that he is home in time for the game, but I'll be happy just knowing that my dad's team has won".

"Well, let's get to Mexico then", I exclaim, my excitement levels having reached an all-time high by now. It is less than thirty minutes drive to the border terminal, and as soon as we arrive, Zara, Darcy and me immediately vacate the car, leaving poor old Daniel to roast-alive in the driver's seat, as he waits patiently in the massive stream of vehicles that are making their way into the huge parking lot. The air temperature is now hovering around the eighty-degree mark, and so we are all feeling a little drained with the heat. After fifteen minutes, or so, Dexter has rejoined us, and he is telling Zara that the cost of parking the car was either $7 for an hour or $7 for a day –and so he chose the latter. (*Try and work that one out if you can?*) The whole area is swarming with people, most of whom are making their way *into* Mexico, and so it is simply a case of *follow the crowds*, as we make our way through a set of very rusty gates, before rattling our bodies around a chronically dilapidated turnstile into the city of Tijuana and the country of Mexico.

Tijuana, which roughly translates as 'Aunt Jane' in Spanish, is not only the largest city in the state of Baja California, but it is also one of the fastest growing cities in the whole of Mexico. Sitting adjacent to the city of San Diego, Tijuana, which was founded on July 11th 1889, is Mexico's most westernmost city, and it claims to be the most visited border city on the planet, with over fifty million people crossing its border every year. Apparently there are over a quarter of a million border crossings made both in and out of the city, seven days a week, 365 days a year, hence it has been nicknamed the 'Gateway to Mexico'. The 'Californian Land Boom' which occurred towards the end of the 19th century, was the beginning of a major tourist industry for Tijuana, as the Californian's flocked to the city for both trade and entertainment. The Panama-California Exposition in 1915 also brought in many tourists, who came to visit the 'Typical Mexican Fair', which included such delights as native Mexican cuisine and thermal baths, along with some of the most popular sports at the time, such as horse racing and boxing.

During the years of prohibition (1920-1933), the lure of legal drinking and gambling saw many thousands of Californian's pour into Tijuana to party their nights away, many of them staying at the 'Hotel Caesar's', which was actually the birthplace of the *Caesar Salad*. In 1928, the 'Agua Caliente' ('Hot Water') Tourist Complex opened, which not only included a hotel, spa and gambling casino, but also a dog track, a golf course, and even its own private airport. A year later a racetrack would be added to its sports facilities, and over the next eight years, famous Hollywood starts, such as Rita Hayworth, who was actually discovered at the complex, along with several major gangsters, would frequent this *den of iniquity* on a regular basis. However, all of this came to an abrupt end in 1935, when President Cardenas outlawed all gambling activities throughout the whole of Baja California. By the early 1950's, Tijuana had completely restructured its tourist industry, by offering more family-orientated attractions to its visitors, and the rest, as they say, is history.

A large sign, stretching right across the width of the road says:
"Bienvenidos - Welcome to Tijuana –the most visited city in the world".

There is an abundance of taxi-drivers swarming around the entrance, and each one of them is trying to coax the *gullible tourists* into taking a lift into the main town centre, which is not that far away from where we are now standing. Daniel tells me that their prices are normally quite reasonable "So long as you make sure that the price is agreed upon before you get into the car –and always check that you have the right amount of change in your pocket, or just a little over, if you want to tip the driver, otherwise there is a good chance that once they have taken your ten or twenty dollar bill, they will claim that they have no change –and so you will end up giving them a very large tip", he added. It was sound advice indeed, and I wondered how many of the dozens of people, who were now piling into several taxis, would be getting ripped-off before the day was out? Walking a little further down the road, I was confronted with a sight that I had half-expected to stumble-upon once we had crossed the border into Mexico, but it still managed to churn my stomach.

Hoards of beggars, most of them draped in nothing but rags, and a few of them with various limbs missing, surrounded me in every direction, but the worst thing of all, was the fact that the majority of them were very young children. This awful scene brought back memories of the poor children in Tunis, as yet another paper cup is thrust into my face, its pitiful owner, a young boy, not so much older than my own son, Liam, gazing up at me with the most pathetic look I have ever seen in my life. I am heart-broken. Upon seeing the look of sheer helplessness on my face, Daniel immediately put his arm around my shoulders, saying that he and Zara had felt exactly the same way when they first visited Tijuana, but since that initial shock they had learned to harden themselves to the harsh reality of life in third world countries. "Of course we care Shaun", he added, "But there are only certain things an individual can do... you cannot help everyone in this world my friend, and you know that, so just try and come to terms with it buddy" he added in a sympathetic tone.

Unfortunately, I was not in the mood for listening to any advice, even-though I knew deep down inside that Daniel was right, of course, and so

I just said "Can we go for a beer somewhere mate?" Ordering a couple of bottles of Budweiser for us two, along with a two Fanta orange drinks for Zara and Darcy, Daniel warns me not to drink out of the glasses that are provided with each bottle, "Because they will have been washed by *piped* water, whereas the bottles will have been sterilized at the factory, and so they will be a lot safer to drink out of", he added, like the proverbial protector, looking-after the well-being of his uneducated protégé. Every five minutes I am learning something new about this country, thanks to my companions, and I know that I will be eternally grateful to them for all that they have done for me. As we begin our shopping in earnest, my *fairy godmother* (Daniel) took on his most-important roll yet –that of saving me money. Having earned my living by selling cars in the UK I was well-accustomed to wheeling-and-dealing, but I had to take a back seat to Daniel when it came to bartering for goods.

After arguing the price of just-about anything and everything (to the point of being quite obnoxious) right down to the last dime, he would simply turn-around and walk out of the shops (or away from the street stalls, as the case may be) stating quite categorically that he could buy the item for a lot less money (even though he never had a clue whether he could or not, of course) from another shop just down the road. If the shop assistant did not chase him out into the street and agree the price right there and then, Daniel would simply walk on to the next shop and start the procedure all over again. If all-else failed then Daniel would return to the first shop and accost a fellow worker to do battle with, until all three of them finally came to an amicable price, which would invariably be a lot less than the price quoted when he first stormed out of the place. Thanks to Daniel's assistance I was able to buy several presents for Caryl and the boys, but just as I was walking out of the last shop (having finished my shopping for the holiday) someone suddenly grabbed my right arm.

"Come and see us again hombre, only next time leave your friend at home, eh", said the shop owner, his smarmy grin telling me that he had made very little profit from *this tourist*. Just for his insolence I immediately handed over two dollars to Daniel in front of him, saying "Sure thing buddy –the money he saves me on buying goods I only have to pay it to him anyway". Although I did feel a smidgen of guilt, Daniel assured me that most of the tourists *give-in at first-base,* thus contributing a healthy profit-margin for the shop owners, and so I wasn't to give it a second thought. On the way back to the parking lot we passed a trio of young children, who are all strumming on various banjos, whilst doing their uttermost to wail-out a song in harmony, even though they are currently failing miserably. However, I figured that they deserved a small donation from their latest admirers, who had stopped for a just a few minutes to watch them do what they do best, and that is 'surviving'.

It was a good job that I had made that trip to the house this morning, because while Daniel had walked on safely through the border checkpoint carrying Darcy, Zara and I had been stopped *as a couple* coming back through customs. With my dark tan and Zara's Latin-American looks we were prime targets I suppose. However, my ten year British passport soon came to the rescue, and so I was immediately waved-on by the security guard. Unfortunately, Zara wasn't so lucky, as the guard was convinced that the poor woman was an illegal immigrant, who was trying to cross the border, as he kept asking her over and over again if her I.D. card was genuine? Thankfully, Daniel was on hand (once again) to come to her rescue, and within minutes all four of us were back in the *civilized* world.

Upon leaving Mexico I knew that I had barely scratched the surface of a country that had so much more to offer, and so I vowed that one day I would return to this land of eternal sunshine. Hopefully I would also be able to help the poor and starving with a much larger donation...with all the money that my movie was about to make me! *(That's it -keep up that positive thinking Shauney-boy!)* As we drove away from the parking lot, Zara said that to see the true beauty of Mexico, one must drive across the vast open-plains, and all the way over to the Yucatan Peninsula "Especially if one wants to swim in the glorious Caribbean waters which surround the southern tip of the country", she added. My mouth began watering at the sheer thought of it.

By the time we arrived back to the *palatial caravan,* Dexter was already sprawled-out in front of the television, watching his all-important football game, which was by now in full swing. Complete with matching tee-shirt and baseball cap, both of which were in his teams' colours, of course, Dexter shuffled-around the carpet, whistling and bellowing-out various sets of numbers and commands –far louder than any of the fans were shouting on the TV, and easily drowning-out the orders from the coaches of both teams. While Zara and Darcy joined Sophie in one of the bedrooms, where she was happily enjoying a bit of peace and quiet from her somewhat rowdy husband, Daniel announced that he was going to watch the match from the kitchen, as he was about to prepare an 'Abalone' meal for us all for this evening. Claiming to be an absolute connoisseur of oceanic food, Dexter reckons that Abalone is "Categorically the best tasting fish in the world". Being used to *cod and chips* for tea, or *sardines on toast* for dinner, I was in no position to argue with this *master-of-the-seas*.

About halfway through the match, Sophie announces that she is popping over to the warehouse, to buy a few Christmas gifts, and so Daniel suggests that I go along with her, as they apparently have lots of excellent bargains, which according to Daniel, are "Simply too good to miss". Everything that one could ever wish to purchase for Christmas, including decorations, food, drink, presents and costumes, is contained within this enormous prefabricated building, and the prices really are a steal compared to anywhere else. Knowing that it was currently unavailable in the UK, I bought a video of 'The Jungle Book' for my two sons, along with a book on baseball heroes for Ted's son, Wills, as I know that he is a great fan of the game. (*I will have to ask Daniel to post it to Ted for me?*) By six o'clock we are back at the house, sitting around the dining table and waiting for our abalone to be served up by Daniel. The sheer aroma that is emanating from the kitchen has conjured-up a huge appetite on my part, and I believe that my fellow diners are feeling the same way too.

Apart from being a self-acclaimed expert on fish, Dexter also claims to be a wine connoisseur, and he has a rack of the finest wines along with many expensive types of claret in his collection just to prove it. After sampling a few of the *lesser-vintage* white and rose wines, both Daniel and I are ready to stay for a second night, but Zara has to be in work in the morning, and so she has insisted on driving us all home this evening.

This is actually more beneficial to me than staying the night, as not only will I be eighty-odd miles closer to San-Francisco, but it also means that I won't have to set off so early in the morning. However, it does seem such a shame that the evening has to end so early, especially after enjoying such a superb meal, which had certainly lived-up to its reputation. Our glorious hosts had also surpassed all expectations, and I will never be able to thank Dexter and Sophie enough for all their help and hospitality, for they had both been truly wonderful. As we make our way northwards, Zara says that she will drop Daniel and I off at one of the local bars, in order that we can have a farewell drink together, but by the time we get to the outskirts of Irvine the pair of us are feeling too tired to go out, and so we agree to settle for a nightcap in the house instead. However, that is if we can get into the house first, because I have just informed Daniel that I have left his front door key on his father's key-ring!

Luckily, Zara has her own key, and so my life is spared –but only just!

CHAPTER 10

The Long Journey Home

Baron's *dawn chorus* has woken me early on my last full day in America, but as I have around four hundred miles to traverse today I suppose I should be grateful for his *rise-and-shine* squawking. Bringing the car around to the front of the house, I begin packing-up the boot (or should I say *trunk*) for the final time, as the rest of the family come hurtling down the stairs and out into the street to say their goodbye's. Breakfast will have to be eaten en-route, as time is of the essence today, because I want to get to a hotel as close as possible to San Francisco Airport before nightfall, in order to have a very short journey for my early flight tomorrow morning. Daniel initially advised me to take the same roads northwards as I came in on, but Zara then intervened, warning me that I would hit all of the weekend traffic, which will be making its way to and from Santa Barbara if I do so, and so she advised me to take the inland route up to Bakersfield, before heading directly to San Francisco. As I always enjoy trying out alternative routes wherever possible, I decided to take Zara's advice, and so after a flurry of hugs and kisses I headed-off on my way to the '605' freeway.

The good news is that it didn't take me long to find the intersection, but the bad news is that I've taken the wrong exit off the junction, and so now I am heading southwards, instead of northwards...oops! Thankfully, there is a turn-off only two miles up the road, and so the problem is soon rectified. It wasn't very long before I had to change freeways, but according to Zara that would be my last turn-off for the next three-hundred miles. According to my map Highway '5' runs right through the centre of the

'Los Angeles National Park', and I have to say that the mountainous scenery was just as breathtaking as the coastal route, except there were no signs of water in any direction. It was yet another gorgeous sunny day, and there was very little traffic about, and so, suffice it to say that driving along at a steady pace whilst listening to my *Love songs* album on the Walkman (whilst keeping a sharp eye out for any police cars, of course) was an absolute pleasure.

With no signs of any towns for miles I pull over at the side of the road, before clambering down an embankment to do what comes naturally, when suddenly I notice something glinting through the undergrowth. Being as nosey as ever I begin clearing away a pile of branches which are currently obscuring my view of what is lying beneath them? "Good lord, it's a hub-cap" I say quite spontaneously, as I wrench it out from its entanglement, the sunlight almost blinding me, as it deflects off the gleaming chrome-work. The hub-cap (or *wheel-trim*, if one prefers) is around two feet in diameter, and so it must have come from a truck (or even a 'monster-truck'), as it is way too big for even the largest four-by-four. As the cap looks almost brand new, and it is surprisingly light in weight, I decide to take it home with me, with the intention of having it engraved as my 'Wheel of Fortune' as and when the movie becomes a *monster hit. (Thankfully my suitcase is the square, flat type, rather than being long and slim, and so I am hoping that the trim will fit comfortably inside of it).*

I had already been warned that beyond the mountains lies nothing but desert, thousands of square miles of the stuff, with only one straight road running right through the centre of it, and now that I have finally left the mountainous terrain behind me I find myself surrounded by millions of tons of rough sandy soil, with only the occasional greenbelt, or cluster of uniformly planted trees popping-up in the distance, to break up the monotony. Mile after mile, hour after hour, it continues, as my odometer clocks up another one hundred miles, and the needle on my fuel gauge moves one digit closer to that ominous red zone. To my great surprise the speed limit along several stretches of this desert road is 65mph (I had always thought that the speed limit was 55 mph throughout the whole of the US) and so I am able to push up my speed accordingly, thus dropping a few more valuable minutes off my journey-time. Having taken a break from the Walkman I sit listening to what I am sure is my *one millionth* commercial on the radio, only this time I actually take note of what the guy is saying:

"Come to beautiful San Francisco and have the time of your life"

"I'm on my way old buddy" I shout out in a jovial manner.

"And why not visit world-famous Chinatown while you are here", he adds, making me think "Yes, 'why not' visit Chinatown while I am here, after-all I may never get another chance to do so —what a bloody-good idea my friend".

I had originally planned to work on the script this evening, but now I have decided to do a spot of sight-seeing instead. I have been making good time, thus putting me ahead of my original schedule, and so I am hoping to be able to park-up the car long before dusk. However one never knows what lies up-ahead, which is what makes travelling so exciting and so much fun... the unknown? For the first time in several hours I am negotiating a long, sweeping bend, which is actually a pleasant change from watching one endless road continue-on for hours-on-end. However, what is not so pleasant is this ungodly *pong* which is now wafting-in through my open windows, thus forcing me to wind them up as quickly as possible. The smell seemed very familiar to me, and yet I knew that it wasn't the smell of a skunk, as I had suffered a whiff of their aroma at Dexter's place, when one of the little critters scuttled across the porch-way, before disappearing into the darkness...and so what the hell could it be?

As I rounded the tail end of this almighty bend, my answer was at hand. Grazing on acres of pasture were thousands-upon-thousands of cows —more than I had seen in my entire lifetime, all together at once. No wonder I had recognised the smell —it was pure cow-shit, or 'dung' if one prefers...a thousand tons of it, I expect, evenly distributed along the rolling hillsides and plains for the entire world to savour its disgusting odour —yuck! After stopping to fill up with *juice* at the next petrol station (which was an absolute necessity, as they were *few-and-far-between* out here in the wilderness) I roared-off once more, full of confidence that tonight I would be back at that same Holiday Inn long before the sun had finally disappeared over the horizon. Highway '5' runs all the way into Sacramento, but I have taken the '580' turn-off, before heading due west. By 4pm I have already passed Livermore, and now I am on my way to Hayward, on the final stretch, before turning-off into *the streets of San-Francisco.*

Battling against an early sunset as I cross the San-Mateo Bridge, which is apparently over seven miles long, I congratulate myself on the completion of another great journey on the road, as yet another invaluable chapter in my life comes almost to a close. Into the big city I roll, and directly in front of me I can see a huge sign for the 'Holiday Inn', and I feel as though it is welcoming me back to where this great adventure all began, less than two weeks ago. Without any photographs to commemorate my initial visit to San Francisco, I decide to use the last six frames in the camera for some colourful shots of Chinatown by night. Pulling up outside the main entrance to the hotel, I enter the building through the small lobby, before making my way over to the reception desk, where I am about to find out if they have a room for me for the night? Standing beside me is a man in a jogging suit (*or 'Shell-suit', if one prefers*) who is probably in his early forties, and he looks a bit of a hippy to me, as he is brandishing a mane of long, scraggy blonde hair, which looks as though it could do with a good wash –and a darn good brushing.

As we are standing so close to each other, I can overhear his conversation. He is asking for the cheapest room available, and he is also very reluctant in paying upfront for his room, before he has even set foot in the place, and so he is giving the poor receptionist a really hard time. Thankfully, the manager is soon at hand, and so the proverbial *trouble-maker* is duly given a very polite ultimatum "Take-it-or-leave-it –it is entirely up to you sir?" Reluctantly, the guy hands over sixty-six dollars of his precious *green-backs*, before walking away with his door key clutched firmly in hand. Upon returning to my car (having paid my dues without the slightest of fuss, I might add) I notice that this same man is standing beside a car which is parked next to mine, and he is fumbling through his pockets, as if looking for a set of keys? Suddenly an idea strikes me, and so, as forward as this may seem, I walk over to him and ask him if he would consider sharing a taxi ride with me into Chinatown this evening?

He tells me that he has to drop his hire car off at the airport first, after he has off-loaded his luggage into his room, of course, and so we arrange to meet in the hotel bar in an hour to discuss things from there. I then do likewise, by carrying my gear (which I have to do in two separate runs) up to my room for safe keeping. Having practically lived out of a suitcase for the past fortnight, my clothing is in total disarray –and having also accumulated various gifts and mementos along the way, means that I have to repack everything a lot more meticulously than I had done before coming out here. Unfortunately the wheel trim is too big to fit inside my suitcase, and so I am going to have to carry it separately. As soon as that somewhat laborious task has been completed, I decide to freshen-up my body by *luxuriating* in a long, hot bath for half-an-hour, in order to let the aches and pains of the day soak-away into oblivion. Thankfully I still have one remaining *clean* outfit left in my suitcase, and so after wiping the last of the steamy soapsuds off my body I suitably attire myself for the evening.

Before meeting up with my new-found friend in the bar, I decide to take one last stroll in the warm evening air, as it will be many months before I can even consider enjoying the delights of any kind of humidity in the UK. "Bloody weather" –there, I've said it! Walking around this vibrant and welcoming city, I can easily understand why Frank Sinatra left his heart here –indeed I think that I may have left mine too? With my lungs now full of that fresh Californian air I make my way over to the bar, where my *hippy* friend is patiently waiting for me. After kindly ordering me a beer, he introduces himself to me as 'Brynley', adding that he is from Saskatchewan in Canada. (*I had heard of the place beforehand, but to this day, I still haven't got a clue how to spell the name?*) I reciprocate of course, before Brynley continues our conversation by telling me that he has just spent a week in America on a working holiday, which included spending a few days in the mountains, as he loves skiing. This time I do not reciprocate, by telling Brynley all about my great adventure, otherwise it might sound like I am boasting, and that is the very last thing that I want to do.

Chinatown is approximately eight miles from the hotel, and so Brynley suggests that I take my car this evening, as booking a taxi each way will be way too expensive "And my budget is almost exhausted", he exclaims, pleading poverty to me, as if he was some kind of pauper, which, I have to say, cheesed-me-off a bit. I tell Brynley that I am a little concerned about finding my way into Chinatown in the dark, without getting totally lost, but he says that he has been there on two previous occasions, and so he will act as navigator for me. "You can then drop your car off at the hire place tonight, rather than waiting until the morning, and then we can both catch the Holiday-Inn free-bus back to the hotel for a night-cap...what do you say to that Shaun?" His idea sounded ideal, apart from the fact that my car might get damaged or broken into –or even stolen (heaven forbid) while we are wandering around Chinatown, whereas if I left it in the hotel car park overnight, I could rest-assured that it would be safe and sound until the morning.

However, having always been a little-bit of a gambler, I decide to chance my luck, and so the pair of us finished our drinks, before heading off into the night. My first port-of-call is the nearest petrol station, as the fuel-tank was full to the brim when I collected the car a fortnight ago, and so it has to likewise be *full* upon returning it, otherwise Budget will charge me double for the replacement fuel, simply because of the inconvenience of having to fill the tank themselves. There are signs everywhere for Chinatown, as it is a major tourist attraction, of course, and so within no time at all Brynley and I are cruising through the packed streets, in an effort to find a parking spot anywhere? After going around in circles for almost half-an-hour, we finally come to a halt in a dark alleyway, where the walls are plastered with signs threatening to tow-away any illegally parked vehicles –great! However, as it seems that we have no choice in the matter, I double-check all of the door locks, before abandoning my trusty steed to the mercy of back-street vandals, mindless hoodlums and opportunist car-thieves –let-alone the police and the traffic wardens, who could be anywhere out there?

Brynley assures me that the car will be fine, and I would love to share his optimism, but then why should he worry; his hire car is now safely locked-up in a compound, and so it is no longer his concern, whereas I am shouldering the burden of worry that everything could go horribly-wrong at the eleventh hour?

The main Chinatown in San Francisco, which was established way back in 1848, is centred on Stockton Street and Grant Avenue. Not only is this the oldest of all four Chinatown's that are situated in San Francisco, but it is also the oldest Chinatown in the whole of the United States. Chinatown is a major tourist attraction for the city, bringing in more visitors each year than even the Golden Gate Bridge. The area of Chinatown itself is approximately half a mile wide in all directions, encompassing a total of twenty-four blocks. The majority of the area is awash with trinket stalls, food stores, mini-shopping malls and restaurants which primarily cater to the needs of the thousands of daily tourists that flock to this oriental *mini-city*. There is also a war memorial to Chinese war veterans, along with a bronze replica of the 'Goddess of Democracy', which was used in the Tiananmen Square protest of 1989.

One of the most popular areas in Chinatown is Portsmouth Square, which is always a hive of activity, as men, women and children of all ages practice Tai Chi in the main centre, while others just sit peacefully at small tables, playing numerous games of Chinese Chequers, whilst drinking several pots of tea. During the 1960's a great number of Chinese immigrants arrived from Hong Kong, setting up homes in Chinatown, and by the mid 1970's the area was one of the most densely populated areas in the whole of San Francisco. During the California Gold Rush (1848-1855), a handful of female Chinese prostitutes began selling their wares in Chinatown, and over the next thirty years their numbers would increase to almost two thousand, with almost three quarters of the female population working in the sex industry. By the turn of the century Chinatown's reputation as a *den of iniquity*, where all sexual fantasies could be fulfilled, brought in even more tourists from all over the world, including many middle-class businessmen, who not only wanted to sample Chinese culture, but they also wanted to indulge themselves in *the pleasures of the flesh*.

As the pair of us wandered aimlessly through the brightly-lit streets, me taking the occasional photograph when anything of interest caught my eye, Brynley is suddenly accosted by this rather buxom woman, who is trying to cajole him into seeing a free striptease show, which is currently going on behind the darkened doorway which she has just appeared from. As Brynley politely declines her offer, the woman immediately shoves open the door, revealing a scantily-clad young lady, who is parading herself up and down a small stage, while a dozen or so goggle-eyed men sit cheering and throwing dollar bills of various denominations onto the stage, in gratefulness for her *obvious talents* –a thirty-six 'Double D' cup being my guess?

For a second time Brynley rebuffs her attentions, before retreating to where I am standing, perhaps trying to convince her that we are gay lovers, in the hope that she will go away? It must have worked, for the woman soon disappeared into a cloud of smoke, never to be seen again by either of us –thank goodness. As the pair of us continued-on walking, past more seedy nightclubs, various strip shows and numerous porn shops, Brynley jokingly quipped that had he been on his own then he might well have taken the lady up on her offer, but he could see that I had no interest whatsoever in what she was selling. Finally we stumbled-upon what looked like a *normal* bar, and so the pair of us parked our bums for a beer. Over the course of the next hour Brynley and I told each other all about our families back home, which included sharing a few photos from our wallets, of our nearest and dearest, of course. I then tell Brynley all about the demise of the British car industry, and how I had been made redundant by my local car dealership, before adding that I will be starting work as a regional manager for Budget Rent-a-Car after Christmas. Telling *porky-pies* is certainly 'not' something I relish in doing, but I was too embarrassed to tell Brynley that I was currently 'On the dole'.

Brynley says that things are not so bad in Canada, before admitting that he actually owned his own car dealership back in Saskatchewan. (*No wonder he was such a cheapskate at the hotel —most of the business people I know are so tight-fisted, it is unreal*). He then asked me if I had heard of the company called 'General Motor's', to whit I replied that *I had*, of course (*let's face it, who hasn't heard of them*) and so he followed-up his question by saying that he actually owned Canada's share of the company! What am I supposed to say to that? As Brynley has *suitably impressed* me with his statement I figure that now might be the right time to impress him, albeit on a much lower level, of course, by telling him all about my movie idea —and also about my liaison with Ted Simon, as I figure that a little bit of *name-dropping* will not go amiss here. It must have done the trick, because Brynley immediately asked me to write the title of Ted's book on the back of one of his business cards, before adding that he actually owns a Kawasaki 1100cc motorcycle, although with all of his business commitments, he gets very little time to make good use of it.

"As soon as I bought the bike I immediately dug out my favourite *bikers* tape to play in the built-in cassette-player. It is called 'Bat out of Hell' by Meat-Loaf —do you know the album Shaun? At that precise moment I could have been knocked down with the proverbial feather. This uncanny coincidence astounded the pair of us, but the biggest surprise (well, for me anyway) was yet to come. Handing me one of his business cards, Brynley calmly said: If you need any financial backing for the movie, then just give me a call my friend". THE DREAM GETS CLOSER. (*Right now I felt like I could get back to Cardiff without the aid of an aeroplane, I was flying so high*). Having said just about all that we could say to each other, Brynley and I spent the next half-an-hour chatting to the barmaid, a very informal conversation which also turned out to be somewhat beneficial for both Brynley and me, as she knew the best Chinese restaurant in town... well, according to her it was anyway.

From the outside the place looked more like a transport cafe —and from the inside it looked like a Kung-Fu school from a Bruce Lee movie, because there were so many Chinese people inside. However, I have to say that the food was exquisite, and the friendly service was excellent too. In fact, Brynley and I were the last two people to leave the place when it closed at ten-thirty. Now all we had to do was find my blessed car again, and hope that it was still in one piece where we had *abandoned* it? Mercifully, the car was just as we had left it —in other words *without a scratch on it*, which, in all fairness, Brynley had assured me it would be. Brynley then kindly offered to drive the car back to the Budget offices, as he knew his way around Chinatown a lot better than I did, and so I willingly accepted his proposal. Back at the office the formalities were swiftly dealt with, and my $128 deposit was duly refunded to me in full. The free-bus then took us both back to the airport, where Brynley immediately rang the Holiday Inn, who sent out their free-bus to collect us.

By eleven-thirty we were back in the bar, hoping to have that final night-cap we had spoken about earlier. However, the pair of us had forgotten that today was Sunday, and so the bar had closed early —*bugger*! A last shake of the hands was in order, as we said our final-farewell's to each other, and then it was 'up the wooden hill' (as my dad used to say) to bed for yours truly. It has gone eight o'clock by the time I wake up in the morning, and I am supposed to be in the airport for nine, and so it's a case of 'all hands on deck' as I drag my *somewhat heavier* suitcase across the courtyard to the rear entrance of the hotel. Trying to look as composed as possible, I slip into the breakfast room, where I quickly grab a bowl of corn-flakes, along with two croissants, before devouring the lot at a startling pace. The free-bus is waiting to leave as I hand over my keys to the receptionist, before duly collecting my deposit, while one of the porters kindly loaded all of my gear onto the back seat of the bus.

At 9.45am our bus pulls in at the British Airways Terminal, and by 10.15am I have checked-in my luggage. Immediately I head for the departure lounge, as my plane is due to fly out to Houston, Texas at 11.08am, and it is right on schedule. After boarding the plane and sitting in my allocated seat, I stare out of the little square window which is overlooking the left wing, but my eyes can see nothing except the pictures in my mind of golden memories. I am oblivious to everything and everyone around me, as I sit their staring into space...remembering. Flying over California in the cool light of day is a fantastic experience, as the scenery below is so diverse, changing every few minutes, as our *big old bird* climbs high into the sky. Large coastal cities disappear into the haze, only to be replaced by small inland towns –and flat, open deserts are soon overshadowed by huge mountain ranges and endless valleys, before the next big city appears, and endless skyscrapers invade the air-space once more. The flight to Houston took just over three hours, thus allowing me a good hour to have a meander around the shops, before my connecting flight to the UK. (*Because of the time zones I had crossed, I had almost landed before I had left!*) Knowing that we would be landing in England around ten o'clock in the morning (local time) I decide to try and get a little shut-eye before even thinking about watching a few movies.

Apart from the norm, the rest of the flight was quite uneventful really, and so all I can say is that our plane touched-down on time, only to be greeted by cold winds and driving rain. Most of the cabin crew was looking rather bored and restless, and I daresay that it had been a very long flight for them, but the captain was obviously in good spirits, because he offered to carry the wheel trim through customs for me. With nothing to declare I headed straight to the Budget Rent-a-car desk, where my papers were all in order, and my car was waiting to be collected from the parking lot. After phoning Caryl to announce my safe arrival in the UK I headed for home. The roads were still very wet, but at least the driving rain had ceased by the time I got onto the M4, and by the time I reached Reading the sun was shining high in the sky. From then on I had virtually clear skies all the way back to Cardiff, thus giving me the perfect ending to an incredible adventure.

As for the dream, well this part had finally come to an end, but like Confucius once said "With each ending comes a new beginning".

CHAPTER 11

Meeting The Experts

Being reunited with one's family is an emotional experience, and one that I, personally, will always treasure. Liam and Carl thought that Christmas had come early, as I gave them their numerous presents, but disappointment soon followed, as we tried to play the Jungle Book video, having not realised that the frequency in America, which the film was recorded on, is not compatible with the European VHF system? I then returned to the car, in order to get Caryl her *small* present, along with a *large* bundle of washing from out of the boot of the car. However, just as I was closing the boot, my *ever inquisitive* son, Liam, asked me what had happened to the front of the car? Knowing that little boys do not ask questions like that for nothing, I hurried around to where Liam was now standing, noticing along the way that the bonnet of the car had been released. I was horrified to find that both the offside bumper and the grill were quite badly damaged. I knew that I hadn't hit anything whilst I was driving, but then who is going to believe my story?

I cursed myself for not checking-over the car before leaving the parking lot, and I began asking myself all sorts of questions. Was the car purposely facing the wall when I collected it from the car park, and did the assistant offer to put my briefcase on the passenger seat for me, just in case I happened to notice the damage? However, the biggest question I kept asking myself was "How could I have driven nearly two hundred miles with the bonnet open, without noticing it at any stage of the journey?" No-doubt these will be the questions which Budget Rent-a-Car will be asking me when I return the car to their depot. Immediately I rang Jules, and we arranged to meet at his office within the hour. I told Jules all that had happened regarding my stay with Ted Simon in California, and he seemed very pleased with the outcome of my trip. I then confessed to him what had happened to the car, pleading my innocence in all honesty, but Jules just told me not to concern myself with the issue, as the car was fully insured, and therefore the lads in the garage would have it looking as good as new again in no time at all.

The first thing I did when I arrived at the office, was to give Jules the forty dollars that I had left from my trip, along with a pile of receipts and a hand written *spreadsheet*, listing every single purchase I had made from day one. Jules glanced over the sheet of paper for a couple of seconds, before tossing everything -except the forty dollars, of course, into his waste-paper basket. He was obviously not concerned about checking-out my honesty, and so I knew right there and then that he trusted me implicitly. Upon returning to the house, I decided to start the ball rolling by following-up on my one and only lead from Ted Simon, and so I rang the BBC Studios in Cardiff. The receptionist put me straight through to Mr. Gerwyn, who was very friendly towards me, especially after I told him that my brother, Gary, worked in the scenery department of the BBC, and that my cousin, Steve (whom James knew) was a video-tape editor. James sounded quite anxious to arrange a meeting as soon as possible, in order to discuss my proposals.

At first we agreed to meet on the following Wednesday, but after double-checking his diary, James saw that he had another appointment for that same afternoon, and so as he thought that this might force us to shorten our *little chat,* he duly asked me if I wouldn't mind changing our appointment to Thursday afternoon instead? Even though I would make myself available any day of the week, no-matter what time of the day or night it was, I purposely hesitated, before saying quite calmly "Yes, I am sure that I can fit a meeting in on Thursday". For the next two days I worked solidly on my *sales pitch,* rehearsing in front of the door mirror on the bathroom cabinet, as if I were talking to Mr. Gerwyn face-to-face. Expressing my true feelings about the movie in a pure, honest-to-goodness way, and knowing full-well that I would not have to exaggerate my beliefs in the film being a huge success, nor bullshit James about my commitment and conviction to the project, I felt confident that the meeting would be a resounding success, just as it had been with Ted.

As the morning dawned I asked Caryl to go through a *practice presentation* with me, knowing only too-well that I dare not fail at this point, otherwise all of my efforts so far will have been in vain. Right from the very first day I knew that there would be no turning-back, once I had begun my quest, and that I must carry it through to the bitter end, no matter what the consequences might be –good-or-bad…success –or failure? As I am driving through the centre *of Whitchurch Village*, a thought suddenly occurs to me "What if Mr. Gerwyn wants to hear some of the music from the 'Bat out of Hell' album?" I no longer possessed a copy of the tape, having given it to Ted, and there are no record shops in the area, where I can purchase a copy –and so what the hell do I do now? "So-much for 'planning and preparation' Shaun", I say to myself, cursing my crass stupidity for not thinking about this a lot sooner than right now.

I am now within half-a-mile of the BBC studios, and I am running smack on schedule, when suddenly I notice that the needle on my petrol gauge is now pointing to the 'E', which (in my eyes) means *'Even though you think that you are going to make it, you will soon be walking pal!'* I had passed a petrol station only a few minutes ago, which I daresay had subliminally reminded me to check my fuel gauge, but my delay in doing so would now be the difference between me being early, or a few minutes late for my appointment? The burning question now is: "Do I turn-around or not?" It is one of those tricky decisions in life which we all have to make at one time or another, only I have about two seconds in which to make up my mind, because right in front of me is the mini-roundabout which will either lead me to my *final destination*, or give me the opportunity to turn around and get some fuel? "Never leave anything to chance son", that is what my father told me as a young lad, and so, heeding his words of wisdom, as I always do, I went *full-circle* around the roundabout, before driving directly to the petrol station.

Filling the tank as rapidly as the pump would allow me, I then rushed inside the kiosk, where a small line of people were patiently waiting to pay their dues. Looking at my watch, I can see that I am already late for my appointment, and so I begin cursing myself (along with my dad's advice, of course) for not waiting until after the meeting was over? As soon as I reached the counter I immediately handed over a twenty pound note to the woman behind the glass partition, waiting *impatiently* for my change, along with a receipt for the fuel. As a handful of coins is pushed gently under the screen, the attendant asks me politely to make way for the customer behind me –but her words have fallen on deaf ears, for my mind is currently undergoing some sort of hypnotic trance.

Hanging on the wall behind the attendant is a rack of the latest music cassettes, and staring me boldly in the face is Meat-loaf's 'Re-vamped Bat out of Hell' album. (*The re-vamped album also includes the song 'Dead-ringer for love', which Meat-loaf sang as a duet with the incredible 'Cher', but as far as I was concerned, this song would not be included in my soundtrack for the film*). "It's a miracle", I suddenly exclaim, much to everyone else's astonishment. Normally I would have deemed this uncanny stroke of luck as nothing more than a coincidence, but since meeting Jules I have become a firm believer in 'Fate', and with everything that has happened to me in the last two months, I feel sure that new doors are going to be opening for me, and that a great future for both me and my family lies ahead. By the time I have managed to find a parking space in the *somewhat cramped* courtyard, I am over five minutes late, and so I leap up the steps to the main entrance, inevitably tripping over the last one and falling flat on my face!

I have no doubts that my *accidental acrobatics* gave everyone a bloody good laugh, including the security guard, who, although he is immediately on hand to help me up, is probably giggling-like-a-fool under his breath... and who could blame him? The receptionist rang through to Mr. Gerwyn's office, to announce my 'arrival at H.Q.' (*it sounds like a James Bond movie*) and shortly afterwards James' secretary arrived on the scene. Escorting me through the front courtyard and back out into the street, the pair of us then crossed the main road together, before entering a second building on the opposite side of the road. The lady then had to tap a few digits into a code box outside the main entrance which released the locks on the front door, and so I could see that this was obviously quite a high-security establishment. After walking through a very small foyer, the pair of us then ascended a set of stairs to the first floor, where the main offices were located. The lady then led me through a set of double doors, and into a small office, which is situated on one side of a long corridor, where Mr. Gerwyn is waiting patiently for his *non-punctual* guest.

Immediately my host rises from his chair, before welcoming me with a warm smile, and a firm handshake. I can see that this man is around the same age as me, although he is slim and has a good head of hair, whereas I am somewhat overweight and my hairline is already beginning to recede. James is also dressed quite casually, in a pair of smart trousers and a nice shirt, but no tie or suit jacket to accompany them, whereas I am fully *suited-and-booted*, even wearing a nice pair of gold cufflinks and a tiepin to match. Beckoning me to take a seat, Mr. Gerwyn immediately relaxes back in his chair, whilst at the same time requesting his secretary to make us a cup of coffee. Immediately I open my briefcase, before handing James a packet of photographs, which included various pictures I had taken of Ted and me on his farm. Coming across a picture of the 'Mendocino Hotel', James has a quick chuckle to himself, before pointing out to me that this was the hotel that he had stayed at for one night when he went to visit Ted back in 1986. I offer the photo to James as a keepsake, which he duly accepts, before asking me for the address of Ted's place in Elk, saying that he would like to add it to his personal file.

I then showed James the photos of the dome, along with a few pictures of Ted and Wills, and he was just as amazed as I was when I first saw the place for real, before adding that he had never met Ted's son. Now that I had filled James's mind with what were obviously fond memories of the past, I decided that now might be the right time to *spread my wares*. Opening-up my briefcase once more, I pulled out the script which I had written for the movie, before handing it over the table to James. As he began flicking through the pages I went into overdrive, meticulously explaining that the script contains a detailed analysis of Ted's journey, from beginning to end, with each event broken down into individual scenes. I also told James that I had worked on location areas for the filming, along with music sequences, complete with explanatory notes on how each of the Meat-loaf songs will represent their respective scenes —and that I had even put together an imaginary two minute film-clip to promote the movie, along with drawing-up a full size poster for both cinemas and billboards.

"Besides all of this, I have also written an imaginary scene, showing how Harvey Edwards of the Sunday Times may have reacted upon receiving an article from Ted whilst he was on the journey", I added, finishing off my little spiel with a humungous amount of positivity. Studying my work very carefully for several minutes, James then closes the folder and pauses for a few seconds, before looking me straight in the eyes and saying: "It is good, but you must know that your back is right up against the wall with a project like this Shaun". Staring straight back at James, knowing that he is obviously impressed with my work, I calmly reply "Well, I know that we are talking six figures".

"Maybe more" James replies, glancing over the top of his spectacles "That one-hour documentary I produced back in 1986 cost over fifty thousand pounds in total, and that was five years ago!"

Shrugging his shoulders and lighting-up his face with a very broad grin, James then added "Still, had you told me three months ago that you intended achieving what you already have, then I would have simply said "Impossible".

As James handed me a video recording of his 'Riding Home' documentary, which he said I could ask my cousin, Steve, to copy for me, I knew that our meeting had come to an end, and a small part of me also feared that this could also be the end of the line for my project? Reaching over to collect-up my script, James gently places his hand on the folder, before asking me if he can borrow the manuscript over the weekend, to whit I gladly oblige, of course, my thoughts somewhat in turmoil, as I wondered what was going through James's mind at this precise moment? To put my mind at ease, James then explained to me that because he only produces drama's and documentaries, and has therefore never done a movie, as such, he wanted to make contact with a few people over the weekend who were *in-the-know*, to try and obtain further information on the best way to go about making the movie?

James then asked me to meet him for lunch on Monday, which I willingly accepted, of course, before humbly suggesting that we dine at 'The Bank' restaurant, in the centre of Cardiff, knowing full-well that one of Jules Gilmores' business partners owned the place, and therefore he would probably set up a special menu for us. As we shook hands I have to say that I felt very confident and content with the outcome of today's meeting. (*Incidentally, the 'Bat out of Hell' tape was not needed after-all, but at least I now had a copy of the films' soundtrack, should it be required at a later date*). As soon as I arrived home I rang Jules Gilmore, primarily to ask him if he wanted to join James and me for lunch on Monday, but he politely declined my offer, saying that he wanted to stay in the background at present, "As you seem to be doing rather well on your own", he added. Jules then advised me to suggest a more *up-market* restaurant to Mr. Gerwyn for our rendezvous, as (in Jules' own words) "The lunchtime menu is not that brilliant". This I did the following day, and so our meeting place would now be 'Champers'.

The weekend that followed turned out to be the longest that I can ever recall, but now Monday was here, and my mind was full of positive thoughts as I entered the small, Spanish-styled restaurant, which is located in the centre of Cardiff. I was the first to arrive, but just as I was placing my backside on one of the barstools, Mr. Gerwyn came walking through the door. The pair of us then sat at one of the small tables, James tentatively placing my script on the tablecloth in front of me without uttering a single word. He then proceeded to bury his head into one of the menu-cards. Unfortunately, I have always been a man of very little patience, and by now I was already on tenterhooks, wanting to shout-out "Well, any news then James?" but, of course I didn't. After a minute or two of deadly silence, James *broke the ice* by saying "Are you going to have a steak Shaun?", but before I even had a chance to answer his question, he followed it up with a second question "Have you ever been here before?"

I sat there in total disbelief of our conversation, my impatience obviously having gotten the better of me, and so in total confusion I simply nodded my head to James's first question, before shaking my head from side to side in answer to his second question. I then said quite calmly "So how did your weekend go then Mr. Gerwyn?" meaning from 'my' point-of-view, of course. Unfortunately, the waiters' *bad* timing was impeccable, as he asked James and I for our orders, just as James was about to give me the answer to my burning question. James gave his order first, and by the time that I had given the waiter mine, James had disappeared over to the salad counter, thus leaving me very much on edge, as I desperately wanted to know if my learned friend had a few positive answers for me? My only recourse was to follow my *mentor* wherever he went, and so like a news reporter, who is determined to get his (or her) story in the end, I duly made my way over to the salad counter as well.

Returning to the table, each of us carrying a glass of wine in one hand, and a wooden bowl, filled to the brim with multi-coloured vegetables in the other, the pair of us placed our food on our respective placemats and our drinks on the accompanying beer-mats in unison, before parking our backsides once more on our seats –and at that point in time our meeting began in earnest.

"Would you mind if I gave you some friendly advice Shaun?" was James' first words to me, and I remember jokingly thinking to myself "If he advises me to have my steak done *rare*, instead of *medium-to-well-done* (the latter being how I like my steaks cooked) then I am simply gonna scream!" James did not suggest anything of the sort, of course, but he did say that I needed to know more about the film industry, before politely suggesting that I make enquiries about enrolling in a lecture course on film making, adding that it would give me a much better insight into script writing, along with teaching me all there is to know about putting movies together, including plots, angles, dialogues, timing, and so on. James then (quite unwittingly, I am sure) dropped the proverbial bombshell, by telling me that the course would only cost me around one hundred pounds, a figure which was currently way beyond my means. However, James' next statement did help to rejuvenate my belief in myself, as I was currently suffering from a severe lack of confidence right now -something which many people have to live with for the best part of their lives, even though we all hate to admit it, of course.

"As for inventiveness and imagination, well you have already surpassed anything that a producer's course could teach you; indeed, I know for a fact that a lot of producers that I have come into contact with over the years, could not have created and produced a manuscript as well as this one has been put together". By now James is vigorously shaking my script in his right hand, while the waiter is finding great difficulty in avoiding James's flailing arms, as he tries his best to set-out the cutlery. By the end of our meal I knew that Mr. Gerwyn could not do anything more for me personally, but he had promised to make further enquiries, in order to try and find someone who might be able to assist me in my quest for recognition, and so all I could do now was wait –and hope for the best. A week went by, and during that time I had heard nothing from anyone, and so I could only assume that I had been forgotten? I so much wanted to pick up the phone and ring Mr. Gerwyn's office, but I knew that this would be so unprofessional –and probably a waste of time anyway.

More importantly, I had to look at the cost of making that phone call during peak times (9am to 5pm Monday to Friday being the only times that I might be lucky-enough to catch James at his desk) as the monetary situation in the *Donovan household* was becoming very tight right now. Unpaid bills were slowly, but surely mounting-up in the kitchen drawer, having been hidden well-away by me, for fear of reminding Caryl and I that our personal recession was worsening by the minute. One morning, whilst I was busy making the beds, and also tidying Liam and Carl's bedroom, the phone rang. As Caryl was currently out shopping (*whenever my wife was at home she would always get to the phone first*) I took the call on the extension line in our bedroom. It was James Gerwyn. Assuming that he was going to ask me about the *Riding Home* video he had loaned me, I began our conversation by pouring-out a volley of excuses as to why my cousin had not had time to copy the programme for me, before apologising profusely for not having returned James' copy to him.

James interrupted my *grovelling* by telling me that the reason he was calling me had nothing to do with the video, before asking me if I had a biro handy. Being totally unprepared for such a call, I immediately pulled out the top drawer of the dressing table, before tipping out its entire contents onto our (freshly washed) continental quilt, in my search for any kind of implement which might be capable of producing legible digits. As one of Caryl's eyebrow pencils rolled off the bed and onto the carpet, I counted my blessings, even-though I would have written on the bedroom wall with one of Caryl's lipsticks, had I needed to. Slightly *shaken* –and most definitely *stirred*, I sat on the edge of the bed, pencil and paper (well, an unused Christmas card actually) clutched firmly in hand, whilst waiting with *bated breath* to hear what James had to say?

"I have managed to contact somebody whom I believe might be able to help you with your movie. Normally he would not entertain talking to anyone who did not have a completed transcript, but I have told him about our meeting, and he is looking forward to your call. 'Robert Staines' is the gentleman's name, and he owns a company called 'Perfecta Film Development'. His office is located on the first floor of the 'Castle Arcade' in the city centre. Give him the same *pitch* as you gave me and you should be okay".

James then kindly gave me the office number, which I immediately rang, without even replacing the handset. The receptionist on the other end of the line asked me "Who's calling please?" -as they always do, before putting me through to Mr. Staines' extension number. Thankfully I am only holding on the line for a few seconds before being connected: "So you are Shaun Donovan; I have been waiting for your call -what can I do to help you Shaun?" Even though Mr. Staines' greeting is a very pleasant opening to our conversation, butterflies are amassing in my stomach by the bucket-load, as I briefly explain what it is all about —at the same time praying that I get a positive response from this man?

"Right then, let us arrange a meeting", Robert (we are on first-name terms by now) says, sounding very interested indeed with my proposed movie.

"Shall we say Friday at two o'clock then, at my office in the Castle Arcade in town...is that fine with you Shaun?"

I feel like shouting-out "Bloody brilliant", but manage to compose myself at the very last second, before casually answering "That's fine by me Robert —I'll see you then". It felt so good to be back on the trail once more, and so once again my best suit came out of the closet for an airing, my black shoes were polished to the highest standard, and my finest shirt was duly washed and ironed to perfection. The days soon passed and it is now 1.45pm on this *very auspicious* Friday afternoon. I am standing just inside the entrance to the Castle Arcade, and I can see the 'Perfecta Film Development' office on the balcony above my head —but how on earth do I get up to the first floor? Walking along the arcade and passing a few small shops in the process, I finally stumble upon a bi-folding door, which is slightly ajar, and so I open it fully, thus revealing a steep, swirling staircase, which obviously leads up to the balcony level.

As soon as I reach the top I am standing directly opposite the *Perfecta Film Development* office, and looking through a *sliding window* hatch, which is adjacent to the main entrance, I can see a woman sitting at a desk, who I naturally assume is Robert's Secretary. Rather than tapping the glass, which might seem a little unprofessional, seeing as it is my first visit, I knock gently on the door and wait to be called in by the lady in question. Upon entering the office I am told that Mr. Stains is currently tied-up on the telephone in his private office, before being kindly offered a seat while I wait for him to finish his conversation. I dislike having to do this, as waiting-around for people, especially when it is someone whom I have never set eyes upon before, always makes me a little nervous, to say the very least. Also, sitting outside a closed door and waiting for someone in authority to emerge, brings back a few *not-so-fond memories* of waiting outside the headmaster's office as a schoolboy, waiting to be caned mercilessly by *'Basher-Beynon'*, the unforgiving head teacher of lower school, who never believed in the word 'leniency!'

I imagine that it is not so easy for the poor secretary either, having to continue-on doing her normal days' work, such as typing-out important letters, sorting-out piles of official documentation, and answering a volley of never-ending calls on the telephone, knowing full-well that her every move is being *monitored* by a pair of beady eyes from an unknown stranger. After a few minutes Mr. Staines emerges, and once again I am greeted very courteously, along with being offered a choice of liquid refreshments, as Robert invited me into his personal chamber. Just like Mr. Gerwyn, Mr. Stains is about my age, maybe a couple of years younger; it was difficult to tell really, but to be brutally honest this man's date of birth was the least of my concerns right now. The office was actually a lot bigger on the inside than it looked from the outside, encompassing two large wooden desks, one of which was smothered with a dozen different manuscripts, most of them as thick as your average *Yellow Pages* phone directory, along with a pile of files and folders that ranged from being only a few millimetres in thickness, to ones that would take two hands just to lift them off the desk.

In total contrast to its counterpart, the second desk was virtually clear, barring only a few small stationary items, and so I headed straight for the latter. Spreading my wares out in front of Robert, I began *plugging; pushing and promoting* my project for all that I was worth, barely stopping to catch my breath for a second. Within the first hour Robert had barely said a word, apart from asking the odd question here and there, while I continued bombarding him with *facts, figures and formula's* for my proposed production. I had originally been granted a one hour audience with Mr Staines, but I had already been with him for twice that long, before Robert finally took over the conversation –probably because he was sick of the sound of my voice! Robert says that the idea is excellent in theory "But you must get a script together in its entirety", he added, rubbing his clean-shaven chin from side to side with the palm of his right hand. Robert then paused for a short breath, while I sat in total silence, *watching, waiting and wanting* him to continue-on with his speech.

"You see Shaun, my job is to take completed works to film festivals, where I meet prospective sponsors, who will hopefully instruct me to go-ahead and make a movie out of one or two of them. You are obviously nowhere-near that stage as yet, but just to get the ball rolling, there is nothing stopping you from contacting the 'Mechanical Copyright Protection Society' which is based in London, in order to find out if the music from the 'Bat out of Hell' album is available, and if so, then to also find out how much it would cost you to be able to use the individual songs as the soundtrack to your movie."

As I begin copying down the address and telephone number of the aforementioned company out of Mr. Staines' address book, Robert stands up and walks over to the main window, which overlooks the city centre. Gazing directly across the street and staring at the main entrance into Cardiff Castle, as if this historic landmark was some kind of inspiration to his train of thought, he paused for a few seconds, before continuing on with his speech:

"In order to produce a decent script, you are going to require the assistance of a professional script-writer, and so I am going to give you the name and telephone number of one of the best script-writers in Cardiff".

Flipping-over a few pages of his address book, which I have left open on the desk, Mr. Staines pointed to the name of 'Josef Zagrodnic', saying that Mr. Zagrodnic is a Polish gentleman, who works mostly from his home, in a place called Radyr, which is situated on the outskirts of Cardiff. As my pen goes into overdrive, Robert informs me that this man has already made numerous documentary films about motor sports, even turning one of them, on the Mexican Car Rally, into a movie. Robert then added that he is quite confident that Mr. Zagrodnic has already read 'Jupiter's Travels', and so he should be very interested about the idea of turning Ted's journey around the world into a movie. To round-off our *three-and-a-half hour* conversation, Robert said that there are two quotes which totally sum-up movie production, the first of which was said by Mr. Alfred Hitchcock many years ago:

"There are three things that make a movie good –'THE SCRIPT, THE SCRIPT, THE SCRIPT', because without a good script the film is worthless".

That quote certainly made perfect sense to me, but I wasn't so sure about the second one, which, incidentally, Robert could not recall who had said it?

"Any movie which cannot be fully explained and totally expressed in five straight-forward sentences is not going to be worth making!"

Nevertheless I would take heed of this warning, before I attempted to convince Mr. Zagrodnic of the merit of my movie. It was almost dark by the time I finally left the confines of Robert's office, but I had learned so much from this man's 'free' advice, that my time could definitely be counted as 'well-spent', after-all, Robert's time is 'money' –whereas mine is not.

CHAPTER 12

Best In The Business

I rang Mr. Zagrodnic that evening, and sure-enough, I was invited to a half-an-hour chat at his home in Radyr on the following Wednesday morning. However, the next morning, whilst I was out at the park with Liam and Carl, Mr. Zagrodnic rang the house, in order to make a slight alteration to our rendezvous, the details of which he duly gave to Caryl. My new instructions were to meet him at 9am on that same morning, only not at his house, but in the car park of the local Mercedes Benz car dealership in Pentwyn, which was fine by me, as the area of Pentwyn is considerably closer to my home than Radyr. Apparently, his car had already been booked in for a service on that day, and so I would be doing him a favour if I could pick him up at the showroom, as he would be leaving his car there until sometime in the afternoon. I was pleased to do this man a kindness, but absolutely horrified at the thought of him having to travel in my *beat-up* old cavalier, which, but for the grace of God, would have been rusting-away in the local scrap-yard a long time ago.

My only recourse was to ring Jules, who immediately offered to loan me his Ford Sierra for the day. If Mr. Zagrodnic had already read Jupiter's Travel's, then he would know how exciting the storyline is, but then a thought occurred to me; what if he had read the book many years ago, and now he only had vague recollections of the many events which took place on the journey? I had to prepare something which would instantly ignite a spark in his memory; some kind of *visual aid* that would immediately create an impression in his mind about how diverse and exciting this movie could

129

be. Using s poster-sized sheet of graph paper I drew-up a list of the all of the major events in the order that they occurred throughout the entire journey. From a scale of 'minus 100' to 'plus 100', I then meticulously plotted the levels of the 'Good times' against the levels of the 'Bad times' –well, how I saw them anyway. Finally I joined-up the dots, so that it ended-up looking like an over-sized cardiogram, rather than a *movie outline,* but so long as it painted the desired picture, then as far as I was concerned (and hopefully, Mr. Zagrodnic too) that was all that mattered.

Even-though I am still a great believer of my own, *self-confessed* adage of "Why use 'ten' words, when a 'thousand' will do", I remembered Mr. Staines' *words of wisdom,* and so I wrote-out '5' (rather lengthy) sentences underneath the graph, explaining the movie in full. I then rolled-up the poster, before sliding an elastic band around the middle of it, and not wanting my beloved children to get hold of my latest masterpiece, I then placed it safely on top of the wardrobe in our bedroom. On Wednesday morning I drove the five miles to the Budget Rent-a-car offices in Collingdon Road, where my great friend, and *proverbial business-partner*, Jules, was waiting for me. After leaving my *junk-heap* outside the main building and collecting Jules' car, I then fought my way through the rush-hour traffic, before *flying* down the A48 at speeds which my car could only ever dream about! Enjoying the comforts of a brand new car again was wonderful, especially when one is used to driving-around in a cronk!

To think that it was only three months ago that I was in possession of a brand-new 'H' registration 1.4 litre Ford Orion car –my company vehicle which I had the pleasure of driving myself and my family wherever we wanted to go, but sadly that *rather large* luxury in my life got withdrawn on 21st September 1991 –the day that I lost my job! However, it was no good reminiscing on the past –I had to look to the future. "One day I will be back on top," I said *positively* to myself, before screaming-up the exit road. I knew that I had to believe in myself, otherwise how could I expect anyone else to believe in me –and above all else I must keep faith in what I am doing if I am going to make it work. "And why shouldn't it work –everyone I have spoken to so far thinks it is a marvellous idea –of course it will work", I whispered reassuringly, before bringing the car to a screeching halt in the Mercedes Benz car park.

As I walked over to the workshops I could see only one man standing in the *Service Reception* office, and as it had already turned nine o'clock, I gathered that he must be the man I am looking for. Wanting to break down any barriers that there may have been between us (i.e. Mr. Zagrodnic being a 'true professional' –and me being a *meagre, low-life amateur*) I decided to try out his sense of humour. In my briefcase was a big red folder, which contained a pile of notes, letters and general information regarding my entire family, as I had been working-on a surprise 'This is Your Life' presentation for my mother's 60th birthday in December, and so I decided to use my *prop* to play a harmless little prank on the man, in the hope that he would see the funny side of it, and that the pair of us would end-up laughing like fools together. "It would certainly make a change from the normal introductory procedures", I said to myself, as I removed the said folder from out of my briefcase, and as so many showbiz actors of stage and screen are always saying "Make 'em laugh" I thought that I would give it a whirl.

Walking up to the gentleman in question, who is now leaning on the end of the main counter, casually flicking his way through one of the accessory catalogues, I politely enquire if his name is Josef Zagrodnic, at the same time making sure that the folder which I am holding behind my back, is completely obscured from his view. As he gave me the affirmative I immediately broke into my rendition of those *immortal words*, made famous by Eamonn Andrews.

"Well Mr. Zagrodnic, you thought that you were here today to talk to me about a movie production, but that is not the case..." (At this precise moment I brought the folder around to the front, so that he could plainly see the four words printed in gold on the cover) "...for I am here to say that 'Josef Zagrodnic', script-writer, movie producer -and renowned film director...'This *is* Your Life'".

As I stood facing Mr. Zagrodnic with a big smile on my face, he only gave me a polite smirk in return, making me feel a little uneasy. The joke wasn't meant to be hilarious, but his quite nonchalant reaction had fallen far short of what I had expected, and immediately I regretted my actions.

Maybe he thought that I was trying to ridicule him, or worse still, he may well have believed that I was serious for a moment, and that my phone call was just a way of entrapping him...surely not? I decided that the least said about this little episode, the better, although I did apologise to Mr. Zagrodnic, as we walked over to my car together, just in case I had somehow managed to hurt his feelings in any way, shape or form? Contrary to my *somewhat negative* thoughts, Josef only replied how grateful he was to me for giving him a lift at such short notice. As we drove away from the dealership, I immediately launched into my *sales pitch*, which caught Mr. Zagrodnic somewhat by surprise, as we had barely had time to say "Hello" or "How are you" to one-another? (*Sometimes my impatience knows no bounds?*)

Nonetheless, our conversation continued for the rest of the twenty-minute journey through Llandaff Village, and along the Llantrisant Road, before turning-off into the relatively small village of Radyr, and finally coming to a halt about halfway along the main street. After crossing the road together, we stopped at one of the terraced houses, which lined the main avenue, where Mr. Zagrodnic began rummaging in his pocket for the key to his front door. Perhaps I had seen too many episodes of 'Dallas', for only now did I realise that my expectations of this man living in some kind of *exorbitantly expensive* mansion, surrounded by acres of farmland pasture, were more than outlandish –indeed they were downright ridiculous. Once inside the house, Josef immediately introduced me to his lovely wife, who took great pleasure in showing me her incredible 'model house', which she had been working on for several years. Constructed at the base of a chimney breast in one of the back rooms, it stood almost five feet in height, and was approximately four feet across.

With over a dozen windows, spread between four storey's but only one large door, which was set in the centre of the ground floor, it looked something like the house I had expected this couple to be living in. It was truly a work of art, and the most magnificent 'fireplace' I had ever seen in my life. Even the three separate rooftops had been brightly coloured in shades of red and blue to enhance its beauty and sheer elegance. However, the best was yet to come. Releasing a tiny hook on the left-hand side of the house, Mrs. Zagrodnic proceeded to open the entire frontage of the model into the living-room space, subsequently revealing literally hundreds of miniature furniture pieces, along with half-a-dozen mini-statuettes of the houses' *imaginary* occupants. Every single piece had been hand-made to scale, before being painted in the most vibrant colours –I simply had to applaud her efforts. The three of us sat drinking coffee and chatting together for around ten minutes, before Josef asked me to follow him back down the main hallway, which leads to the front of the house. Both the middle and the front rooms had been knocked through into one large office-cum-study, and over in one corner of the room was another doorway, which led to a very small room, not much bigger than your average closet.

Shelf-loads of video cassettes, along with an endless number of books, adorned the one wall of the main room, and underneath these sat a large television set, along with a rather plush-looking video-recorder. Whilst Mr. Zagrodnic positioned himself on one corner of the sofa, which was situated in the well of the bay window, I immediately plonked my *backside* in the armchair directly opposite him, in readiness to begin pleading my cause once again. Josef had already told me in the car on the way over here, that he had read Jupiter's Travel's, and so with all the enthusiasm in the world, I immediately unravelled my posters, before meticulously going through my script, whilst playing a few songs from the Bat out of Hell album on *his* music system, all the time knowing that I only had around half-an-hour to convince this *very valuable* man about the credibility of my project. Josef listened intensively to what I had to say, before taking a *pink-covered* paperback book off one the shelves, which was entitled 'Zen and the art of motor-cycle maintenance' -a book which I had heard of beforehand, but had never read.

Josef said that it was a very encapsulating book, which delved deeply into the innermost thoughts running through the author's mind as he worked on his beloved motor-cycle in his workshop, and also as he rode his bike through the streets of his home town. Josef then added that whereas Ted wrote about a 'one-off' journey, as opposed to lots of small trips on a bike —and even though Mr. Simon had also revealed his innermost thoughts along the way, just as the other author had done, that was where the similarity between both books ended. After hearing this, I knew that I must read 'Zen and the art of motor-cycle maintenance' if I had any hope of fully understanding what Josef was talking about, and I kind of hoped that my *host* might offer me his copy to read, but sadly my luck was out.

In the days that followed I obtained a copy of the aforementioned book, and all I can say is that it 'completely blew my mind' —in more ways than one. The totally unexpected ending revealed everything about what the author had written in the book, and indeed 'why' he had written it. Obviously I am not going to spoil it for those of you who might also want to read this book, but if you are considering doing so, then all I can say is that you are going to need every ounce of your wits about you, and that you had better have a box of Kleenex at the ready.

Josef said that he had once thought about putting together a documentary based on 'Zen and the art of motor-cycle maintenance', but as for making a movie out of Jupiter's Travel's, well that idea had never entered his head. However, I had obviously succeeded in opening his mind at the prospect of turning Ted's journey into a film, because the answers to my many questions came next.

"The filming would take about six to eight weeks, to complete all of the location work, as many of the scenes would be shot on a studio set in this country, in order to minimise the cost. As for the music, well you could be looking up to as much as sixty thousand pounds, which would take a hefty chunk out of your overall budget, which I estimate to be around the four hundred thousand pounds mark –or possibly even half-a-million?"

Josef then went on to talk about the numerous problems he had already encountered in obtaining filming rights, and how the varying laws in many of the countries he had previously filmed in, had caused him major upsets in the past, which brought home to me the reality of what I might be letting myself in for should this great movie idea of mine ever reach fruition. He also added that whenever a difficulty arose in the past, it would always turn out to be very expensive to stabilise it, and that volatile governments in various countries, may not take Ted's views on them too kindly, and so filming in any of these countries would be a definite *no-no*.

This had easily been my most informative meeting to date, even if a lot of what I was hearing was veering on the negative side. However, I was undeterred, and even more determined to make it work. My half-an-hour allowance had already turned into one and a half hours, and Mr. Zagrodnik still showed no signs of *letting-up*, as he continued on with my extremely enjoyable –and absolutely invaluable, lesson.

"The hardest part would be condensing something in the region of twelve to fourteen hours of good documented scenes into a two-hour production, without racing through the journey, thus spoiling the effect and the impact which you will obviously need to create. At the same time you must also make sure that the audience does not lose track of where Ted actually is on each stage of the trip, and so wherever possible you must give each scene its own genuine purpose"

My brain was beginning to boil over with a mixture of excitement and bewilderment. Trying to soak-up all of Josef's knowledge of the industry in one fell swoop was one thing, but being able to put his professional words of wisdom into practice would be an entirely different matter. However, I need not have concerned myself with such *trivial matters*, for my most harrowing question was just about to be answered.

"The first thing you must do Shaun is to get yourself a professional scriptwriter. He will then have to spend at least two or three days with Ted Simon, before he can begin setting out the scenes and dialogue, etc. From this point on you could expect a finished product within five to six weeks. A round figure for his expenses would be approximately five thousand pounds, give-or-take a buck".

I should have known that when Ted told me that he did not have the skill to write a script for a movie, the chances of me writing one on my own were very remote, but now I humbly accepted the fact that I would have to find five grand 'just for openers' —and then I could start worrying about the other costs as and when they reared their ugly heads.

Whilst we were getting-down to the *nitty-gritty* -and the *do's and don'ts of* movie-making, Josef veered onto the lines of 'constructive criticism', which I was only too-willing to listen to, of course, especially from the likes of people who knew what they were talking about, such as Mr. Zagrodnic.

"Meat-loaf's music is basically "Rock 'n' roll, and I have no-doubts that it will work fine for the scenes in North America and Australia —and also for the opening and closing credit scenes in Europe, but do you really think that it will be suitable for a South-American prison setting? Personally, I think that some local music, such as a Latin-American song, would be more appropriate?"

On this occasion, I had to disagree with Josef, but not before I had given him my reasons as to why I thought that Meat Loaf's music would be perfect.

"If the song was one of Meat Loaf's livelier tunes, then I agree that it would certainly not be suitable for a prison scene, but this is a very emotional ballad, with both the lyrics and the music aptly portraying

Mr. Simon thoughts when he is at his weakest and most vulnerable –and even Ted, himself, admitted that the words of the song depicted his actual thoughts, along with the psychological traumas that he was going through during this horrendous ordeal, and so what could be more appropriate than hearing it from the horses' mouth?"

Although we had *begged to disagree* on that last issue, Mr. Zagrodnic and I were still getting on famously, and our meeting still showed no signs of termination, as Josef turned on his television set, before placing a tape into the cassette slot on his video-recorder. Josef then asked me to watch a programme which he had made for a major network station, which had apparently been shown on television only this week. When Josef told me that the documentary was entitled 'A bit on the side', I wondered what to expect at first, but all would soon be revealed, as the picture came into focus on the screen. It was all about motorcycle and side-car combinations *through-the-ages*, and the keen enthusiasts, who had spent a great deal of their lives collecting and working on them. Some of the characters that had been interviewed for the programme had tremendous personalities, and the variance in the sizes and shapes of some of the side-cars (compared to your *run-of-the-mill* 'torpedo-shape') were simply unbelievable. As our meeting drew to a close I felt confident that Josef was on my side, but now I needed to *take the bull-by-the-horns* and see how he would react to my offer:

"You said that five thousand pounds would be a round figure to cover the writer's costs, and so I am asking you –if I can raise that amount, would you be prepared to take-on the contract?"

"Sorry Shaun, but had you asked me that question a year ago, I may well have taken you up on your offer, but unfortunately I am fully committed to other projects at the moment."

Leading me into his small office, Josef then pulled a small chalkboard off the wall, which had three titles listed in numerical sequence on it. Pointing to each one in turn, he said that it was going to take him the best part of a year to complete all three of them, and so he would simply not have the time to fit in a fourth project. I then asked Josef if he knew of any other scriptwriters who might be willing to take on the task, but he said that every scriptwriter he knew was currently *chock-a-bloc* with work. "Robert Staines sees far more script-writers than I do –why not ask him for a couple of names; I am sure that he will have somebody on his books who is looking for work," he added.

Thanking Josef profusely, not only for his time and his hospitality, but also for his wealth of knowledge, I then offered to pick him up this afternoon, and take him back to the dealership, in order to retrieve his car, which he immediately accepted. Driving straight to the *Budget* offices, I informed Jules about the outcome of this morning's meeting, before asking him if I could borrow his car for a few more hours, simply to save me the embarrassment of having to pick up Josef in my car. "Talking of embarrassment Jules..." I say quite sheepishly, knowing that I am just about to ask him my most awkward question to date "Is there any chance that you would be able to help with the money for the scriptwriter?" Unfortunately Jules said that he would be unable to do so right now, as the majority of his money was currently tied-up in setting-up the new Budget Rent-a-car franchise in Swansea.

My meeting with Jules had gone on a lot longer that I had expected it to, and so I arrived at Josef's house about half an hour later than we had agreed upon. Josef had arranged to pick-up his car from the dealership at 2pm, and so as it was now two-thirty, his wife told me that he had already taken-up the offer of a lift with someone else -shit! I was very angry and duly cursed myself for letting him down after all the assistance he had given me this morning, but it was too late to do anything about it now. Anyway, I had far-more important issues to deal with at this moment in time, such as finding a *generous* sponsor, who would be willing to come up with five G's?

CHAPTER 13

This Is Your Life

Staying awake until way past midnight, I rung through to the international operator in Canada, who kindly connected me through to the home of Brynley Graham –the gentleman I had met in San Francisco a few weeks ago. Brynley's wife answered the telephone, and so I immediately introduced myself to her, telling her how Brynley and I had met at the Holiday Inn Hotel –and also how we had ventured into Chinatown together for a meal on our last night in *Frisco*. (I did not mention his generous offer to me, just in case Brynley had omitted to relay that to his wife when he returned home). She told me that Brynley was still at work, as it was only around 6.30pm in Saskatchewan, adding that he rarely returns home before seven-thirty in the evening, but that she would get him to call me the minute that he walked in through the front door. I was certainly *not* looking forward to staying up until the early hours of the morning, in the hope that Brynley would call me back, but "When needs must..." as they say.

However, within ten minutes the phone was ringing –it was Brynley on the line. He said that his wife had contacted him in work immediately after she had spoken to me, and so, rather than keeping me up half the night, he had decided to call me directly from his office. Brynley then went on to say that he had told numerous friends, along with several business associates about our chance meeting, and also our little chat regarding my proposed production, before adding that every one of them had been enthralled with my idea. This was great news, of course, but 'saying' and 'doing' are two entirely different things, and so I had to find out whether any of these people were genuinely interested in becoming involved in my movie production, in other words, willing to *put their money where their mouth is*, so to speak. Initially, I personally offered Brynley the opportunity of becoming a partner in this great venture, as he was the person who had made the original offer of investing in the project.

I then said to Brynley that if any of his business acquaintances were genuinely interested, and wanted to know more about my proposition, then I would be willing to fly over to Canada, to give them a full presentation of the movie concept –and also to expand on my terms and conditions of possible *joint-venture* contracts. *Thankfully, Jules had offered to sponsor that trip should the need arise?* Brynley said that he already had money tied up in different ventures, before adding that he was also going off on a skiing holiday this weekend, but that he would put the word out to his fellow associates in the next few days, and upon his return he would assess his own financial situation, to see if he was in a position to invest a proportion of his own personal capital into the venture? I had not expected an immediate answer, and so when Brynley offered to call me back with possible proposals within the next seven to ten days, I was more than satisfied with the outcome of our discussion. The following day I rang Mr. Staines to see if he could help me with a writer, but he said that he only saw scriptwriters as and when they had a completed script to present to him, before adding that, as far as he was aware, all of his regular writers were currently inundated with work.

However, after a short pause, Robert said that he had another idea which might work "But don't build-up your hopes too much Shaun", he added commandingly. In my current position I was willing to take on anything that gave me a glimmer of hope, and so I almost *begged* Robert to let me in on his idea, "Andrew Coles is the man who *may* be able to help you" he answered, before explaining that this gentleman is the head of the Drama Department at the H.T.V. Television Studios in Culverhouse Cross, which is located on the western outskirts of Cardiff. "This man has the power to commission writers Shaun, and although I have great difficulty in getting an audience with Mr. Coles, you seem to have the *knack* of meeting the *right* people at the *right* time, and so he has got to be worth a try". Robert seems genuinely concerned with my plight, telling me to come back to him if I have no luck with Mr. Coles, and so I know that he is not just *passing-the-buck*.

Also, like everyone else I have spoken with to date, Robert has not asked me for a single penny for the time he has given me, or for all of the invaluable information and the contacts he has passed on to me, and so I will be forever indebted to these people for their assistance in helping me realise my dream. (*In fact if any of the aforementioned people had placed any kind of fee in front of me for services rendered, then the whole project would have been brought swiftly to an end*). Managing to get through almost straight away to Mr. Coles' office, I am duly informed by one of his producers that 'Andrew' is an extremely busy man, and therefore he only accepts written applications for an interview. "Unless you have been introduced personally by a fellow producer, of course", the gentleman added in a somewhat patronising tone, and basically telling me that without being as 'important' as Mr. Coles obviously is, then my chances of getting an audience with him were pretty slim.

In all truthfulness it was the response that I had expected all along, but it was better than nothing I suppose, and so I duly wrote a two page letter to Mr. Cole, outlining my intentions, before popping it in the post to him that very same day. The price of a first class postage stamp is quite nominal, and the cost of my petrol and parking fees to and from my various meetings I have managed to cope with quite well, but upon receipt of this morning's telephone bill I nearly died! It is approximately three times the amount of our normal *quarterly* bill, and far in excess of one hundred pounds. I cannot afford to be without a telephone line at this stage of the game, and Caryl certainly couldn't cope without having a *lifeline* to her family, and so we have no choice but to go cap-in-hand to her parents for a *relatively small* handout, so that we will be able to pay our bill. The borrowing has started, and it will only get worse, unless something positive turns up really soon? With a very high mortgage to pay, the only jobs that are offering a large enough salary to encompass all of our monthly outgoings, whilst still leaving enough money to survive on, require a string of letters after one's name, and so with nothing more than a handful of CSE's to my credit, my chances of getting a decent job are quite minimal.

There are always the sales jobs to go for, of course, with stupendous O.T.E's. (On Target Earnings), but the basic salaries are so pathetically low, or even non-existent in some cases, that I have to skip on them, as 'paying to go to work' (i.e. transportation, clothing costs, food, etc.) is certainly not an option in my current situation. "I wonder if I am the first person on the dole to suffer with 'executive stress'", I joke to myself, trying to make light of a horrendous situation. However, as I have not invested thousands of pounds into this venture of mine, such as renting a business premises, or purchasing office furniture and a company vehicle, I suppose I should consider myself quite fortunate, for if everything does fall flat on its face, then at least I have nothing tangible to lose. In the past I have read and heard about so many entrepreneurs who have risked everything when setting up their new business, only to end-up losing the lot. Also, the people who have succeeded in building-up businesses, and amassing huge profits over the years, are currently suffering greater losses than ever, which they simply cannot cover, and so I often ask myself "Who is really worse-off in life —me or them?"

It is pointless continuing-on with my quest without having any doubts whatsoever, otherwise if everything did go *belly-up* in the end then the strain would be too much to bear. So far I have not been given a single guarantee that anything will come of my efforts, but so long as there is a possible chance that my dream could become a reality one day, then I must keep at it one hundred percent. To satisfy my curiosity regarding the cost of the main soundtrack for the film, I rang the Mechanical Copyright Protection Society in London, who duly informed me that I must contact the 'Carling Music Corporation' —and so I did. The gentleman I spoke to, a Mr. Drew Carter told me that they would have to apply to the United States for permission to use virtually a whole album in one film production, adding that the company had never received this kind of request beforehand. Drew said that he would ring me as soon as he had received his answer from America, which should be within the next seven to ten days.

The following morning I received a *Dear John* letter from Mr. Coles, saying that unfortunately he would not be able to spare me any time for a chat, which, as you have probably guessed by now, is exactly what I had expected from the off. However, he personally wished me "All the best" with my venture, adding that the H.T.V. Studios were currently overrun with programme scripts, which would see them well into 1992 anyway, and so having a meeting at this point in time would be fruitless...point taken Andrew. To make matters worse, I then received a call from Mr. Carter in the afternoon, telling me that unless I could give them specific budget figures, along with determining whether the movie would be shown in the cinema, or on television then they would be unable to apply for realistic and accurate figures for the proposed soundtrack –double shit! As I did not have the answers to either of Mr. Carter's questions at this precise moment in time, I said that I would get back to him at a later date. By now I was seriously *clutching at straws*, and as much as it really greaves me to admit it...even to myself, I really didn't know which way to turn?

In all fairness to Mr. Staines, he had done all that he could for me, and so going back to him for help would probably result in one more dead end. Also, the last time we spoke together, it was as 'good friends', and so if I rang him again it would be as a 'pest' in his eyes, I am sure, and that is the very last thing that I would want. Mr. Gerwyn had told me to give him a call if I managed to raise the £1million it would take to make the movie, which was just about '1 million' miles away right now, and Mr. Zagrodnic had already turned me down once, and so the only real hope I had left was to hear from my Canadian friend, Brynley. Unfortunately I had heard nothing from him in almost a week, and so I was beginning to seriously doubt his sincerity in my project. Slowly, but surely I began sinking into the realms of depression, the light that once flickered at the end of my personal tunnel, having seriously dwindled to nothing more than a tiny spec in the distance. Things were not looking good at all.

I could always ring Brynley and simply ask him the pertinent question, of course, but my pride had gotten the better of me. "Surely there is someone out there who would give their right arm to have a piece of the action I am selling", I whinged out loud...but no-one was listening, except possibly the cat, who was currently curled-up in a ball on my lap, purring-away to his hearts' content. "Oh how I wish I was a cat sometimes; I might even be able to sleep a lot better than I have been doing of late", I grizzled even more to my ever-so confused, and *totally battered* brain.

However, as strange as it may seem, Jules had not lost any faith in either the project, or more importantly I suppose, in me. When we met up a few days later, he told me that when he had read my letters of reference all those weeks ago, each of which had stated quite categorically that I always put 100% of myself into any task that I undertake, he was quite sceptical at first. "However, since we have been working together, I would say that it is more like 150%" Jules added. As for the project, well Jules insisted that 'fate' would play a great part in it being a success in the end, before adding that that it would only be a matter of time before things would begin falling into place. I only wished that I could share in Jules' cheerful optimism, but I was currently feeling *all-burned-out* inside, even though I knew that I would have to *snap-out-of-it* pretty soon, otherwise there would simply be no hope for me?

"Remember the day that I purchased your flight ticket to California Shaun, and you reciprocated by treating me to a copy of Jupiter's Travel's, saying that I could read all about Ted's journey whilst you were away", Jules said jovially, his face beaming with positivity. I will never forget that day as long as I live, of course, but unfortunately I was not in the same frame of mind as Jules right now, and so all I could answer him with was a *stone-faced* nod of the head, my *miserable attitude* just wanting him to get to the point of his story as quickly as possible.

"Well, a few days after you had flown-off to America, my car was broken into right outside Budget's main office, and my briefcase was duly stolen from off the back seat. I was absolutely horrified when I found out that the thieving bastards had smashed one of the front windows, in order to gain access to the car, thus leaving hundreds of fragments of glass splattered all over the seats –and I was even more upset about losing my briefcase, which had been a birthday present to me from my wife, and was therefore of great sentimental value to me. But the thought of losing that precious paperback which I had been reading well into the early hours of each morning -and loving every minute of it, had devastated me beyond belief.

In total despondency I walked around the car, simply to check if there was any other superficial damage, but thankfully the thieves had only been interested in ransacking the inside –and so, after brushing the majority of the broken glass off the driver's seat, I reversed the car into the nearby workshop, in order to have it repaired. Returning to my parking spot, with a view to sweeping-up the glass from the floor, I could not believe my eyes –for there, lying on the ground in front of me, was my 'Jupiter's Travel's' paperback, Ted Simon's face on the front cover staring directly up at me, as if he was saying "You didn't think that I would let those fuckers ruin everything did you Jules?" Even though I had not heard from you as yet, I knew that this was my sign, telling me that your meeting with Mr. Simon had been a total success".

Christmas was approaching rapidly, and as always Caryl was entering into the true spirit of things, but all I could think about was how I was going to get up-and-running again? We had managed to buy a few presents for Liam and Carl, with the meagre amount of spare cash that we had left in our cash reserve fund. This was primarily the money that we had been saving for all of those huge winter bills, which would inevitably be upon us after Christmas. However, with our children's happiness being far more important to us, we had decided to worry about them as and when they arrived? Talking of arrivals, well our third child had still not decided to enter this world of ours as yet, even though Caryl's due date was yesterday, and so whether we would be buying one or two additional presents for a baby boy or a baby girl was still uncertain right now?

The decorations had certainly brightened-up the inside of our house, and the fairy lights glistened and sparkled on the Christmas tree, but to me the world was still a very dark place. However, I refused to allow myself to stray near the *boredom border-line*, and so I worked tirelessly on my mother's 'This is Your Life' surprise party, which was due for perpetration on Sunday 22nd December, which is two days after mum's official 60th birthday on the twentieth. Working in collaboration with all of one's relatives, both home and abroad, can be quite daunting, especially when some of them live as far away as Kano in Nigeria, but in comparison with the amount of work that was necessary for my other project the task of contacting everybody via letter and telephone, was a relatively simple one.

Whilst I was out collecting the guest's *speech* forms (*these forms contained individual stories about my mother, which the members of my family would tell on stage*) from relatives who lived in and around Cardiff, my father-in-law had rung the house, leaving a message with Caryl for me to ring him upon my return. As Donald (my father-in-law) owned a tyre and motor-factors company in Cardiff, I simply assumed that he wanted me to book our car in for its annual service –but I couldn't have been more wrong. Donald told me that the son of one of his long-standing customers, had called into the garage this morning, in order to have a few new tyres fitted to his Mercedes. Whilst he was waiting in the reception area he and Donald had struck-up a conversation, in which Donald had told him all about my trip to the US, along with my subsequent meetings with the *powers-that-be* in the film and television industry. Donald then added that as the gentleman is a television producer, himself, he showed great interest in my story, immediately giving Donald his personal business card, before asking him to ask me, to contact him at my earliest convenience.

"The gentleman's name is Ryan Willis, and he works for a company called 'Agenda'", Donald said, before reading Ryan's direct telephone number to me off his business card. Immediately I rang Mr. Willis, who told me that although he lives in Cardiff, he works in Swansea (which is located about 40 miles / 60km due west of Cardiff) adding that the two main television companies he deals with are Channel 4 and S4C (*S4C is a counterpart company of Channel 4, that deals primarily with Welsh-speaking programmes*). After briefly enlightening Ryan on the outcome of my meeting with Ted, along with telling him about all of the 'positive' things which had happened to date, we duly arranged a meeting at the company's offices in Cardiff for this coming Friday, at 3pm in the afternoon. As soon as I had replaced the handset, I immediately jumped to my feet, before screaming-out at the top of my voice "I am alive again".

However, things didn't end there, because when I rang Jules to give him the good news, he also told me that an old acquaintance of his, who had slipped Jules' mind up until now, worked as a set director at Pinewood Studios in London. Things were getting better by the minute. Jules told me that his name was Terry Atwell, adding that he would try and make contact with the man after Christmas, with a view to arranging a meeting between us. All systems were definitely on 'GO', and my mind was on fire, as I parked my car outside 'Grosvenor House' (the name on the plaque outside the building) which is located in the centre of Cardiff. I was so pleased that Ryan had suggested us meeting in Cardiff, rather than having to go all the way to Swansea, which would once again have necessitated the use of Jules' car, as my *rust-bucket* would simply *not* have made it. Having finally learned what *time-management* is all about (which, in all truthfulness, it has taken me the best part of twenty years to master), I am over half-an-hour early for our appointment, which I am sure will impress Mr. Willis no-end.

However, my *grand-entrance* was short-lived, as his secretary graciously informs me that Ryan has just popped out to the bank, before adding that he will probably be back in around fifteen to twenty-minutes time. Leading me up a very long staircase, and then across a small landing into the main reception area, I sit watching the minutes tick away on the wall clock, whilst gently sipping a cup of boiling hot coffee, which the lovely lady has kindly made for me. At 2.55pm Ryan enters the room, immediately introducing himself to me with a firm handshake. Ryan is built like a rugby player from Wales' front row, with a huge chest and a pair of very broad shoulders to match. He is also donning a pair of dark tinted glasses, which assisted in maximising his *somewhat brutal* appearance. However, what I noticed more than anything else was the fact that he was wearing a black suit –which just so happened to be what I was dressed in, only in a much smaller size, of course.

We both laughed in unison as I expressed to Ryan how lucky we were that we were not of the female agenda, otherwise one of us would have had to go home and get changed –'colour-clashes' and all that *female* stuff! For some inexplicable reason I did not feel nervous at all, as the pair of us walked back down the stairs together to the main meeting room. Maybe it was the fact that Ryan was an old friend of my father-in-law's that I felt totally at ease with this man's company, or perhaps it was because my little joke had immediately broken the ice between us? Or could it be because I had finally overcome this massive inferiority complex, which I have always suffered from, with persons I consider to be far more intellectual than me? Whatever the reason, I was feeling on top of the world right now, and I couldn't wait to get the meeting underway. As we entered the office together I felt as if we were walking into a basement flat, because the room was so dark.

There were no lights on at all, and natural daylight could only gain access through a narrow, scullery window, which was situated in the far corner of the room. Unfortunately, as today was a typically dull, December afternoon, the dark grey skies outside only assisted in darkening it further. Ryan began our meeting by saying that even-though I had given him a general overview of the movie over the telephone, he now wanted me to tell him the whole story to date, in order that he could fully understand my intentions. This I did willingly, of course, and although Ryan had originally known nothing at all about Ted's journey, prior to our telephone conversation, along with the fact that he had only ever heard snippets of Meat-loaf's Bat out of Hell album, but by the time I had finished my *spiel*, he had resigned himself to the fact that not only would he have to read Jupiter's Travel's as soon as possible, but that he would also have to purchase a copy of Meat-loaf's 'Bat out of hell' album, before listening carefully to the words of every song from start to finish.

By now Ryan's enthusiasm was almost on a level-par with my own, as he told me emphatically how he would love to work with me on a television series about Ted's journey. Ironically, Mr. Zagrodnic had suggested the very same thing to me during our *rather lengthy* conversation at his home, even advising me at the time to contact a Mr. Gerald Wakefield at the BBC in London, who was apparently the commissioning editor for documentary programmes. However, Ryan could see that I had my heart set on making a movie, and so he said that he would give me the name of a television producer in Cardiff who actually owned his own film production company, adding that if I were to fail in my current direction with the movie, then I was to get back in touch with him immediately. Knowing that I had something to fall back on, should all else fail, was certainly a benefit, although I was still determined to achieve what I had set out to do in the first place, and that was making an epic movie.

'Jim Howells' is the name of the next gentleman, who will have to *suffer the pleasures* of being *hounded and pounded to death* by yours truly. Jim is the Managing Director of a company called Ffilm Cymraeg, which is situated in Cardiff, although Ryan has told me that he will not give me the whereabouts of Jim's offices until he has spoken to Jim first, in order to introduce my name to him, along with giving Jim a brief insight to my project.

"If he likes what he hears, and I certainly think that he will, then I will give you his telephone number, so that you can ring his office and arrange a meeting with Jim via his secretary. She will give you a specific date and an exact time for you to meet up with Jim, as he is a very busy man...oh, and she will also give you the location of his offices, of course".

Jim sounded like the man I had been waiting to speak to all along, although our meeting would have to wait until after Christmas, because apart from it being the festive holiday, I also had a more important issue to concern myself with, and that was the perpetration of my mum's 'This is your life' surprise *bash* –an *all-out party* that I had arranged for her 60ᵗʰ birthday, which will be taking place in only a few days time.

My two brothers, Gary and Paul, had arranged to take my mother out for a couple of drinks on the Sunday evening, whereas I had used Caryl's *overdue* pregnancy as an excuse not to be with them, promising my mother instead that Caryl and I would take her out for a *slap-up meal* after the baby had been born. *Mum had already made me promise her faithfully that I wouldn't do anything 'over-the-top' for her special birthday, as she knows what I am like, and so I was sure that this little ruse of ours –i.e. me being unable to make it, would work a treat –and sure-enough, it did, because she didn't suspect a thing.*

On the night in question I drove around to Gary's house first, before moving on to Paul's place, collecting-up five of Kitty's (*my mothers'*) grand-children, before dropping them off at around 6.30pm at the Pendragon Pub –our local *watering-hole* in Thornill, Cardiff, where Caryl, Liam and Carl were waiting to greet them.

The landlord of the pub had put together a magnificent buffet –a mouth-watering spread which now adorned two large tables in one corner of the Function Suite. He had also set up a row of seats at the front of the stage area, along with several rows of chairs behind them, all of which had been set in an arc formation, from one side of the room to the other (with a small aisle in the centre) and so it really did look like an original 'This is Your Life' stage set. In one corner of the room stood a tripod stand, and sitting on top of this was a very expensive looking video camera, ready and loaded to capture this evening's entertainment for life. The stage had been set, and the invited audience were starting to build up in the lounge area of the pub, while the participating 'story-tellers' hid way out of sight in the bar, only to emerge when our budding *Michael Aspel* (my second-cousin, 'Glyn') gave them their pre-arranged cues to enter the auditorium. At this point they would tell their comical story about 'Kitty', which would invariably leave my mother (as well as the audience) in fits of laughter, or in certain cases as red as a beetroot with embarrassment.

Glyn, the man armed with the 'Big Red Book', also had several letters from those family members who had been unable to attend this evening's bash for various reasons, which he would read out in turn, as the story of my mother's life unfolded from her initial birth (apparently weighing-in at an incredible '14' pounds) to the *four-feet, eleven inches tall* woman we all know and love today. As the taxi arrived, the whole place came under a deathly silence, as my mother was led into the function suite by my two sister-in-laws, Lorraine and Pat. Glyn was immediately on hand to welcome the ladies, before turning to my mum and uttering those immortal words: "Kitty Donovan, mother of Paul, Gary and Shaun, 'nanny' of Marisa, Kara and David, Diane and Scott –and Liam and Carl –with another grandchild on the way, I might add; you thought that you had come here tonight for a quiet drink with your immediate family, but that is not the case, because I am here to say that tonight 'Kitty Donovan' –'This is Your Life'.

The look on my mother's face as dozens of family members, along with all of her closest friends, came pouring-in through the doorway, was something that words could not express. All I can say is that it made all of the hard work which I had put into the surprise well worthwhile. However, if the initial shock wasn't enough to *pop-my-mum-off*, then the embarrassment that followed from the hilarious testimonials, along with a dirty 'mistaken-identity' trick, which I also perpetrated on my dear, unsuspecting mum, would surely be enough to make sure that she never showed her face in public again –jokingly, of course. At the end of the presentation (45 minutes in all) I organised a 'guess the baby's weight' –and also a 'guess the time of birth' raffle, which ended-up raising over £100 in total. This we duly gave to my mother as an additional birthday present, in order that she could buy herself a portable colour TV, as her old one had *ceased to function* only two days earlier.

My child had not been born yet, and already he or she was making more money than his / her father. Whilst our family partied the night away, which included drinking the pub almost dry, (the landlord only had cider left at the end of the night) I kept reminiscing on how much pleasure and satisfaction the evening had given me as a person, and it made me forget all about the numerous problems in my life, which would inevitably be waiting for me when I woke-up in the morning.

CHAPTER 14

'Hayley Donovan' Has Arrived

The night after mums party Caryl's labour pains started in earnest, and so the pair of us immediately got ready to go to the hospital. This being our third child, we knew exactly what to do, and so while Caryl began putting together her sleepover bag, I drove to my brother's house to collect my niece, Diane, who had kindly offered to babysit for us, as and when our new baby was ready to come into this world. By the time I got back to the house Caryl had already phoned the hospital to let them know that we would be on our way very shortly. As soon as we reached the hospital the nurse took all of the usual tests, whilst asking Caryl the pertinent questions about the timing of the contractions, and whether or not they were becoming more painful, while I sat in one of the armchairs, as instructed by the nurse, just in case I might be needed for anything, but in reality I was only there to offer moral support to my wife.

According to the nurse, Caryl's cervix had only dilated two centimetres, of which, if I remember correctly (please forgive me if I am wrong ladies) it would need to reach around five centimetres, before going into the delivery room —and ten centimetres, before being able to give birth to our baby. Having been given this information, I knew that this was going to be a very long night. Because of her current situation, Caryl was duly sent down to the labour ward by the midwife, to try and get a couple of hour's rest, whilst I waited in the day room for anything to *take shape*, so to speak. The clock struck midnight, and Christmas Eve was now upon us. Caryl and I were still undecided upon names, regardless of whether our new off-spring would end up being a boy or a girl, and so it looked as though the poor little *mite* would have to enter this world nameless...unless the pair of us could finally come to an agreement within the next couple of hours, of course?

In all truthfulness, we had gone through the same scenario when both Liam and Carl were born, and so why should this time be any different? At 3am in the morning a nurse came into the day room to inform me that the contractions had settled-down, adding that Caryl had asked her to ask me to go home, adding that she would phone me if anything happened. It was nine o'clock the following morning when Caryl finally phoned me, but only to ask me to collect her as soon as I was ready, for nothing else had materialised during the night. Christmas Eve turned out to be like every other Christmas Eve, with Caryl doing all of the last minute rushing around before the 'Big Day'. There were also lots of jobs for me to do in and around the house, of course, which was nothing new. Like most parents, we sent our children to bed early on Christmas Eve, and while I spent the rest of the evening sorting out and wrapping-up the boys presents, before neatly placing them around the base of the Christmas tree, Caryl busied herself by filling-up Liam and Carl's Christmas stockings with all kinds of sweets and chocolates, along with lots of mini-toys and puzzles for them to play with in the morning.

The final chore of the evening was to bring down Liam's brand new bike from upstairs. Caryl and I had hidden it at the base of the fitted wardrobes in our bedroom about a week ago, before covering it with a blanket, in readiness for the *grand unveiling ceremony* tomorrow morning... and the pair of us couldn't wait to see the look on our little boys' face when he saw his big surprise. As I climbed into bed, having left Caryl to finish-off her final *bits and pieces* in the bathroom, I suddenly heard the sound of fast-running water, and so I immediately vacated the bed and returned to the bathroom, in order to see what was going on? As I reached the doorway I saw Caryl carefully stepping into the bath, and so it did not take a genius to realise what was happening. Being my usual *unsympathetic* self, I jokingly blurted-out "I hope this is not another false-alarm young lady?" Caryl's sister, Sheila, had kindly offered to look after the children for us, should anything happen tonight, as she lives less than a mile from our house, and therefore she could be over within minutes, and so I immediately *got on the blower* to her.

I then rang the hospital to inform them of our imminent arrival. It was now ten-thirty in the evening, and so our *early-to-bed, early-to-rise,* Christmas Eve / Christmas morning scenario had just gone right out of the proverbial window. As if that wasn't enough to dislodge our quiet evening, Caryl then asked me to put all of the presents, including Liam's bike, into the attic, as there was no way that she was going to miss her beloved boys opening their Christmas presents –hopefully on Christmas Day! As soon as we arrived at the hospital the same nurse who had been on duty last night, did all of the usual preliminaries, which she had done the previous night, and once again the results were the same. However, this time Caryl was allocated a bed in one of the small side wards, which could only accommodate four patients at a time, and as luck would have it, the other three beds were empty anyway. As there was no one else in the room, Caryl and I spent the next hour or so discussing names, until we finally agreed upon both of the names for either a boy or a girl.

If our *unborn child* turned out to be a girl then we had originally thought of calling her 'Holly', as we both love the name, but as there was now a strong possibility that our daughter could be born on Christmas Day, we decided against the idea. School children can be rather cruel if they can conjure-up any reason to take the rise out of a fellow pupil, and the last thing we wanted was for our little girl to be ridiculed, or possibly even bullied, because of her name. And so we decided on the name 'Hayley' instead, not only because it is a lovely name, but also because I had always had a *mega-crush* on 'Hayley Mills' (*award-winning actress and daughter of the legendary actor, 'Sir John Mills'*) ever since I was a young lad in school, after seeing her in the classic movie *'Whistle-down-the-wind'*. I also wanted to use the name 'Michelle' for Hayley's middle name, as not only does *Hayley-Michelle* roll off the tongue quite nicely, but also because (once again) I had always had a *soft-spot* for the gorgeous actress, *Michelle Dotrice*, ever since I first saw her in the comedy classic series, 'Some mothers do 'ave 'em', starring the incomparable Michael Crawford.

Unfortunately Caryl overruled me on that one, saying that as she had never been given a middle name by her parents, she would be quite happy just calling our new born baby 'Hayley', without giving her a second name. (*Many years later, my daughter regressed me severely for not 'sticking by my guns', insisting that she would have loved to have been christened*

'Hayley-Michelle', rather than just plain old 'Hayley!') If the baby turned out to be a boy, then after much deliberation, the pair of us finally agreed on the name 'Callum', not only because it is a lovely name, but also because it represented a mixture of his two brother's names 'Carl' and 'Liam', which I personally thought was rather a nice touch. With that little problem now solved, I duly left Caryl for her to have some well earned rest, exiting the ward and heading off in the direction of the day room, with a view to watching a few festive programmes on the television.

As I passed through the main reception area, the place was completely deserted; not a soul to be seen anywhere, and so I gathered that the poor staff who had been called in to do the Christmas shift, had scuttled off to have a little Christmas party of their own. "And why not" I whispered softly under my breath, my vocal chords not wishing to break up this deathly silence that now encompassed me. The day room was just as deserted as everywhere else, which was not a bad thing really, as it meant that I could now choose whatever channel I wanted to watch on the television. Having settled myself in, in readiness to watch a late-night movie, my peaceful bliss is suddenly interrupted by the sound of a woman's voice? It is the ward sister, and she is asking me why I am sitting in the day room? I explain to her that my wife is in the side ward, expecting a baby, naturally assuming that this answer would suffice.

However, I was swiftly informed that the room was for 'patients only', before being assured that my wife would not give birth to our baby this evening, and therefore I would have no other choice but to leave the hospital. And so, begrudgingly I said my goodbye's to my wife, before making my way out of the building. However, this turned out to be a task that was going to be *easier said than done*, because the nurses had obviously locked all of the doors leading into the ward, in order that they could have a little privacy, and so in the end I had to push open a fire door, not really caring whether I would accidentally set off all of the hospital alarms by doing so, before clambering over a five-feet high perimeter wall, in order to get to the multi-storey car park and retrieve my blessed vehicle –what a palaver!

"At least I can now have a decent nights' sleep" I jokingly said to Sheila, after returning to the house, and releasing her from her Christmas Eve duties. However, I could not have been more wrong, for only a few hours later (as the clock struck 4am, to be precise) the telephone rang, and so I picked it up, simply expecting to be told by one of the nurses that I would now have to rush back to the hospital as soon as possible, as my wife was about to go into labour. However, it wasn't one of the nurses –it was Caryl. "Well, you've got what you always wanted Shaun...your baby daughter, Hayley Donovan, is here, and the pair of us are both doing fine", she said quite nonchalantly, before adding that she would like to get some sleep right now, and that she would ring me again in the morning, after the doctor had been to check on both her and the baby. I was simply ecstatic (although a part of me was deeply saddened that I had missed the birth of my baby daughter, of course) and I knew that there would be no point in trying to get back to sleep after hearing such news, and so I made myself a cup of coffee, before sitting back on the sofa, silently revelling in my excitement, whilst letting all of my worries disappear into oblivion.

My two beloved sons continued sleeping in their beds, the pair of them unaware that they now had a baby sister to play with, and that from this day forward Christmas would never be the same again for the *Donovan* family. I desperately wanted to tell someone -or perhaps I should say 'everyone', my wonderful news, but I could not consider waking anybody at this time of the morning on a normal day, let-alone on Christmas morning. Then an idea struck me; it may be the early hours of the morning over here in the UK (by now it was 4.40am), but in California it was 8.40pm in the evening (California time being eight hours behind *good-old G.M.T.*) and so I decided to give my great friend Ted Simon a call. Unfortunately I was met with his usual answer-phone message, apologising for him being unable to take my call right now, before adding that he would get back to the caller upon his return. For a moment I hesitated, my heart really wanting me to express my good fortune to a fellow human being, rather than a piece of celluloid, but then my head told me to just say what I had to say, and get off the damn phone, because this call was costing me a small fortune.

Undeterred by my failure to communicate directly with Ted, I then decided to ring my two other great friends in California, Daniel and Zara, and this time I *struck it lucky*. These loving parents were doing all the things for their beloved daughter, Darcy, which Caryl and I had been doing for our two boys less than eight hours ago, and they were both highly-delighted with my news, telling me to send all of their love and best wishes to Caryl and Hayley when I see them in a few hour's time. *After replacing the handset, I suddenly realised that I had just phoned my friends on 'Christmas Eve', to tell them that my daughter had been born on 'Christmas Day'—how weird is that?* I also had to admit to myself that it was nice waking-up on Christmas morning without the proverbial hangover, which was something I have suffered from on many previous occasions, after getting slaughtered at the annual 'Christmas-Eve' office party. This was in the days when I was a 'working-class' citizen, of course, and not a sponging parasite, who had been claiming benefits on the old 'rock 'n' roll' for the last four months. My situation was starting to eat me up inside, and I was beginning to feel like a 'pimple on the arse of humanity'.

Realising that my emotions were obviously all over the place right now, I decided to switch on the multi-coloured fairy lights which abounded our extensively decorated Christmas tree, before tentatively tearing apart one of the Christmas crackers (with the children still asleep I dare not let it 'bang') and wish myself a "Happy Christmas". I also contemplated opening the bottle of champagne which Caryl and I had bought in readiness to celebrate the birth of our third child, and pour myself a small glass of bubbly, before proposing a toast to our latest *bundle of love*, but then I thought that that moment would be better shared with my wife, rather than doing it all alone. As the clock struck 6am, I began preparing the breakfast table for me and my boys; purely because I had nothing better to do, of course, when suddenly the phone rang. It was Ted Simon, returning my call and congratulating both Caryl and me on the birth of our precious new daughter, which I graciously thanked him for.

However, as excited as I was about our new addition to the family, I wanted to *talk shop* with Ted, and so I told him all about the various meetings I had had with James Gerwyn, Robert Staines and Josef Zagrodnic. I also told him about Ryan's offer for a documentary series, which Ted liked very much, saying that it certainly sounded a more practical idea, as it would allow us five or six hours of coverage, rather than squeezing the whole journey into a two hour production. Knowing that this call was costing Ted an arm and a leg, I said that I would send him a letter in the post, giving him a full update of everything that had happened. (*Sadly, I didn't have the luxury of e-mails in those days. How much easier my life would have been had I known then what I know now about the internet and the World*

Wide Web). By eight o'clock the boys were up, but as they were both too young to fully appreciate what day it was, I just told them over breakfast that they now had a little sister, who we would all be going to see for the very first time, in the next couple of hours.

Shortly afterwards, Caryl rang the house to say that we could come into the hospital any time after ten o'clock, once the doctor had *done his rounds*, and so it was a case of *all hands on deck,* as I washed and dressed my two little boys, before taking them to the hospital to meet our precious new arrival... 'Hayley Donovan'. I have to say that Caryl looked rather sprightly, especially considering that she had gone through the horrendous ordeal of childbirth less than seven hours earlier. As for my new baby girl, well she looked absolutely beautiful, lying there in her little cot, her innocent mind totally unaware of the *big, bad* world outside –and *hopefully* the exciting life that lay ahead of her. Wrapping my hands around her tiny body ever so gently and lifting her slowly out of the cot, I held my beloved daughter in my arms for the first time, and looking deep into those beautiful blue eyes of hers, I swore right there and then to my little girl that I would give her everything wonderful that this world has to offer her, and that I would love her forevermore.

I then kissed her forehead, as daddies do so that she could feel the love of her father next to her soft skin...it was a 'very special' moment. Caryl's parents arrived at the hospital shortly afterwards, and by twelve o'clock all seven of us were sitting in the lounge area of their house, Caryl's mum having prepared *the Christmas Lunch to end all Christmas lunches.* During the afternoon, Caryl's Sister, Sheila, and her family, along with Caryl's brother, Ian, and his family, popped in to join us for a few festive drinks, and also to share a few celebratory toasts to the *new addition* to our family. The handing out of presents came next, as the grandchildren (i.e. Liam and Carl, along with several of their cousins) began furiously tearing away at dozens of sheets of Christmas wrapping paper, before screaming with tumultuous glee at the presents inside. After the usual onslaught of various after-dinner puddings, the *Donovan clan* said their fond farewell's to the rest of the family, before returning home, only to begin the unwrapping ceremony of an array of presents all over again. (*Liam absolutely loved his new bike, by the way*).

As the best Christmas Day of our lives finally came to a close, I truly believed that Caryl and I could not have been happier, but then my wife was never one to express her innermost feelings, and so I suppose that I can only speak for myself. It is now Boxing Day, and Caryl and I are both fully aware that our *precious little bundle* will be the centre of attention for many days to come, as the house became swamped with excited family members from all over Cardiff, along with numerous friends and neighbours, most of whom wanted nothing more than to catch a glimpse of our little baby whilst she was awake. However, one or two of them also asked if they could possibly pinch a quick cuddle –a humble request which was also granted without question, of course. A journalist from the local newspaper also contacted us, asking if he could call up to the house and take a few photographs of Hayley, as apparently she was the *second* baby to be born in Cardiff on Christmas Day this year (1991), the first one having been born just after midnight.

Naturally, I accepted his request, and when he arrived the following day Caryl and I proudly presented our little bundle to him for her very first photo shoot. (*My daughter was only a few days old, and she had already attained more recognition in our community that I had achieved in a frigging lifetime!*) After giving him the general details about Hayley's birth weight, along with the exact time that she came into this world, I joked to the reporter about the raffle which I had created at my mother's 'This is Your Life' bash. To my great surprise he loved the story, and before I knew it my dear old mum was making front page headline news -alongside Hayley, of course. They say that mixing *business* with *pleasure* is not the *done-thing*, but I made good friends with the reporter, telling him that I may well have another great story for him to cover in the not too distant future, but for now this was Hayley's moment, and so I said that I would discuss the matter with him at a later date.

It is now New Year's Eve 1991, and what has been a very eventful year for our family will soon be drawing to a close. Tomorrow a new year will begin –a year of hope for billions of people around the world, but tonight there will be merriment and dancing in the streets of hundreds of countries, and thousands of cities across the planet, as the human race prepares to 'ring-out the old', and 'ring-in the new'. As for our little close in the village of Thornhill, well my fellow friend and neighbour, Graham, has managed to talk Caryl into letting me out for a couple of pints with him in the local pub later on in the evening, so long as I am back in time for the meal which she has *painstakingly prepared* this morning. Graham says that he will call for me around seven o'clock, and we can then take a slow walk up to the Pendragon Pub together. (*This is the same pub where my mum's 'This is Your Life' presentation had taken place just over a week ago*). Graham was as good as his word, gently tapping on the front window of our house just as the clock on the wall struck 7pm.

However, this was only minutes after everything had gone horribly wrong for me. Fate, it seems, would be against me tonight, as a string of seemingly innocent events culminated into me having one of the worst accidents in my life –indeed, one that could so easily have been fatal. At around six-thirty Caryl had decided to take Liam and Carl upstairs for their nightly bath together, leaving Hayley sleeping soundly on the sofa, and so I decided to square-up the kitchen, which was now cluttered with a handful of dirty dishes from the boys teatime meal. Having washed the *pots*, I then wiped over the worktops with a dishcloth, before scooping-up the crumbs into my right hand, with a view to throwing them into the pedal-bin, which was located under the sink unit. However, upon opening-up the cupboard door, I could see that the bin was already overflowing, and so I decided to empty the contents of the pedal bin into the large dustbin, which was situated outside in the garden. Unfortunately that bin was also full to bursting, and so my next step was to take the black bin liner out of the dustbin and tie it up, in readiness for the dustmen to take it away in a few days time. (*This was long before the days of green bags for recycling waste products, and food dispenser bins, of course*).

As it was a bitter cold evening, I thought it best to bring the bag inside the house, before resting it on the internal door mat, which was sitting just inside the patio doors in the dining area. Taking one of the tie wraps from out of the kitchen drawer, I began flattening the contents of the bag as much as possible, in order to be able to tip the rubbish from the pedal bin into the bin bag as well. Putting all of my weight onto the top of the bag, I managed to squash down the majority of the rubbish, but something seemed to be stopping it from flattening any further, and so with one almighty *heave-ho* I slammed my palms down onto the top of the bag, and that is when things went horribly wrong! Suddenly there was a very loud 'popping' sound, just like a bottle of champagne being corked, and all I could see was this huge red fountain of what looked like blood, spraying all over the dining-room curtains and across the ceiling directly above my head. Looking down at the bin bag, where both of my hands were still resting, I could see that the aforementioned *blood* was spurting from the main artery in my right wrist —and it was pumping out at a rate of knots.

Immediately I grabbed one of the tea-towels off the towel rack, before wrapping it firmly around my wrist. I then shouted up to Caryl, asking her to throw down one of the hand towels from out of the airing cupboard, as my *temporary bandage* was already nearing saturation point. Having had no response to his gentle taps on the window, Graham decided to ring the door bell instead, and by this time I was on hand to open the door —with my left hand, of course, as my right hand was feeling very weak, and pretty useless right now. Upon seeing my upper-torso, which was completely splattered with blood stains, and the sodden tea-towel hanging from my right wrist, which was also dripping blood everywhere, Graham looked more shocked than I had been a few minutes earlier. Immediately he offered to drive me to the hospital, before running across the double driveway, which separates our two houses and unlocking the door to his garage. By now Caryl had joined me downstairs, after wondering what on earth had been going on in her absence.

Unfortunately, there was only time to give Caryl a brief outline as to what had happened, as Graham was now reversing his car out of the garage, whilst at the same time tooting his horn several times for me to *hurry-up* and join him. As we drove to the hospital at a frantic pace, I explained to Graham what had happened, even though at this point I was still totally in the dark as to what implement had actually managed to slice through my wrist like a hot knife through butter. The hospital was empty when we arrived, apart from an abundance of nurses and ancillary staff, most of them having been drafted in to cope with the aftermath of drunken violence, and the victims of car-accidents, invariably caused by intoxicated people who had insisted on driving their vehicles home. To *rub salt into the wound (if you'll pardon the pun)* the desk clerk congratulated me on being the first *casualty* of the evening, before jokingly remarking that my explanation of my injuries was "A likely story!" Clipping my x-rays to a big white screen (*an event I had seen a hundred times before, back in my biking days, when I managed to break more bones in my body than your average dog could eat in a lifetime*) the doctor duly informed me that no bones had been broken on this occasion, although I had severed the main artery, which was cause enough for concern.

He then praised me for immediately applying pressure to the wound, which is the correct procedure whenever a vein or an artery has been severed. However, he followed up his 'compliment' by chastising me for strapping a tourniquet around my right upper-arm (*this was the hand-towel from Caryl, which I had asked Graham to tie in a knot between my elbow joint and my bicep, before setting-off for the hospital*) saying that this had effectively cut off the blood supply from my heart to my arm, which apparently was certainly 'not' the right thing to do. A local anaesthetic was then administered, before being given the pleasure of watching the trainee nurse thread seven stitches in through one side and out through the other side of this gaping hole in my forearm. Knowing that the initial danger had now passed, and that I wasn't about to *croak-it* on New Year's Eve after all, I began to see the funny side of this whole *somewhat traumatic* episode.

In fact when the nurse gave me a prescription for a course of antibiotics, I simply asked her politely if it would be okay for me to start the course tomorrow morning, as I had my own source of *painkillers* for this evening –medication that came in the form of a golden coloured liquid, full of nerve-numbing chemicals, which I assured her would see me through safely to the following day. By eight-thirty the catastrophe was finally at an end, and so as painful as my hand was right now, the evening went ahead as planned, with me and Graham having a couple of pints in the local pub, before returning home to our respective partners (a little later than we had originally planned, of course) for a right good *nosh-up*. As for the cause of the accident, well the culprit turned out to be an old coffee percolating pot. Caryl had been washing it earlier on in the week, when suddenly a chunk of the glass suddenly broke off in her hands, but thankfully without hurting her.

Being as safety conscious as ever, Caryl immediately placed the slice of glass inside the pot, before packing it out with a load of grocery bags, in order to stop any more pieces from breaking off. She then placed the said pot in the middle of the black bin bag, before surrounded it on all sides with a mass of empty food tins, along with a handful of cardboard boxes and all sorts of household refuse, not thinking for one minute that I would go slamming my arms down on top of the bag, just to get a few extra pieces of rubbish into the bin. It would be a week before I could drive again, and nearly a month went by before I could even consider putting pen to paper. Those four weeks also turned out to be the hardest of Caryl's life, for not only did she have to look after two very young children, along with catering to the every need of a new-born baby, but she also had to cope with a veritable invalid –a man who couldn't even pick-up his precious little girl for a cuddle, let alone, feed or wash her, or even change her *shitty* nappy. (*If truth be known I actually became very good at changing nappies, along with making up bottles, bathing and dressing the children, and I was always on hand whenever they wanted to have a big cwtch (cuddle) whilst drinking their bottles of milk).*

At the end of January Ryan phoned me, simply to say that he had been unable to get in touch with Mr. Howells, and so he had decided to give me his telephone number anyway, to see if I would have better luck in contacting him? After several attempts, I finally got to speak with Mr. Howells over the telephone, but all he was interested in right now was seeing Mr. Gerwyn's 'Riding Home' video, before committing himself to any kind of interview, and so he duly gave me the address of his offices in Cardiff (*which only turned-out to be in the same arcade as* Robert Staines' *office*), where I could drop off the tape. This I did that same afternoon, of course, and so all I could do now was to sit back and wait for that all-important phone-call. It took over a week for that call to come, but now Mr. Howells' secretary was duly informing me that I had been granted an audience with this 'very-important' gentleman the following week. Unfortunately, this meeting turned out to be one of several which would end up being cancelled on the day prior to our appointed slot.

Whilst waiting for that all-important meeting to materialize, I duly rang the Sunday Times offices in London, to see if it would be possible to obtain copies of Ted's articles, which he had written for the newspaper during the first two years of his journey, to whit I was duly informed that I would have to pay a visit to the 'Newspaper Library', which was located in Colindale, North London, if I wanted to obtain any references to archived material. However, Rory Clifton, the gentleman I spoke to on the phone, said that the Sunday Times would be seriously interested in doing a story if the film *was* made, before asking me to keep him personally informed of the movies' development.

Ironically, my friend from the South Wales Echo (*the gentleman who had written the article on Hayley's birth*) contacted me the following day, to see if I was ready to do an article with them, but I told him to *hang-fire* for a month or so, as I was hoping for some major developments to transpire over the forthcoming weeks.

CHAPTER 15

Men, Movies And Moguls

On the tenth of February 1992 I decided that I would make *that* journey to the Newspaper Library in London, having put the trip off on several previous occasions. Ted had already told me that he had written around seventeen or eighteen articles during the first two years of his travels, before he stopped writing them altogether, after the Sunday Times interest in the project had somewhat deteriorated in the third year. I did not have a clue how long it would take me to find the original transcripts, or indeed if I would manage to find any at all in only one day, but the challenge was on, and by now you should know what a sucker I am for challenges. As I was still unsure whether I would be catching the bus or taking a train from Cardiff to London, before leaving the house I rang the Newspaper Library to find out their opening times, and also to ask them which was their nearest underground station, or which bus I must take once I get into London, in order to get to the library as quickly as possible.

As Caryl pulled-up at the entrance to Cardiff General Station, so the National Express Bus drove out of the exit lane, on its way to London –shit! It was now almost eleven o'clock and according to the bus timetables it would be over an hour before the next bus to London was due to call in at Cardiff Station. According to the young lady in the ticket office, the journey to London would take approximately three to three and a half hours to complete, and therefore I would be getting into London at around three-thirty in the afternoon, which would simply be too late for me to do what I needed to do. And so I decided to check on the trains instead. The journey by train would take only two hours, and the next train would be leaving for Paddington Station in less than twenty minutes time, and so my choice was obvious. Unfortunately, the fare was £30 for a return ticket, compared to only £18 on the bus, and this was also on the condition that I did not return on any of the trains that would be leaving Paddington Station between 4.30pm and 6.30pm (rush-hour) this evening.

Although twelve pounds was a lot of money to me right now, I knew that I had no choice but to take the train, and as for me returning to Cardiff at around 9pm this evening, because of time restrictions, well that was something that I could easily live with. At 1.25pm I am walking around England's capital city, having already spent a further two pounds on my underground ticket, which would see me through to the 'Colindale' terminal. However, it would still take me three more train rides and another hour before I finally reached my destination, and so I only hoped that all of this running around would be worth it in the end. Thankfully, the library is only about a hundred yards from the tube station, and so by 2,30pm I am making my way inside the building. Using my passport as identification (*having been told to bring it with me only this morning*) I was then instructed to leave my briefcase with two burly security guards, who were standing next to the reception desk, on the understanding that I would be able to retrieve it again on my way out...how security-conscious is that?

Upon entering the main auditorium, I am duly handed a 'request form', which I must fill in completely, including the name of the newspaper that I wish to research, along with every individual month, in each corresponding year, that I wish to search through, as apparently each binder contains issues over a period of six months. The maximum amount of binders I can have at any one time is four. Unfortunately I am unable to determine which 'heading' the articles would come under, such as 'Sport' 'Politics' 'Financial Sector' etc and so the receptionist humbly suggests that I look through the *Library Index* first, before I begin my in-depth, and very time-consuming search. After choosing one of the desks at the back of the room, I duly leave my notepad and pen in the centre of the table-top, as if staking-a-claim to *'my desk'*, as I can see that there are very few available tables left to sit at, before shuffling-off to have a closer look at the *index files*.

The room is alive with the sound of flickering pages and scribbling biros, as the students and businessmen, who have collared most of the tables, finally find the articles that they have been looking for, their ballpoints obliterating the once clean sheets of paper in a matter of minutes. Upon returning to my seat I can see a note on my desk, stating that the articles I requested are no longer in 'book' form, but they have been recorder on microfilm, and that these films are now waiting for me at the counter. Working in the confines of a small, dark cubicle and glaring into a microfiche for hours on end is certainly 'not' my idea of fun, but once my eyes had adjusted to the light, and I had mastered the controls then it was *bearable*. After ten minutes of searching I came across my first article, which is dated '13th January 1974' and describes Ted's worst week of the journey, since starting it only three months ago. I know that this cannot be the first article that Ted will have sent all those years ago, because it only covers Southern Sudan and Ethiopia, with no mention of Europe or Egypt, which he obviously will have encountered first.

Any articles which Ted would have written prior to this would be on an earlier micro-film, covering the last six months of 1973, which unfortunately, I did not order, simply because Ted only started the journey on the 6th October, and so I had wrongly assumed that he would not have written any articles within the first three months. Undeterred by this I decided to press-on, fastidiously searching every inch of every page, looking for Ted's next article, but from the beginning of February right through until the end of June there is nothing? After changing over the spools I begin searching through July and August, but still there is nothing, and just as I enter the month of September, I am being informed by the security guard that the place will be closing shortly. Almost two hours of solid searching, and all I had to show for my efforts was one blessed article! I decided that racing against the clock must have been my downfall, as Ted must have written several articles in 1974 –and in my hurry to uncover them I had missed the blessed things altogether.

There was no doubt in my mind that I should have been here when the doors opened first thing this morning, and that I should not have arrived here halfway through the day, expecting to find everything in a couple of hours –how stupid was I? Now I was paying the price for my stupidity, having no more than one article to my name. However, before

I can have a copy of the said article, I must purchase a 'photo-copying disc' for one pound Sterling, which actually allows the owner to have two copies. Unfortunately, my first copy is so bad that I end up throwing it in the bin, before adjusting the brightness on the screen and repeating the printing procedure. The second copy is far from perfect, but it will have to do, as apart from using up my quota of copies, I am also out of time. As I exited the building I conceded to the fact that today's little jaunt would have to be put down to experience, as I am totally dissatisfied with the outcome of my efforts.

Returning home with only *one* photocopy of *one* newspaper article which has cost me around £40 to acquire, is not going to go down too well with my wife at all, especially as we are currently counting the cost of every single penny that we spend. Now it was time to begin the long journey home. After boarding my first tube-train, which is bound for Euston Station, and having accepted the fact that today was *not* going to be one of my most positive days, I decided to read Ted's article, to help me pass the time away.

Even though I am engrossed in Ted's write-up, I manage to grab the tail-end of an announcement, which has just informed the passengers on our train that due to smoke in the tunnels, our tube is being diverted to Charring-Cross Station. Quickly checking the underground map, which is located directly above my head, I can see that I must get to Kings Cross Station, before getting onto the main line, which leads to Paddington Station. 'Warren Street' is the next stop, and so I vacate the train as soon as the sliding doors open, before running to the nearest porter for help. My *learned friend* informs me that I must go back out onto the main street, and walk down to the traffic lights, before crossing the road into 'Euston Square Station', where I will be able to pick up the main line to Paddington Station. The streets are heaving with commuters who have been caught up in this rush-hour *emergency*, and the traffic has been brought to a virtual standstill, as swarms of people continue crossing at the pedestrian lights, even-though the amber light has stopped flashing ages ago, and the light above it is now clearly showing 'red'.

"Do you know that you can use your train ticket on the bus in this situation", I hear a teenage girl say to her friend, but I am in too much of a hurry to adhere to her advice, and so I just keep on rushing around with the crowds. When I finally reach the station, two porters are evacuating everyone out of the place, but after battling against a great surge of bodies, and explaining to these guys what one of their colleagues had told me, one of them immediately pulled me to one side, before allowing me onto the main line platform, where a tube had just pulled into the station. By the time I get back to Paddington Station it has just turned 6.30pm –and the next *Inter-city 125* train to Cardiff Central Station will be leaving at 6.55pm, and so I have time to chill out with a steaming hot cup of coffee before boarding the train. The return journey was a lot smoother than my outward bound train, and so I also enjoyed the pleasures of a little snooze on the way home, waking-up just as the train was pulling-in to Bristol, Temple-Meads Station.

When we finally arrive in Cardiff at around 9pm, I immediately pull my ticket out of my pocket, in readiness to give it to the ticket collector. As I am walking down the platform I begin reading the small print on the reverse of it, only to discover that my train ticket entitled me to free travel on the London underground, and so I cursed myself bitterly for having wasted four pounds on buying additional tube tickets. The following day I rang Mr. Staines to tell him how I had *struck-out* in my attempts at meeting-up with Andrew Coles, along with the good news that I would soon be meeting-up with Jim Howells. Robert knew Jim very well, saying that besides being the Managing Director of Ffilm Cymraeg, he also had a very high standing in the BBC. Now that certainly was *music to my ears*. I then told Robert about Ryan's offer to make a series for television, either in a documentary form, by returning Ted to places he had already visited almost twenty years ago, or as a drama, by re-enacting the journey in full, but using actors instead. Robert congratulated me on my continued success with the powers that be, adding that he sees Josef quite often, and so he will keep him fully *up-to-speed* with what is going-on.

With everything looking so promising, my curiosity had gotten the better of me, and so I wanted to know whether Brynley Graham had simply ignored my offer, or if indeed, he was as good as his word, but had simply not had the time to sort out his finances as yet? It is now 1.45pm in the afternoon in the UK, and so, with Saskatchewan being six hours behind good old GMT, it means that it will be 7.45am in the morning over there, and therefore probably a good time to catch Brynley before he goes off to work. I was right, as Brynley answered the phone within a couple of rings. Immediately he apologised for not getting back to me, as he had promised, adding that although everyone still liked the idea, and would love to become involved in the project, nobody had any spare money to invest in an industry which they knew absolutely nothing about. In other words it was purely a 'trust' issue, and nothing else. Brynley also went on to say how Canada was now feeling the effects of the worldwide recession, before enlightening me on a few of the problems that it was causing him personally, and so, much to his dismay, he had no choice but to turn down my offer of being a part of this great venture of mine.

I had heard from a very reliable source that Jim Howells had the power to commission movies up to a maximum budget of £1.25million, and so when Mr. Howells' secretary rang me to say that the meeting was 'all systems go' for next week, I knew that this was going to be 'the big one', and so I prepared myself for the *fight of my life.* Having kept a photographic journal of my journey to America, which I had aptly entitled 'The Dream Begins', along with a follow-up album entitled 'Follow The Dream', which tells the story of my endeavours to realise that dream, and which also included photos of all of the people who had helped me along the way, I now purchased a third album and entitled it 'Dream to Reality', with the intention of filling it with photographs and stories of the 'making of a movie', thus proving to Mr. Howells that nothing was going to stand in my way of completing what I had started less than five months ago.

The big day had finally arrived, and I was more than ready for it. The last time that I had called into Mr. Howells' office, I had simply dropped off Mr. Gerwyn's video with Jims secretary, but now I was here to do battle (*in the politest sense of the word, of course*) with a person whom I had barely spoken a few words to, and to a man whom I had never even met before, and so I only hoped that he would be as nice in person, as he had been over the telephone. From the outside, the office, with its glass fronted door, which had been obscured at the top by a pair of very colourful hanging baskets, looked more like a quaint little tea-house, or a Welsh craft shop, rather than a theatrical film studio, but the sign above my head is telling me that I am at the right place, and so here I go, into the *valley of the shadow of death* —or maybe a brand new life ...who knows? Unfortunately, I am stopped in my tracks, as the door is locked, but thankfully Mr. Howells' receptionist, whom I had met on my last visit, immediately recognises me through the window, and so after depressing a button under her desk the lock on the door is instantly released.

After confirming my name with the young lady, she duly rang Mr. Howells on his extension line, before leading me through a small doorway and up a flight of stairs onto a landing, where several movie posters adorned the walls, all of them showing Mr. Howells' name as Executive Producer in bold letters. Directly opposite me is an open doorway that leads into a large study, and as I walk into the room I can see a tall, slim figure, looking out of one of the widows which overlooks the interior of the arcade. However, the gentleman in question is standing with his back to me, and so I begin to wonder whether he even knows that I am here? Jims secretary follows swiftly behind me, and as the young lady kindly asks if either of us would like a cup of tea or coffee, the man turns around to answer her, and instantly I know that I have finally come face-to-face with the 'He-bull' himself –Mr. Jim Howells...the most *elusive* man that I have ever had the pleasure of shaking hands with. Dressed in a denim shirt and a pair of corduroy trousers, the man looked so relaxed that he made me feel the same way –and yet here was I, all *suited and booted* in my best pin-stripe suit.

As the pair of us made ourselves comfortable, Jim sitting in an armchair, and me on a sofa (both of which are on opposite sides of the coffee table) I begin my plea of recognition, my literary begging-bowl clasped firmly in both hands, as I began spreading my wares on the table. Jim then spent the next few minutes browsing through my 'Dream' albums, before sifting through my three motor-cycle travel journals that I had written back in the early eighties, and which I had also brought with me today.

After returning each album to me in turn, Mr. Howells then relaxed back in his armchair, while I showed him Ted Simon's inscription, which he had written inside the cover of my 'Jupiter's Travel's' paperback, confirming his complete backing of my movie project. I then opened out the 'Freedom Run' movie poster, which had enthralled Ted at our initial meeting back at his *dome-home* in Elk. Jim says that he admires the way that I have pre-empted the outcome of my efforts, by calling the third album 'Dream to Reality', before asking me if I write *'off-the-cuff'*, or from previous notes, to whit I honestly reply "A bit of both". Jim's next question kind of caught me on the hop?

"So how do you see this film Shaun; a completely true story; based on a true story; an idea by....or what?"

I know what I really want to say, but because of my naivety in the industry, I am afraid of making a fool of myself, and so I simply decide to pull out the script, and show Jim the movie outline, along with my four-way split idea, including the opening scene, which seemed to impress Mr. Gerwyn so much.

Mulling over the script for a few minutes, Jim then slammed it shut, before telling me, quite honestly, that it read more like a documentary project to him, rather than a movie, and if that is the case then he cannot help me, as he only deals with movies. Mr. Gerwyn had warned me that my back would be *up-against the wall* with my project -but now I felt as if I had been well and truly pinned down on the floor. It would take a karate expert to get out of such a position, and so, even though I am only dealing with it in a metaphorical sense, I decided to rely on my old sensei's words of wisdom to get me out of this hole. "Attack is the best form of defence", he always told me, and so I must use this analogy to my best advantage. "Then pray tell me how you would go about turning this documentary-styled script into a movie project then Jim?" I say quite abruptly, but not in an offensive tone, in the vain hope that he will answer his own question. It is obvious to Jim that I was stumped for an answer, and that I had to succumb to drastic tactics, in order to get myself out of the mire, but instead of *shooting-me-down-in-flames*, Jim just paused for a few seconds, before springing into action:

"Okay, what we have here is a great storyline, with superb accompanying music, but let us take a look at it from Mr. Simon's angle. To copy the contents of the book, word for word, would only insult Ted, and yet if we change the film entirely, then we may not do the book justice, and so what we have to do is to find ourselves a level where everyone is satisfied. Also, having read Ted's comments on your project, I don't think that there is much chance of him relinquishing his rights to Jupiter's Travel's, which would have to be done before we could even consider working on a movie project...are you with me so far Shaun?"

"Please continue Jim, I am listening and learning", I say quite nonchalantly, before crossing my legs, folding my arms and leaning back on the sofa as if I owned the place.

"We definitely have something here Shaun, but now I want you to look at it from a completely different angle. I watched James Gerwyn's video, Riding Home, the other night, and to be quite frank with you I think that the journeys you have done and the journal's you have written are every bit as exciting as Ted's journey".

This was something that I had not expected Jim to say, and I am flattered to the point of embarrassment, and so, rather sheepishly, I thank Jim for his kind comments. However the best was yet to come.

"What's more, is that I find you as interesting a person to talk to as anyone on that tape –hell, you have only been speaking to me for half-an-hour, and already I am getting all wound-up, talking about blessed motor-cycles, of all things. Do you still have a motor cycle Shaun?" However, before I had a chance of replying to his question, Jim is asking me another one:

"I'll bet that you would love to do Ted's journey, wouldn't you Shaun", Jim says, glaring deep into my eyes, as if beckoning me for a positive reply.

"If you mean would I like to leave my family for four years, to go on the journey of a lifetime, then the answer is a resounding 'No' Jim", I say quite emphatically, my mind now wondering where this conversation is leading to? "However, if I could do the journey in separate stages, such as dropping me off in North Africa, and then flying me home from South Africa three months later, before doing a similar thing in South America, Australia, The Far East and so on, then I suppose I would be interested", I said quite unexpectedly –both from Jim's point of view –and also from mine!

Jim then asked me if I had mentioned this idea to James Gerwyn, but as the thought had never occurred to me before today, I simply shook my head in the usual negative manner. By now I am trying to ascertain what Jim is getting at, as I watch his mind ticking away in front of me, like a time bomb waiting to go off, and knowing full-well that if it does explode, then there will be nothing that I can do to stop it!

"Before I make my final decision, I would like to have a word with Mr. Gerwyn –and also Ryan", Jim announces, before asking me if I would mind him doing this —as if I would? I am just thrilled to be here, knowing that in the not-too-distant future I will either be making a movie or, worst case scenario, a documentary series for television. I also feel as though a great weight has been lifted off my shoulders; as if I have done all that I can do with the project, and now it will be up to someone else to make all of the important decisions. At this point the formal part of our meeting was over, and so the pair of us talked freely about our friends and family, both home and abroad, as if we were old buddies who had decided to meet up for a quick chat in the local pub together. I told Jim how reading Ted's book had made me feel as though I already knew the man intimately, and how, after spending less than a week together at his homes in California, I felt as though I had known the man all of my life.

Jim reciprocated, by saying that he shares that same kind of relationship with a friend in New York, before telling me how they initially got together, and how their friendship had grown over the years. What had started out as purely a business meeting has once again, turned into an informal friendly conversation, as Jim now offers me a small scotch to round off our *little chat*. Unfortunately, the security guard has started locking up both of the entrances to the arcade, and so we will have to take a rain-check on our little tipple, as Jim confesses that he does not possess a key to the main padlock, and so unless we leave now, we are likely to get locked in overnight. Again Jim's politeness shines through, as he asks me if I would be good enough to meet up with him again in two week's time, at which point he will be happy to confirm his proposals. I cannot agree quickly enough, and as sad as it may seem, I begin wishing-away the next fourteen days of my life.

CHAPTER 16

The 'Mean-Machine'

I have now been out of work for five months, and the effects have been devastating. Our collection of unpaid bills has turned into a mountain of reminders and final demands for payment. The *polite* letters have also turned into 'last warnings', and the companies who could not give me enough credit a year ago, were now curtailing my accounts and forbidding me to use their facilities until I returned to full time employment. Leaving school at fifteen and walking straight into a job as a painter and decorator, before turning to office work a year later, I had been fully employed for over half of my life. I had only ever spent four weeks on the dole previously, and during that time I knew that I had another job to look forward to. This crippling recession had taken its toll on my whole family, and it had affected each one of us individually. Caryl no longer went on shopping trips into town with her mother and sister, Liam's special treats of small toys and endless bags of mixed sweets had been curtailed, and Carl had no choice but to wear hand-me-downs, including Liam's old shoes, which if not taken off on a regular basis, would rub his feet, thus making them quite sore after a while.

Hayley would only get pretty new dresses when other members of our family bought them for her, and I had long-since lost the delights of a brand new car! However, much worse than this was the fact that I had lost my *average* status in the community, and my pride had gone with it. I had become a 'bum', a redundant dole outcast, who could no longer tell stories about 'what happened in work today' to his mates over a pint of beer in the local pub. The days were long, and the nights were even longer. Highlights of the day included washing and wiping the dishes, pegging out the clean laundry, and making-up the beds. I became an expert at changing nappies and dressing the children, and my culinary skills in the kitchen were second to none, as I revelled in cooking and preparing at least two, and on many occasions, *three* meals a day for my family.

Mind you, when it came down to putting the vacuum cleaner over the carpet, polishing the woodwork, cleaning the windows, and doing any kind of ironing, then I would either make myself very scarce, or end up doing these *very boring* chores under duress. However, I would always be more than willing to do the weekly shopping, as this gave me a good excuse to get out of the house for an hour. I also drew-the-line at *mother and toddler* group sessions, although I spent many hours playing over the park with Liam and Carl, along with taking Hayley for endless walks, in order that Caryl could get on with her many *spring-cleaning* chores. Having spent almost half a year as a 'house-husband', I have to admit that my sympathy for the hundreds of thousands of men who were currently out of work took a *back seat* to every common housewife in the country. Never again would I express to Caryl what a 'hard day' I had endured in the sales showroom, after experiencing her role in life.

With only four days to go before that all-important meeting, tensions were mounting for both Jules and me, and the pair of us desperately needed something to deter our minds away from the horrendous pressure we had put ourselves under, and so that evening I rang Jules to see if he fancied taking a trip up to the 'Museum of British Transport' in Coventry tomorrow morning. Ted's Triumph is a permanent exhibit in the motorcycle section, and to see that *mean-machine* 'in the flesh', so to speak, would be a dream come true for me. Jules said that he would also love to see the bike "in all its glory" (Jules' words) and so we arranged to meet at Jules' flat at 10am the following morning. Before leaving the house I rang the museum, to find out its opening and closing times, along with asking a few directions on getting to the place once Jules and I were in Coventry? Arriving at Jules' flat just before ten o'clock I was invited to wait in the front room for *ten minutes*, as Jules was currently discussing a business proposal (as always) with one of his many associates.

Unfortunately the 'ten minutes' soon turned into an hour, as Jules kept popping in to tell me that *'something else'* had suddenly cropped-up, which required his *undivided attention*. It had just turned eleven o'clock by the time that we finally *hit-the–road*, Jules apologising profusely about the unavoidable delay due to unforeseen circumstances, even though I assured him that it really wasn't necessary. Thankfully, the roads were dry and the traffic was sparse, and I had also decided to prepare a packed lunch for the pair of us this morning, so that we wouldn't have to stop off for something to eat along the way. However, what I had omitted to pack was my *Atlas of Great Britain*, a veritable bible for travelling long distances in the UK. (This was long-before *Satellite navigation systems* had been invented, of course). However, as the car we were using was part of Budget's hire fleet, Jules knew that there would be a road map in the glove compartment, and so while Jules kept his eyes on the road ahead, I planned our travel route from the M50 turn-off (which we were now rapidly approaching) all the way to the city of Coventry.

In-between listening to the music on the radio, Jules expressed to me how much he was enjoying reading Jupiter's Travels, saying that he was now up to the chapter where Ted is in India, and how "Utterly fascinating" the whole adventure was, before adding that he still found it hard trying to put the book down for more than a few hours at a time –much to his *good lady's* dismay! This was just the invitation that I had been waiting for, and so I told John that it was about time that he heard some of Meat-Loaf's music, before pulling out the tape from my coat pocket and popping it into the cassette player. As each song played in turn I carefully explained the relative scenes to Jules, in an attempt to paint a visual picture in his mind of what had actually happened to Ted on this amazing journey of his. Although the music was quite loud at times, Jules seemed to enjoy the songs as much as Ted had done, only in Ted's case there was no need to try and conjure-up an imaginary setting, of course, for he already had the true picture of what the songs were portraying in his mind.

After two hours of driving we entered the city of Coventry, and not long after that Jules and I were driving in hypnotic circles, until we finally reached the top floor of a multi-storey car park. As soon as Jules had parked-up the car, the pair of us began tucking into our *goodie-bags*, which, apart from containing various sandwiches, a bag of crisps, a chocolate bar, and a mini-pork pie each, they also included individual cans of diet Pepsi, to help wash down our packed lunch. Having thoroughly satisfied our hunger and thirst pangs, Jules and I duly headed for the escalators, where we immediately descended into the bowels of a huge shopping centre. As we made our way through the humungous crowds of people, who had literally engulfed the entire complex, a strange feeling came over me. It was as though Ted's bike was actually calling to me; as if it had been waiting for us to arrive, and the closer we got to the museum the stronger the vibes became, to the point where I was half-expecting the bike to say to me "About bloody time too" when we finally arrived at the stand.

As we approached the entrance to the museum I took a few photos for the album, in order to commemorate our visit here on this somewhat special day in my life –the day that I finally came face to face with Ted Simon's 'round-the-world' motorcycle. After paying our dues to the cashier, the pair of us entered the main exhibition hall, where we were duly greeted by an abundance of classic cars from all over the world. Jules was in his element as we sauntered around dozens of pre-war and post-war models of 'priceless' (well, in Jules' eyes anyway) vintage cars. Pristinely clean and immaculately polished, these vehicles of yester-year, which once roamed Jules' native streets of London, stood in total silence, nothing more than fond memories to the people who once drove them. There were also several Rolls Royce's and Bentley's, many of them once owned by royalty, a number of ex-racing cars and various land-speed record holders, along with a massive collection of incredible Ferrari's, that would leave racing enthusiasts drooling forever.

Just as Jules and I were about to move on to the next exhibition hall, an announcement was made over the speaker system, saying that they were about to show 'Thrust Two's' magnificent victory over the world land-speed record, a victory which had finally brought the crown back to British shores. According to the narrator of the film, an average speed of '633.468mph' on a two-way stretch (reaching over 650mph at one stage of the run) was achieved on this magnificent day in the history of British motoring. As time was swiftly moving-on, Jules asked a porter if he would kindly direct us to the motorcycle exhibition hall, which turned-out to be on a separate floor to the car exhibits. Within minutes we were standing next to our (or should I say 'Ted's') *mean-machine,* the pair of us paying homage to its majesty, as we read the brief inscription, which was printed on a small card that was sitting on the floor, adjacent to Ted's bike. Jules then insisted that I sit astride this *great beast* so that he could take a few more photographs to add to our ever-increasing collection.

However, as soon as Jules removed one of the restricting chains from around the stand he was immediately pounced upon by one of the security guards, but after explaining to the guy why we were here, our friend decided to turn a blind eye, asking us to be careful with the bike "For all our sakes", he added, before calmly walking away. Now it was me who insisted that Jules pose for a few photos -after-all, it was he who had made the dream possible in the first place. I then checked on the route which Ted, himself, had painted on the rear pannier boxes all those years ago, and sure enough, it outlined his entire journey from start to finish. (*I was even able to open-up one of the boxes –but there was nothing inside, of course*). The speedometer registered nothing like the 64,000 miles which Ted had apparently covered on his travels, and so one could only assume that either the speedometer cable had snapped along the way, or that the whole unit had been replaced at some stage –perhaps both, but only Ted would know the answer to those questions?

Before saying my final farewells to this *illustrious beast*, I stopped and wondered whether we would ever get her back out on the road again? Ted had told me that the bike was officially 'his' possession, and that it was only on loan to the museum, and so there should be no reason why we would not be able to use it in a movie, or for a documentary series, after-all Ted had taken it out of the museum when he did the *Riding Home* documentary for Mr. Gerwyn back in 1986. Descending to an underground level in the building was like stepping back in time, for in the bowels of this amazing museum was a wondrous collection of vintage motorcycles. Standing in one corner of the room was a 1927 *'Round-the-world'* (well, according to the card sitting next to it, it was) bike and sidecar combination. How I would love to read a book about the adventures which these intrepid explorers must have encompassed on their travels over sixty years ago –a time when the majority of the roads in third world countries were nothing more than glorified gravel-tracks, fuel stations were literally hundreds of miles apart, and the availability of spare parts for their vehicle was virtually non-existent. What an incredible achievement.

There was also a great collection of Triumph's, Norton's and BSA's, both post-war and pre-war, along with a selection of old-time classic bikes, such as the infamous 'Ariel-square-four' and the amazing 'Matchless' motorcycle, both of which took pride of place in the centre of the auditorium. I cannot even begin to imagine the total value of all of the exhibits in this place, but I would hate to be the insurance company covering the cost if this museum ever burned down –heaven-forbid. Apart from buying a souvenir catalogue of the museum and its exhibits, I also purchased a dozen postcards, all of which had a picture of Ted's Triumph on the front of the card. I would now send these as my form of communication to Ted in the US, as I dare not ring him anymore after seeing the total amount outstanding on my last telephone bill. Jules and I had seen what we came here to see, along with a lot more than we ever could have imagined, but now it was time to take our leave of the place.

After taking the elevator up to the same floor that we had parked the car on in the multi-storey car park, the pair of walked over to the parking space where we had left the car, but to our shock-horror, there was no sign of our vehicle? For the next half-an-hour or so Jules and I checked out all of the floor levels from top to bottom, but to no avail, and so it looked like we would have no alternative but to phone the police and report the car stolen, before making alternative arrangements to get us home. However, just as we had given up all hope of finding our vehicle, Jules suddenly spotted an identical building to the one that we were now standing in –sitting directly across the street! Needless to say that our little fiesta was standing peacefully where we had left it and that it was us who had lost our bearings from the museum and ended-up going into the wrong lifts! By now it was five o'clock, and our slight oversight was about to cost us dearly, as we hit all of the rush hour traffic coming out of Coventry.

Our journey down the M5 was also a lot slower going south, than it had been when we were travelling northwards earlier on today, as mile-upon-mile of heavily congested traffic, most of which was coming out of Birmingham, clogged the roads as far as the eye could see...and beyond. Nonetheless, the journey had been well worthwhile, and we had both enjoyed an exciting and somewhat eventful day. On the way home Jules asked me to put some 'lighter' music on the cassette player, after openly admitting that, as much as he had enjoyed listening to my 'Bat out of Hell' album earlier, he now wanted to serenade himself with a bit of classical music in the background. Once again that all-important day had arrived, and once again, I was a bundle of nerves. As I climbed the staircase leading up to Mr. Howell's office I had a premonition that the outcome of today's meeting would not be what I had hoped for. Something inside of me was saying that it would not be good, and yet it would not be bad, and so I was neither optimistic, nor pessimistic about what lay in store for me.

All that I do know for sure is that whatever Mr. Howells decides upon, I will have to go along with his decision...period! After waiting patiently for several minutes (which seemed like an eternity) in the reception area, I can see Jim's previous appointment exiting Jim's office, and so I know that I am next in line. Again Mr. Howells is very courteous to me, as we sit together in his private domain gently recapping on our previous discussions for several minutes, before Jim finally takes command of the situation. "I am not going to *beat around the bush* with you Shaun", Jim exclaimed quite commandingly, thus causing my ass to twitch a little bit, and my cup to gently rattle in its saucer.

"The last time we met I told you how interesting you were to converse with, and how well you wrote your storylines, and personally I think that you could talk into a camera equally as good. This morning I spoke at length with Ryan about what I have in mind, and he agrees wholeheartedly with me. I truly believe that you should go on another journey; it does not matter where to, or how far that you want to travel, so long as you can express your innermost feelings along the way, by stating clearly how you are affected by your surroundings".

A cold shiver ran down my spine, as I suddenly realised that Ted Simon is no longer the issue here. I have to admit that after our last meeting I had half-expected Mr. Howells to follow this train of thought, and so I knew that it was now time to pop the ultimate question.

"Are you trying to tell me nicely that *Freedom Run* is out of the window, and that my only option is to do a documentary series for television?"

"I suppose so", Jim replied, obviously sensing my huge disappointment, before trying his best to uplift my dampened spirits, by telling me how easy it is with current technology to do most of the video recordings on portable cameras, and how exciting it would be for me to follow Ted's example, by writing all about my own personal experiences on the road, as well as being able to show the best of my adventures on film. Jim then explained to me why he had chosen to go down this route, saying that the general public are more interested in 'current' and 'future' exploits, rather than delving into the past. Jim then insisted that doing it this way would certainly be more beneficial to me in the long run, which I had to agree with. I fully understood what Jim was saying, and I had to respect his reasoning, after-all, he knows a damn-side more about both the film and television industries than I will ever learn in a lifetime. However, I still could not accept the thought of my fabulous movie being shelved until god-knows when?

Six months ago I would have given my right arm for such an offer, but now that the opportunity is actually staring me right in the face, I am feeling totally confused and my thoughts are all over the place.

"How about Ireland" Jim exclaims "What about doing a journey around the Emerald Isle? With your Irish background, you could even trace your family heritage as you went along?"

"Sorry, but I don't think that would work at all for me Jim, as I want to create a great adventure around the world, rather than simply travelling around one relatively small island", I reply quite adamantly. Why do I feel as though I have reached square number '98' on a *snakes & ladders* board, and that I have just rolled a '1' on the dice, which will land me on the head of the biggest snake on the board, before being thrown back to square number '32'? Talk about *one-step forward —and two steps back*!

"Perhaps it is because I have just returned from Ireland that I am so enthusiastic about the place", Jim says in a kind of sympathetic tone, although I can tell by his mannerism that he is more than a little put-out about my *negative* attitude towards his *positive* idea.

"Never mind; you think of somewhere to go, and then have another word with Robert Staines to see if he can assist you in any way?"

I hated to sound ungrateful for all of the help which Jim has given me to date, but I am now beginning to feel as though I am going around in circles, and getting nowhere fast! Jim also advises me to speak to Josef Zagrodnic again, to see if he would be interested in doing some filming for me, adding that he is a 'wizard' at doing speed scenes and cutting them perfectly. By now I am feeling a little flustered with everything, and I know that it will not be very long before our meeting will come to an end –but what will I do then?

Ryan Willis is the next man to be brought into the equation, as Jim tells me that it will be his job to apply for permission for the air space, and so I must get in touch with him and arrange a meeting at his earliest convenience.

"If it will help then I will also look over the application, and give it my own sanctioning, but don't worry because I have a lot of friends in the business, both here in Cardiff, and also in London, who will be only too willing to help. Always remember that we all work together at the end of the day, Shaun, and so never be afraid to call on any of us".

It was a touching speech from Jim, and I only hoped that he was right, because I was going to need all the help that I could get, from all parties concerned, if any kind of production was ever going to reach fruition. Once again the security guard is preparing to lock up the arcade, only on this occasion we do not have to concern ourselves about being locked in, because since my last visit Jim has had his own key cut, and so we just relax back in our respective armchairs and finish off our drinks –a few wee-drams of *the hard-stuff*, which Jim had kindly poured us earlier on.

The main roller shutter door leading out of the arcade has been put on a latch by the security guard, and so, according to Jim, all I have to do is to pull it tight after exiting the arcade. However, just as I approach the door I can hear a voice bellowing down the arcade "Let me know when you have spoken to Ryan –I want to know all of the details". All I can do is wave my right hand above my head in acknowledgement to Jim's statement, because by now my mind is encapsulated in another world. "No-longer will I be a mere co-producer in a blockbuster movie, or a meagre assistant director in a travel documentary, but a famous 'television celebrity presenter', and master of my own series", I chuckle to myself as I begin my journey home. Suddenly a thought crosses my mind; what about Ted? Whatever Ryan and I decide upon it must include Ted, for I owe so much to the man. And let us not forget my great friend, and veritable sponsor, Jules Gilmore –what is he going to say about this *monumental* change of direction?

When I labelled that third album 'Dream to Reality', I had no idea of the implications of that statement, and yet now by some strange twist of fate, the dream that Ted Simon had created for millions of people, has now

181

become a reality for me. I have been given the opportunity of taking on a similar adventure; to experience what it must be like travelling all alone in the wilderness –man and machine versus the elements...wow! However, getting back to the 'real world' for a moment, there are numerous questions which need to be answered, such as 'what journey am I actually going to undertake and can I really do the trans-Africa project on my own, or should I take on something considerably smaller to start with?' How many weeks will I be allocated to complete the journey, as there is no doubt in my mind that in the end it will all be down to the money? Will I get a full film crew following me as back-up, or will I simply be given a hand-held camera and told to "Get on with it!"

Finally, and most importantly of all, is "What will Caryl say about all of this?" My mind was boggling with uncertainties, as I pulled up on the driveway outside my house. Needless to say that my wife was not at all happy about the latest turn around, but as there was nothing else on the horizon for me to do, work-wise, she reluctantly accepted the fact that I really had no other choice but to go along with the decision. Well, either that, or scrap the damn project altogether and become one of those *no-hopers* in life, that simply wasted their lives away. Unfortunately, the photographs which Jules and I had taken at the Museum of British Transport failed to materialise for some reason, and so instead of having numerous pictures to treasure forever, all we would be left with is our fond memories of that somewhat eventful day, along with a pile of post-cards, of course, one of which I have already sent to Ted, giving him an update of the latest transition appertaining to our project.

It is a good thing that I am not superstitious, for my next meeting with Ryan is scheduled for Friday 13th March. I have already told Ryan over the telephone that I wish to undertake a crossing of the African continent, starting at Alexandria in Egypt, and finishing in Cape-Town, South Africa –a distance of approximately 8,000 miles. My intention is to follow Ted's route as closely as possible, even though I will not be able to start in Tunisia, as he did, because crossing Libya is a definite no-no at present. This is due to the breakdown in diplomatic relations with the country, after the shooting of police woman, Yvonne Fletcher outside the Libyan Embassy in London on 17th April 1984. I also told Ryan that if all goes to plan and I manage to reach Cape-Town in one piece (positive thinking there Shaun) then I would like to fly the bike over to South America for the next stage of the journey.

After crossing over from Latin America into North America, I would then want the bike shipped from California over to Australia for the crossing of that great continent, before finally freighting it to the Far East, where I would begin my long journey home via India, before finishing off my round-the-world extravaganza by travelling from one end of Europe to the other. My intention was to make a six-part documentary series for television entitled 'Steps of Jupiter -20 years on', 'Part 1' being the African crossing, 'Part 2' the 'American Dream', 'Part 3', The Australian crossing, 'Part 4' The Far Eastern adventure, 'Part 5' The Indian sub-continent –and 'Part 6' The crossing of Europe, including the final journey home. Rather than heading for the *unknown* African continent, Caryl wanted me to do the American crossing first, of course, as she thought that this would be a much safer place for me to get used to travelling on my own, but when I suggested this to Ryan at a later date, he was not enamoured with the idea, saying that the vast differences of culture in Africa, far outweighed the small variances of culture in North America, and therefore the African project would have much more appeal, not only to the *powers-that-be*, but also with the general public as well.

CHAPTER 17

Battling With Bureaucracy

Ryan was waiting for me when I arrived this time, and so he immediately led me into the front office downstairs, which was a lot brighter than his room, thanks to the huge bay fronted window, which faced out onto the main street. Ryan was happy for me to take a few pictures of him for my ever increasing photo album, and after doing so we both sat next to a large table in the centre of the room, where we got straight down to business. I estimated that averaging between 100 and 120 miles per day, the journey would take between ten and twelve weeks to complete. Apart from traversing the roadways, there would also be an overnight ferry crossing, and even a train journey to take into account, once inland, for many of the roads are impassable at certain times of the year, primarily during the rainy season. The bike would have to be modified to be able to carry a 15 litre drum of fuel on one side, and the equal amount of drinking water on the other side, to keep the balance in order. If I drink a pint of water, then the equal amount of fuel must be siphoned into the tank if I am to maintain equilibrium throughout the journey.

Both the water and the fuel containers would then have to be replenished at the first opportunity. Ryan is becoming more interested by the minute, as I continue-on *baffling-him-with-science*. "A mountain or desert bike would have to be used because of the terrain, as opposed to using a standard road bike if I was travelling purely on Tarmac surfaces. Also, a tank bag and a set of rear tank panniers, along with a top box, would also have to be fitted, in order to carry my personal luggage", I insisted, as if giving my orders, which must be carried-out to the letter. Ryan had listened enough to me *waffling-on* by now, and so he decided that it was time for him to take the stage.

"Now, about filming sequences, I think that the minimum amount of crew you would require is two. They would consist of a cameraman, along with a sound recordist, both of whom would have to be willing to follow you in a Land Rover or jeep throughout the entire journey. This would also necessitate a driver for the duration, of course. As for Mr. Howell's idea of you using a hand-held video-recorder –forget it".

I was really pleased to hear Ryan's last statement, as going it alone had been worrying me all along. Earlier-on in the week I had called into the A.A. (Automobile Association) offices in Cardiff, with the intention of purchasing a few maps of Africa, but to my great surprise I was informed that the A.A. no longer do any route planning guides, other than in Europe and certain parts of America –that is unless I was willing to spend around twenty pounds on a complete atlas, of course. On my way out I asked the assistant if Penelope Bushen still worked at their head-office branch. *Penelope had shown great interest in my holiday journals back in 1983, and she had even considered publishing them at one stage, and so I hoped that she might be able to assist me with my project.* Unfortunately, the lady I spoke to had never heard of Mrs. Bushen, but she then kindly advised me to speak to their public relations manageress. I told Ryan that her name was Josephine Lewis, and that the assistant had kindly given me a telephone number in Bristol where she could be contacted. Without further ado Ryan tells me to ring her immediately, but my luck was out again, as her colleague informed me that today was her day off, and that she would not be back in the office until Monday morning.

Ryan says that the bike can easily be shipped to Egypt and that I can fly out and pick it up at the ferry port, but I insist on riding the fifteen hundred miles from the UK to Turkey, before boarding a ferry to Cyprus –and then a second one to Alexandria. Apart from enjoying the journey tremendously, I am sure that this trip would also give me a chance to get used to the handling of the bike whilst fully loaded, while at the same time clocking up the mileage I needed to run the bikes' engine in properly, before my *assault* on Africa. *I had emphasized the words 'New Bike' earlier-on in our conversation, just in case Ryan had any notions of trying to borrow a second hand machine.* "In that case" Ryan blurted-out, before pointing the index finger of his right hand directly at my face, whilst at the same time staring up at the ceiling, as though great inspirations were now running amok in his brain "In that case" he repeated once more "Why not have the opening scene showing you and Ted chatting together on the ferry, just as it is docking into port in Egypt —what a great start to the documentary that would make".

I agreed entirely with his idea, as my mind began working overtime on Ryan's suggestion. "Then perhaps Ted and I can meet up again in Mombasa, in Kenya, at the halfway stage of the trip —and finally in Cape-Town, South Africa, where we can talk about the changes that have taken place in Africa over the past two decades." By now the batteries in my brain were on full charge, and so the ideas were flowing through my head like a torrent. "What if we even tell the story of how this whole documentary series came about; we could do a reconstruction of how I first pitched Ted at the dome in Elk, and how Ted showed me the BBC documentary at his farm in Covelo. We could even get an actor to portray Jules Gilmore, and recreate the scene where he first offered to sponsor my trip to the US". My mind had most definitely gone into overdrive at this point. Taking a rather large deviation, I veered onto the subject of charities, suggesting that we raise money for third world countries, to enhance the value of the trip to the general public.

"If we ask companies to sponsor us at one penny per mile from start to finish, then the most they would have to pay out for a ten thousand mile journey would be £100. For that measly sum we could offer to put their company logo on the bike, as well as giving them a mention in the credits —and even a 'Thank-you' in the follow-up book, which I intend writing once the series has been completed". I then suggested that we could stop off in some of the worst hit villages, to film the starving people and ask for donations from the general public as well. At this point Ryan held up his hand, like a policeman on point duty, halting all of the oncoming traffic, immediately ordering me to "Slow down". He said that he fully appreciated my sentiments, before stressing how we could not afford to deviate off the beaten track, as there would be time constraints in place. "However, that is not to say that we *won't* look at this issue, but if we do, then it will be at a much later stage" he added.

Ryan then brought me crashing down to Earth with his next set of statements.

"It will be your responsibility to acquire the loan of a motorcycle, and also to find someone who is willing to donate all of the fuel that will be required to complete the entire journey. Also, can I suggest that you contact either South African Airway's, or perhaps Virgin Airlines for a flight seat from Cape-Town to the UK –and keep working on the A.A. to see if you can talk them into donating a few ferry tickets for the trip. Anything else that you can think of which we might need for the journey, then speak to the relative companies concerned, and always do your best to attain it at minimal cost, or if possible, for free, because the less we have to fork out will keep the budget to a minimum, and the smaller the budget then the better chance we have of selling the project to a sponsor".

I realised now that I would not be sitting back in my chair, whilst Ryan does all of the *donkey work*, as I had so wrongly assumed; indeed it looked as though the real work was just about to begin. Having been under the impression that teams of telesales operatives would be doing all of the ringing around for me, and that the standard *begging* letters would be dispatched in their thousands from offices all around the country, was obviously nothing more than *wishful thinking* on my part. However, while we were on the subject of budgets and costs, I thought that now would be as good a time as any to ask about my own personal finances for the trip, after-all, there would be no way that I would be entitled to claim any of the normal benefits whilst I am out of the country –and the main reason that I am doing all of this is to be able to sign off the old *rock-n-roll*. Ryan assured me that all of my personal expenditures would be taken into account, and so I openly told him that I would have to have a guarantee of this money, otherwise I could not even entertain doing the journey...period.

Ryan then asked me if I had any idea what the cost would be of shipping the motorcycle back to the UK, but I said that I would want to put it into storage in Cape-Town, with the intention of shipping it out to South America, in readiness to begin part two of the journey. According to Ryan it should only take a couple of weeks after applying for air space, to know if the project has been accepted, but before deciding which broadcaster to try first, he said that he would like us both to have a final meeting with Jim Howells. As I intended starting the journey in the autumn of 1992, I knew that a lot of work would have to be done in a very short space of time if I was going to take off on schedule. A company which specialised in Trans-African safaris had sent me brochures over a year ago, but lord-knows where I had put them? However, I would search the house from top-to-bottom until I found them, as the company could be invaluable to our cause. Ted has also intimated that his telephone book was bulging with names and addresses of people throughout Africa whom we could contact along the way.

Ryan said that information such as this would certainly be beneficial to the project, and that any offers of help, in any way, shape or form, would be readily accepted by the purchaser of the series. Before leaving the office, Ryan insisted that I would have to become more official with my letters, as I currently had no headed paper, no logo for the journey, and nothing to print out my pleas with, as every letter had been hand written up to now. I must admit that I was initially hoping that if I listed my requests and requirements on sheets of plain paper, then Ryan would get one of his young ladies in the office to type out everything for me on his own headed paper, but it seems I was wrong again. Ryan simply advised me to purchase a basic word processor, saying that I could pick one up from as little as three-hundred pounds.

Unfortunately I needed that kind of advice like a *hole-in-the-head!* Finding three pounds would be hard enough right now, let alone getting hold of three hundred of those precious greenbacks, so forget it! I was stuck in a *catch-22* situation, and there seemed no way of getting out of it. I desperately needed some money right now, in order to purchase the equipment which would enable me to ask for sponsorship donations in the future. This was certainly a *chicken and egg* scenario, and somehow I would have to find an answer?

If I could only get someone to finance me for a short period of time, until my official contract came through, then I would be okay. I knew that it would be no good going to the bank, for I already had a large loan outstanding with them, and an even bigger overdraft, which increased dramatically with every new quarterly statement I received. With never ending bank charges, ludicrous interest charges, and ridiculous 'fines' for sending me a wad of printed letters, I hardly stood *'a hope-in-hell's'* chance of ever being able to get straight. Approaching finance companies for a personal loan was out of the question, of course, for I was currently out of work, and the fact that my house was now in a *negative equity* situation, thanks to the massive slump in the housing market, meant that even if I was prepared to take out a second charge against my property, there was simply no equity left which I could use as collateral. Every way I turned I just came up against a brick wall.

Dreams of the movie making me wealthy were mostly *pie-in-the-sky*, and no one gets rich from a one off documentary series, but neither of these scenarios seemed to be the point to me right now. I just wanted to concentrate all of my energies on a rewarding, knowledgeable existence; one that would hopefully achieve a much greater standard of living than my current 'shocking' state. Being a sun worshipper does not help either, when one lives in *wet-'n'-windy Wales,* and every day the urge to move to a healthier climate grows stronger within me. However, the chances of that happening at the moment are 'Zero' –and that is with a capital 'Z'. The beauty and warmth, and most of all, the freedom that this blazing star emanates from its bowels in the heavens, has never been so prominent in my mind. It is probably the fact that I have to spend so much time being surrounded by the same four walls in my living room that has caused my feelings to erupt so fiercely against the atrocious weather conditions that this beautiful country so often suffers from.

"I was born and brought-up in Cardiff, so why do I suddenly want to leave my native land, which holds so many wonderful memories for me?" I keep asking myself, without being able to determine an honest answer? I daresay that my ancestors would probably *turn in their graves* if they thought for one minute that I was even contemplating abandoning my relatives in the UK, but as there is nothing I can do about it anyway, I guess that a few harmless thoughts won't make any difference. My working class lifestyle seemed hard enough as it was, with me working an average of sixty-odd hours a week, and my wife going out to work the moment that I set foot inside the house, but like most married couples we learned to live with that scenario and accept the situation, regardless of how difficult it became at times. However, our current predicament has changed both of us completely, the result of which can only be described as 'disastrous'.

Our attitude towards each other has changed completely, as mixed feelings of extreme guilt and embarrassment from my side combine with endless worry and a deep-seated fear for the future from Caryl's side. The pair of us have very little to say to each other, for we are nearly always together, and when anything does go wrong it is always the other person's fault, when in reality, most of the mishaps were simply unavoidable. Friendly discussions soon become heated arguments, and expressing an opinion, which is neither wanted, nor welcomed by one's partner, nearly always turns into a blazing row. Being unable to afford a babysitter also means that on the odd occasion when we do decide to have a break from the house, it is never together. However, considering the current state of our nerves, along with the huge tension which increases by the day, thus preventing us from having any kind of level-headed communication, this is probably a good thing.

Having the pleasure of someone else's company can be a real tonic, especially if they have a good 'listening ear' when one is desperately in need of a shoulder to cry-on. Despite all of the above Caryl and I still love each other very much, and so we always put on a *brave face* for the sake of our friends and family, rather than burden them with our numerous troubles. The children mean everything in the world to us, and so as long as they are happy and healthy, we will manage to get through each day as it comes, standing together as one, and smiling in the face of adversity. Unfortunately, the A.A. are unable to assist me with my plight, but they kindly give me the number of Stanford's in London, who are reputedly the largest map and travel shop in the world. I am so intrigued by their *claim-to-fame* that I just have to visit their shop, to see if it is all that it is cracked-up to be, and Jules has told me that he would also love to come along with me. Today I must start my cold-calling in earnest. It has been a long time since I last practiced any kind of door-to-door sales techniques, and so to say that I was feeling very apprehensive about the outcome of today's efforts would be an understatement. With regards to the loan of the motorcycle, well I would have much preferred contacting the concessionaires directly, but as Ryan wanted to keep this an *all-Welsh* project, I decided to contact one of the local motorcycle dealerships instead.

I had known the owner of the main Kawasaki dealership in Malpas for around fifteen years, and so assuming that he would be my best bet, I set off for the twenty minute drive to the outskirts of Newport. Unfortunately, walking onto his premises dressed in a smart suit and clutching a black briefcase turned out to be a big mistake, for he immediately assumed that I was going to try and sell him a chunk of advertising space, as the last time we spoke I was a sales manager for one of the local car magazines. Thankfully, I managed to grab him just as he was making a dash for the wooden staircase leading up to his office, which overlooks the showroom, vehemently assuring him that I was no longer working for the company, before he had the chance to order a handful of his burly salesmen to throw me off the premises. Knowing that my time would be seriously limited, I quickly explained the situation to him, saying that if he was willing to loan me a bike for three to four months then I would be happy to advertise the name of the dealership on the tank of the motorcycle, thus giving it some serious media coverage.

Unfortunately, *my old friend* said that even if he wanted to loan me a new motorcycle he would be unable to do so, as he now only dealt in second hand bikes -and even if he loaned me the best bike that he had in the showroom, it would still be too risky for such a long journey. "Why don't you try the bike manufacturers themselves; they are always lending bikes to magazines for lengthy periods of time, in order to road test them all over the country. In fact, one of their 'Paris to Dakar' bikes would be ideal for your journey" he added positively. Although I dislike the word 'Defeat' intensely, I felt as though I was hitting my head up against a brick wall, and what made things even worse is that I also believed that I would get the same response from any dealership I tried. I was knackered! As I left the shop a deep sense of urgency ran through me; I needed more help fast, and I knew that there was only one person I could turn to –Jules Gilmore.

Driving straight to the Budget Rent-a-Car offices, I managed to catch Jules just as he was leaving for an important business meeting, but as always he was willing to spare me some of his valuable time. Sitting in the passenger seat of his car, I placed all of my *theoretical cards* on the table, as I had always done with Jules from day '1', and said my piece:

"As you know Jules, one of the journalists from the local newspaper has been hounding me for a story about what we are doing, but each time he has approached me I have held back on telling him anything until we had some definite proposals in writing. If Budget Rent-a-car would allow me to use their brand name, and you could possibly give me the use of one of your typewriters in the office from time to time, then I would immediately order up some letter-headed paper and start sending out 'official' sponsorship letters by the dozen. I would then be more than happy to give Budget a big splash in the paper, of course –what do you think about that my friend?"

Jules calmly replied that he would be going to London with the regional director of the company this afternoon, and so he would be happy to proposition him on my behalf. I then emphasized the fact to Jules that he was the ONLY person who had taken all of the risks so far, as it was he who had put up the entire cost of my trip to California. Also, up until now he was always happy for me to *verbally* use the company name of Budget Rent-a-car as the 'official sponsors' when pitching for further sponsorship, so surely the company would have nothing to lose, and everything to gain by letting me use their name on paper. I was beginning to sound desperate, and perhaps at this stage I was literally *clutching at straws*, but I knew only too well how much work I still had in front of me, and how quickly the time would dwindle away before that final meeting with Ryan and Jim, and so I just had to have something positive to put on the table for that crucial round of talks, otherwise everything that I had done to date would simply *go down the pan*.

That same evening the phone rang, and to my pleasant surprise it was Jules, asking me if I could meet him at eight-thirty in the Holiday Inn, as there would be someone accompanying him whom he would like me to meet. Jules then added that Gareth, who is a sales representative for the company, and someone whom I had already met on a previous occasion, would also be joining us this evening. Naturally I assumed that this 'special person' would be the regional manager of Budget, and so I quickly scribed a letter together, outlining exactly how I intended attracting sponsors for my journey, before dressing accordingly for this 'big' meeting. Driving through the streets of Cardiff I truly believed that things would finally be getting off the ground. With the Budget Rent-a-car management team backing me, I felt that *the 'wheels-within-wheels'* syndrome was about to be put in motion, and that once the 'jungle-drums' had reached maximum pitch, word would expand rapidly, thus bringing more and more influential people into the picture, along with some serious amounts of cash.

As I enter the hotel I can see Jules and Gareth sitting together at a table in one corner of the lounge area, but no signs of a third person, and so I naturally assume that the regional manager hasn't arrived as yet. Making my way up to the bar in order to purchase a drink I am swiftly joined by Jules, who immediately insists on paying for my beverage –as always, whilst at the same time reminding me about this *pub singer* that he had told me about a few weeks ago, which we were planning to see on stage the following evening. Jules then said that he had played one of the singers' music tapes in the car this afternoon, and that he thought his voice was excellent, adding how he couldn't wait to hear more of his music. I take note of Jules' comments, assuming that he is only reconfirming out little outing tomorrow evening, but it is only when I turn around and see this rather thin man, who is dressed in a pair of scruffy jeans and a tee-shirt -and donning a mane of very long brownish hair, talking to Gareth, that I suddenly realise that *this* is the 'special person' whom Jules wants me to meet. I am devastated. Doing my best to hide my disappointment from Jules, I walk over to the table, where Gareth duly introduces the gentleman to me.

His name is Theo, and I would say that he is about the same age as me, although appearances can be deceptive, of course. Anyway, over the next hour or so we become bosom buddies, Gareth singing Theo's praises to the hilt, saying how he has written music for the group 'Aswad' in the past, and how adaptable his voice is for any kind of music. "Apart from using Meat-loaf's songs for specific scenes in the movie, surely you must be thinking of using other kinds of background music throughout the rest of the film Shaun." Gareth blurts-out, immediately putting me right *on-the-proverbial spot!* "That is why I have been brought here tonight, for Theo to sell himself to me!" I thought, the tables having been turned on me for once –and I didn't like it one bit. I had to say "Yes, of course Gareth", as I didn't want to cause any ill-feeling, but the truth is that until I receive the go-ahead to use Meat-loaf's music, along with someone to pay for the soundtrack, then I wouldn't even consider looking at any other songs from additional musicians.

At ten o'clock the four of us wandered over to 'The Bank' night club, as Jules wanted to have a chat with his business partner, who owned the place. By now I was seriously thinking about giving Theo the opportunity to sell his musical talent to me; after-all a handful of people had helped me with my project, so why shouldn't I give someone else a possible break which they could so richly deserve. Once inside the club Jules gives me the best news of the evening, saying that he may well be able to supply me with a motorcycle for the journey, because the regional director's brother used to own a bike dealership, and so he is currently making enquiries about borrowing one from a friend in the motor trade.

"Ring the Budget offices in Cardiff in the morning and ask the girls to arrange a meeting for us with Toby, and he will be happy to fill us in on all the details", Jules added. I was *over-the-moon*. Before I had the chance to make that call, Jules rang me early the next morning to alter my instructions. He said that I must now contact a gentleman called 'Thomas', who is the Sales Director of the 'Almost New Car Company', a subsidiary company of Budget-Rent-a-car. Slightly confused with the change of contact, but as obedient as ever, I ring Thomas and enlighten him on my proposed journey from start to finish, including the type of terrain I anticipate travelling on, the amount of mileage I intend covering, and so on. Thomas sounds very professional over the phone, and he is also quite optimistic about me obtaining a *complimentary* vehicle for the journey, and so I am happy to *leave the ball in his court,* as they say. However, after hearing nothing for over a week I decide to give Thomas a call and see how he is doing with our proposition?

"Sorry Shaun, but I have had no luck at all in contacting my friend –perhaps you would have better luck in getting hold of him. His name is *George Thomas,* and he runs the Caerphilly Motor-cycle Club, which meet regularly at the Goodrich Pub. The landlords name is *Kevin Jacobs*, and so if you cannot get an answer from George's phone, then just leave a message at the pub and Kevin will get hold of him." Thomas then read out George and Kevin's respective telephone numbers, before wishing me luck -with a small sigh of relief, I might add, as if I had now relieved the man of his obligation. "If you want a job done properly, then do it yourself", I say to myself after replacing the handset, confident in my mind that this man's efforts had ceased after the first attempt. Mind you, Thomas was right about one thing –getting hold of George was not an easy task, as I failed on several attempts to catch him at home, eventually resorting to ringing Kevin at the pub, who assured me that he would get hold of George, and duly ask him to ring me as soon as possible.

Unfortunately, but not wholly unexpectedly, those were the very last words that I would hear from either one of them. If things had been moving any slower, then they would have been going backwards. In desperation I arranged a meeting with *Matthew Green*, the reporter who had written the article on Hayley's birth. Although I still had nothing *concrete* to give *Matthew*, I desperately needed to get the ball rolling from any angle whatsoever. However, *Matthew* said that he would rather wait until I received some kind of commitment from a broadcaster, before doing a full interview, adding that he would like to read up on my *diaries*, in preparation for the 'big splash', and that if I could get a charity involved with the project then this would add a lot more credence to the article.

I had mentioned the idea of doing the journey in aid of a good cause –i.e. for charity, (as well as doing it for television, of course) to Jules on one previous occasion, to whit he had suggested the 'Matt Taylor Appeal Fund', which Budget-Rent-a-car were apparently already involved in sponsoring. I had never heard of this particular charity beforehand, but Jules told me that it was initially set up with a view to raising money for a young autistic lad, in order to be able to send him to a specialist school for autism in Boston, USA.

CHAPTER 18

Stanfords Of London

It is now Thursday 26th March, and today I am going to sell myself to 'Stanford's of London'. Unfortunately, Jules has had to decline my offer of joining me on this trip, due to other (more-important, no-doubt) business commitments, but he has kindly arranged to loan me one of Budget's fleet vans for the three hundred mile round trip. Arriving at the depot just after 9am, Jules is already waiting for me, and while he is sorting out the necessary documentation, one of his office clerk's is drafting a letter for me, outlining the fact that 'Budget Rent-a-Car International' is now the 'Official Sponsor' of the '1992 Trans-Africa Solo Motor-Cycle Crossing'. Upon completion of the letter, Jules had several copies printed, before signing each one individually, saying that the more I had in writing the better chance I would have of obtaining sponsorship –and the more companies that offered to be a part of my project, the easier my job would become.

After thanking Jules for the letters, along with his usual *words of wisdom*, I set off into Cardiff's bustling streets, the unfamiliar sound of a diesel engine now rattling in my brain, as I battled with what was left of the rush-hour traffic in the city centre, before hitting the infamous M4 motorway on the outskirts of the city. Thankfully, there was a cassette player in the dashboard, along with a few tapes sitting in the consul, and although the music was not my preferred choice, at least it was loud enough to drown out the vibrating sounds of this somewhat noisy engine, which I was having difficulty in coming to terms with. However, the good news is that it was a really pleasant day, with no signs of rain in the skies at all. In fact the sun was doing its uttermost to break through the thinning grey clouds overhead, and I, for one, was living in hope that that great fireball in the sky would be victorious in battle, and that once again I would be able to feel the warmth of its rays, beaming down on me from the heavens above.

Remembering that the clocks would be going forward an hour this weekend, thus signalling the opening up of the summer months (well, as far as I was concerned, it would anyway) put me in an even better mood, as the endless stream of broken white lines continued being sucked under the bonnet of my van, and the miles disappeared into oblivion. My target was to reach the outskirts of London by midday, and sure enough, as the hands on my watch struck noon, I turned onto the North Circular Road, heading directly into the Earl's Court region. I already knew that finding a free parking space around lunch time in London, would be the equivalent to looking for a needle in the proverbial haystack, but I sauntered on through the streets regardless, whilst secretly living in hope that my guardian angel would be looking down upon me today. "Where's the nearest car parking facility, my friend", I bellow-out through the open window of my driver's door, to an unsuspecting taxi-driver, who is currently parked-up at the side of the road -and right in the middle of collecting his latest fare from his passenger.

"About a mile up the road on your right-hand side", he shouts back at me, his voice barely audible over and above the sound of the horrendous traffic congestion. Raising my thumb high in the air I scream back "Cheers buddy –and what are my chances of being able to park there for the afternoon?" By now I know that I am pushing my luck for a reply in time, because the traffic lights ahead of me have already changed to green, and in my mirror I can see the face of the driver who is sitting directly behind me, becoming somewhat contorted with frustration at my reluctance to move with the flow of traffic until my question had been answered? "Now Bladdy Chance Mate" was his *broad cockney* reply. A huge grin then swept across his face, as he slammed shut the rear door of his Hackney Carriage on his latest *hostage*, who was about to be given an elongated tour of the sights of London, no doubt, before been ejected from the cab miles from where the poor guy wanted to go in the first place!

Feeling a little disappointed with his somewhat sarcastic answer, I immediately roared off through the junction ahead, the car behind me now hot on my tail, for fear of the lights changing to red, no-doubt, when suddenly I spotted my salvation. Flicking on my indicator light, I immediately changed lanes, before taking a sharp left at the next corner and slipping into the small car park of a 'Budget Rent-a-Van' establishment. (*By pure chance I had caught a glimpse of their sign from the main road).* Thankfully, there is a space in one corner of the yard, and so after *abandoning* my van, I make my way into the small *Portacabin type* office, where I am immediately greeted by a very friendly young lady, who asks very politely if she can help me? I tell her that I have an urgent letter from Budget Rent-a-car in Cardiff, which I must deliver to Leicester Square (which happens to be where Stanford's is, of course), before adding that I would like to leave my van in their courtyard for the next couple of hours, my 'pretty-please' face now working overtime, as I stare solemnly into her big brown eyes.

"I will have to check with the manager first", says my *little helper,* and by the soft tone of her voice, I already know that she is on my side. However, not so soft is the tone of the managers' voice, for he has just emerged from a small office at the back of the cabin, having obviously overheard my plea for assistance, and so he is asking for proof of my *so-called* 'emergency letter'? Immediately I show him the sealed envelope, which Jules had given me to hand to the manager of Stanford's, Budget's insignia standing out proud in one corner of the envelope, and the simple words 'Delivered by hand' sitting in the middle of it. Thankfully, this is enough to convince the guy that I *am* on 'official business', on behalf of the company, and so permission is duly granted. However, this was only on the proviso that I retrieve the van before 4.30pm, because between 5pm and 5.30pm is when the company vehicles which are currently out on the road, are returned to the depot by their respective drivers, before being locked up in the compound overnight. I agreed.

With that little item sorted, the next thing on the agenda was to find my way into the heart of London. The nearest tube station to the Budget depot is the 'Olympia' terminal, which is about 100 yards down the road, and even though it is part of London's 'Underground' system the platform is actually 'over-ground'. After purchasing my ticket from the vending machine at the end of the platform, I sat on one of the slatted benches, where I set about rearranging the paperwork in my briefcase, in preparation for my forthcoming sales pitch. The trip to Earl's Court Station should take around twenty minutes (so long as there is no *smoke* in the tunnels, of course) where I will change trains, and also platforms, before boarding my final tube to Leicester Square Station.

At Earls Court Station *chaos and mayhem* (which are my constant travel companions) reigned supreme, as I continued following the hoards of people through a maze of tunnels and seemingly endless corridors, before traversing an escalator that simply *reached for the heavens.* I have to take my hat off to the designer of this huge bowl of concrete spaghetti, and as for the man (or woman) who was given the unenviable task of organizing the thousands of timetables for the hundreds of trains that traverse this labyrinth to end all labyrinths, day in and day out, well in my eyes you are nothing less than a genius. One day I might pop over to the United States, so that I can try out New York's version of Bolognese pasta, but not before I have acquired my gun licence, that is for sure.

Although Stanford's of London is actually in Covent Garden itself, and the main tube station is only just down the road from the shop, the switchboard operator that I spoke to on the telephone this morning, has advised me to get off the tube at Leicester Square Station, which at first I thought was rather odd? However, as the majority of the stations in London's underground network are virtually within a stone's throw of each other, I decided to take the young lady's advice. Asking for directions to Stanford's, and being duly escorted to the front entrance by a very nice middle-aged chap, I soon realise that I have already passed the shop once, without even noticing an eight feet high map adorning the entire front window. I had also managed to miss their name, which was printed in bold letters on a huge sign, hanging above two Roman pillars outside the main entrance. Maybe it was finally time for me to accept the fact that I should get my eyes tested, having already been advised to wear spectacles many times by the school optician...over twenty years ago!

As I enter the shop I am feeling a little nervous, and so I decide to hold over the job in question for half an hour or so and just marvel at my incredible surroundings. The shop has been split into dozens of individual sections, each one covering a selection of countries from all over the world, and the walls have been obliterated with hundreds of shelves, crammed to bursting point with thousands of hardback and paperback books, all telling their own individual stories of wondrous adventures and exciting expeditions. Formal guides bulging with information on every town, village, city and state on the planet are stacked high on centre stages, like New York skyscrapers, towering over the human's around them. There are also endless racks of foldable maps, all standing tall on a raised platform at the back of the shop, along with boxes upon boxes of rolled Ordinance Survey Maps, which have been strategically placed around every pillar in the place. I am also very pleased and impressed to see 'Jupiter's Travels' standing proud on top of its own individual stand, which has been placed only a few feet away from the shops' main entrance.

I could quite easily have spent the rest of the day here, seriously indulging myself in these glorious surroundings, but I had come here to do a job –and do it I will. Sauntering over to the 'African' section I boldly introduce myself to the young lady assistant, who politely confirms that she is the woman I spoke to earlier on the telephone, before offering to assist me in any way that she can. Immediately I produce the letter from Budget Rent-a-car, along with Ted's autographed copy of Jupiter's Travels, before telling her all about my proposed journey from Cardiff to Cape Town. Having now gained the lady's full attention, I ask her if Stanford's can possibly help me out, by sponsoring a selection of maps and books, which I have already chosen, for my cause. Unfortunately, the lady is not in a position to give the go ahead on such matters, and so she duly calls her supervisor to the counter –who likewise does not have the authority to hand out 'freebie's', and so he is soon on the phone to the store manager (who is currently working in the back office) and he tells his colleague that he will pop out and speak to me within the next ten minutes or so.

By the time Mr. Gould, the store manager, arrives on the scene, I have already chosen two books and a handful of maps on the continent of Africa, which I figure should be enough to suffice me...well, for now anyway. My host is very polite, and listens patiently to my story-cum-sales-pitch, but again my request is denied. Jules had already pre-empted the outcome of my efforts, and so he told me that if all else fails then I was to ask the store manager if he would be willing to give the merchandise to me, and then send the invoice directly to Budget rent-a-car for payment, but unfortunately Mr. Gould was having none of that either. "I am very sorry Shaun, but there is no way that you will be leaving this store with any merchandise unless it has been paid for in full and that is my final word on the matter". It was patently obvious to me now that simply mentioning the BBC was never going to be enough to sway the minds of any prospective sponsors, and so my only recourse was to play the old 'take-away' game. "Well that is a shame", I sighed, as if it was going to be Stanford's loss, and not mine.

"You see with Budget Rent-a-car being my official sponsors, I intended spraying the motorcycle in their company's colours, and I was hoping to have the words 'Route planned and prepared by Stanford's of London' embossed on the side panels, with your company logo sitting underneath the wording, but never-mind; I am sure that Waterstone's or W.H.Smith's will be happy to sponsor me with books and maps, and duly have their name splattered all over the television screens in return", I added nonchalantly, before turning away, in readiness to take my leave of the place.

"Oh 'Sponsorship', is that what we are talking about here Shaun –I thought you were just looking for a handful of friendly donations; now where did I put those books and maps we were talking about?"

By now Mr. Gould was fumbling about under the counter, trying to retrieve the array of goods I had brought to the cash desk a few minutes ago. The tables had been completely turned, and I was now feeling very positive about the outcome of my visit here. While I was at it I also gave myself a swift pat on the back for succeeding on my first 'cold-calling' sales pitch in a very long time.

After delicately placing the items in a large Stanford's carrier bag, Bryn (by now we were on 'first name' terms) handed me one of their compliment slips, before pointing out the logo in the top left hand corner of the paper. He then asked me if there were any other books or maps that I might like to take with me, or indeed if there was anything else that he could personally do for me. I was so enthralled by this world of unlimited travel which now engulfed me in all directions, that I could easily have taken everything in eyeshot, before living *happily ever after* in a dream world of fantasy. However, not wanting to overstep the mark, I simply said "Well, a few photos for the album wouldn't go a miss Bryn", before pulling out my camera and explaining to my latest sponsor that one day I intended making a 'documentary about the documentary' in other words: 'The making of: *Steps of Jupiter -20 Years on*', and that is why I wanted photographs of all the people who had been involved with my *little* venture from day one.

In conclusion of my business, and with a view to showing good faith -along with an abundance of confidence, to my new sponsors, I duly offered to sign a receipt for the *donated* merchandise, adding that should the journey not go ahead as planned, for whatever reason, then I would be willing to pay the outstanding amount for the items received, although lord knows how I was going to do that? The two Michelin maps were excellent, to say the very least, and they were probably all I would need for my forthcoming crossing of Africa. The two books I had selected, entitled 'Africa Overland' and 'Through Africa' also seemed to contain every mortal thing that one might need to know about this vast continent. Stepping out into the busy street once again, I am immediately confronted by a large blue 'White Arrow' parcel delivery van, which has been abandoned in the middle of the road, the driver having dived into one of the local shops, no doubt, with a view to dropping off his parcel as swiftly as possible, before returning to his vehicle, and disappearing into the distance, long before the local police arrive, and begin decorating his windscreen with a flurry of parking tickets and court summonses.

Flashbacks from the past begin running amok in my brain, as I recall the many long nights I spent working *the graveyard shift* in White Arrow's *Cardiff depot,* listening to a very noisy printer spewing out the days' parcel figures, before printing out reams of 'round delivery schedules' for the drivers to work with the following day. It was during those twilight hours that I did most of my daydreaming about travelling to far off lands, crossing endless mountains and deserts, and discovering all that there was to know about this great planet of ours. A few years down the line, and I can now admit to everyone, including myself, that it had become an obsession of mine, and even though I believed in my own mind that there was very little chance of me ever being able to do anything about it, I still hated it when people kept telling me that I should forget about ever going travelling again like I used to do in my single days, because I was now a married man, with a family to support and a mortgage to pay.

However, something deep down inside of me kept telling me that this was not the 'end of civilization' as I knew it, nor the curtailment of my travels…and now here I am, planning the journey of a lifetime, and within a few short months I will be living the dream that had haunted me for so long. As I begin walking down the main street, I suddenly stop in my tracks, as I come face to face with the ultimate touring machine. Standing proud at the end of a row of motorcycles, this Honda 1.5 litre 'Gold Wing' overshadowed all of the others with ease. Like a true 'king of kings', or an emperor being followed by his loyal subjects, this majestic looking beast paraded itself at the head of the line for all the world to see. Complete with a full touring fairing, including a dark tinted windscreen, along with a large top box, complete with a set of rear panniers, one can easily see that there is certainly no shortage of luggage space for long distance riding. The 'armchair' type seat also adds comfort and style to this amazing machine, which has been sprayed in a deep metallic silver paint and decorated with gold and red pin-stripe, to give it that finishing touch of class.

Because of its sheer bulk, the Gold Wing would not be suitable for a trans-African crossing, of course, but it is the perfect machine for travelling on motorways, freeways and autobahns, where one can truly believe they are gliding on air. It is also the very first motorcycle to have a reverse gear incorporated in the transmission for easier parking and manoeuvrability. Mind you, weighing in at around one third of a ton, it really needs it! As I make my way through London's underground stations, a dark shadow is suddenly cast over my day. At the foot of one of the staircases sits an attractive young woman, who is no more than twenty five years of age, but she has such an unhappy face; one that bears the burden of a lifetime of misery and suffering. In her right hand she has a plastic cup, which she is holding aloft in the air, in the vain hope that she might hear the sound of a few precious coins jingling in it, as the crowds continue barging passed her, the majority of them totally ignoring her existence, and one or two of them even treading on her bare feet, albeit accidentally, as they scurry off into the distance.

Her other arm is wrapped tightly around a small boy, who is sitting on her knee, and clinging onto her neck for grim death, his eyes transfixed with sheer fright as the latest 'herds of cattle' come charging towards him from all directions, engulfing his world of total fear and apprehension –the only world that he knows. Immediately I think of my own son, Liam, who is around this poor lads' age, as a few solitary tears dribble their way past my cheek bones, before dropping aimlessly to the floor, only to be swallowed-up by the dust of the day. "There, but for the grace of God", I whispered softly, before dropping a flurry of coins into her miniature begging bowl, my heart bleeding with pain for them both, and my conscience forcing me to say "Sorry, but that is all I can afford" to them as I passed them by, for I wanted to give them so much more.

I was not in Tijuana, where begging is rife, nor in Africa, where starvation is common-place, but in London, the capital city of one of the greatest empires the world has ever known, and yet this kind of thing is still going-on! "Where did it all go wrong?" I asked myself, as I trundled off into the crowd, safe in the knowledge that these two people would at least have a meal today, and yet I knew that this was no compensation for what I really felt deep down inside. There was simply no answer to my question. A hop, skip and a jump later, and I am back in the compound, and as I sit in the van, quietly contemplating the universe, and wondering what life ahead has in store for me, my thoughts are suddenly dispersed by the sound of a honking horn. In my rear-view mirror I can see a large transit van, the driver of which is looking none too pleased that I am parked in his space, but after a quick apology, all is well with the world again, and so I say my farewell's to the office staff, before heading out into the wide, blue yonder.

It was now just after 4pm, and so the rush hour traffic was building up by the minute, as I meandered my way through the busy streets, before finally taking the slip road onto the M4 motorway. The journey home was just as uneventful as the outward journey, which is a blessing in disguise if one stop's to consider the amount of things that could go wrong –and indeed, do go wrong, even on the shortest of journeys. Jules was still at Budget, when I finally reached the office at around seven o'clock, although he was literally packing up his desk for the evening, and so while he began locking up the place, I quickly transferred my belongings from out of the van, and into the boot of my car. I then handed Jules the keys, before telling him all about the days' events, in order that he could share in my triumph. However, as soon as I got home I realised that I had left the movie script in the well of the vans' passenger seat. I had taken it out of my briefcase, when I parked up the van at lunch time, knowing that I would not need it at Stanford's, and I had forgotten to replace it when I returned, but it was too late to do anything about it now.

The following morning I rang Jules, who said that the van had already been sent for cleaning, before adding that he would retrieve the scrip, and that I could call in for it at any time, as he would be in the office all day. I had already decided to take more photos for my album this afternoon, and so I told Jules that I would pick it up on my way to Mr. Staines' office around two o'clock. However, it seems that someone else had other plans in mind, because Jules was back on the phone to me within the hour, saying that he had personally checked over the van, and there was no sign of the script anywhere, before adding that the only other person who had been in the van (the driver-cum-cleaner) since I parked it up last night, had also confirmed that nothing whatsoever had been left in the van? Immediately I double checked my car, but there was no sign of the manuscript anywhere? I could remember seeing the folder when I returned to collect the van, and so it had definitely not been stolen whilst I was in Stanford's, but what could one do?

Because he was literally on his way out the door when I got there last night, Jules had simply left the keys in the top drawer of his desk, rather than putting them in the main 'key box', which in turn, was locked away in the garage maintenance office, and as the drawer was not locked, then anyone could have taken the keys I suppose? Jules and I were the last people to leave the building last night, and so we could only assume that someone had taken the keys and opened the van early in the morning, prior to it getting taken away for cleaning, but the chances of someone having seen that the script was in the seat well, and felt it worth risking their jobs (and possibly prosecution) for a 'folder' seemed so remote, that the pair of us were at a total loss for words. Where the script did end up has remained a mystery to this day, and unless I see a movie being released in the future called 'Freedom Run', or Ted rings me one day to say "You'll never believe the documentary I've just been asked to do" then I guess we will never know the answer to this somewhat unusual whodunit?

Feeling a little down, but certainly not 'out' I continued on with my day as planned. Crying over spilled milk was never one of my failings in life, but I did feel very annoyed at the person who perpetrated this crime, for I felt sure that he (or she) would have no use for the script at all, and that it would probably end up getting thrown in someone's dustbin, rotting-away and soaked in the stench of stale food. Actually, as I was now treading a completely different path, the script was no use to me now anyway.

CHAPTER 19

The Sponsors

Next on my list was to pay another visit to Robert Staines, who had kindly agreed to pose for a few photos for my album. The last time I called at his office unannounced, his secretary informed me that he was in the United States hence I had made firm arrangements this time. Expecting to be with him for only a few minutes, I was pleasantly surprised when Robert kindly offered me a choice of liquid refreshments, before insisting that I update him on the whole project. To my even greater surprise, Robert then kindly offered to assist me even more with my cause. Robert said that in a few weeks time he will be attending a major Film Festival in the South of France "But sadly 'NOT' the Cannes Film Festival Shaun" he added rather swiftly –just in case I was getting my hopes up too high. However, there would still be several big film companies, along with a plethora of film promoters (such as Robert) from all over the world, who would be attending the bash.

Included in the above would be a selection of delegates representing a handful of African Film Companies, and according to Robert, they normally have a lot of money to spend on movies and television documentaries. If Jim Howell's approved of Robert pitching the film companies for a joint Anglo-African production, then Robert said that he would be more than happy to act as our representative for the project. I was thrilled to think that Robert was willing to take time out of his busy schedule and do this for us, and even more impressed when he gave me his rendition of how he would plead our cause to the hierarchy of the film industry. Whilst I was taking a few photos, Robert told me that when he was working on the 'Paris to Dakar' race a couple of years ago, one of the African film companies had hired two light aircrafts, which leapfrogged over one another for the whole distance of the route, from start to finish, just in case any of the bike riders or car drivers had broken down, or gotten themselves bogged down in the sand dunes whilst traversing some of the harshest desert terrain on the planet.

"Who knows Shaun, they may even offer to loan you a driver for your Land Rover, who would know the safest, and possibly the shortest route across the desert." Robert then finished off our conversation by ordering me to write up a synopsis of the proposed programme, and to deliver it to his office within the next seven days, in order to give him plenty of time to prepare his spiel. I could not thank Robert enough for his continued support, and for the first time my excitement levels caused my whole body to come out in a rash of goose pimples. Jim Howell's had told me where he lived in one of our meetings, and so that night I drafted him a letter, asking for his permission to allow Robert to approach film companies outside of the UK to become a part of our project, before hand delivering the document to his home. Another week passed as I began organising what would surely be the penultimate meeting between Ryan, Jim and me. Easter came and went, and I had still not sorted out any letter -headed paper, business cards, or compliment slips, let alone setting up an *official* sponsorship form.

Robert had advised me to contact 'Prontaprint', who were located in Cardiff's city centre, adding how he always uses them, and how they are very forthcoming in sponsoring businesses with a free *trial pack,* with a view to acquiring repeat business, along with fellow recommendations of course. And so off into the town centre I ventured, full of confidence that I would sell myself, and also my amazing project to the company, before returning home with one of their starter packs. Unfortunately, life rarely goes the way one wants it to, because according to *Charlie,* who is the sales assistant I spoke to in the shop, the manager happened to be visiting another outlet on this day, and, he wouldn't be back in Cardiff until Wednesday, which was two days away. However, Charlie was kind enough to give me the owners' personal business card, and so the minute I arrived back at the house I duly gave the man a call. Maybe this wasn't the most professional way of doing business, as selling yourself over the phone, as opposed to doing it face to face, can lose a lot of impact, as there is no body language incorporated in the sales pitch, which is something that can so often mean the difference between success or failure.

However, today must have been my lucky day, because the gentleman in question stopped me in my tracks long before I had reached the pinnacle of my verbal presentation. "Stop right there my friend –you have made your point, and not only can I see the merits in what you are doing, but I also like the fact that you are obviously committed to your project, and so the answer is 'Yes', you can have one of our starter packs made up for you free of charge. Do you have your headings and insignia all drawn up and ready for photocopying?"

"Absolutely", I said, telling a little white lie, but not wanting anything to delay this major move forward in my professional approach to the project.

"Then pop in and see Charlie again, with your proofs at the ready, and tell him that I have given the go ahead for the pack; in fact I will ring him now and tell him myself. Good luck with the journey Shaun, you're sure going to need it."

Immediately I drew up a company logo; a simple map of Africa, just showing the outline of the continent, along with the proposed route pencilled in from start to finish. However, the heading was not so simple: **"8,000 Mile solo overland journey from Cardiff to Cape-Town –by motorcycle"**, but at least it would get the message across to prospective sponsors. As soon as I had drawn up the proofs I returned to the shop, where Charlie duly complimented me on winning over the boss over the telephone. "You must be one smart cookie, young man, for our boss does not suffer fools" he added, which I kind of took as a back-handed compliment. Anyway, after quickly scanning over the proofs, Charlie said the pack would be ready for me to collect in two days time, which suited me fine. Over the next 48 hours I prepared a standard 'begging letter', which I would adjust accordingly depending on whom I was writing to, and what I was asking for.

After collecting my pack, I began writing out my first sponsorship letters, before posting them off to their respective companies in brilliant white A4 sized envelopes, in order to look that little bit more professional.

I have always believed that large envelopes get opened before small ones, as not only do they look more *official,* but they also stand out in the pile. One of the letters went to Virgin Airlines...for the 'personal attention of Mr. Richard Branson, of course. In the letter I requested a one-way flight from Cape Town to the UK, to be used in the month of December 1992, as I already knew that Virgin Airlines would be opening new links on that route in the autumn. A second letter I duly sent to the Automobile Association, for the attention of Josephine Lewis, in which I requested a complimentary ticket to cover my Channel ferry crossing. My father-in-law's motor-factor company had kindly offered to supply two sets of chunky tyres for the trip, and Stanford's of London had already supplied me with maps and guide books for the journey, and so my sponsorship list was slowly but surely taking shape.

In the evening I popped into to the Heath Pub in Cardiff, to have a *swift half* with my cousin, Tony, having already ordered a takeaway curry from the local Indian restaurant, which would be ready in around half an hours' time. Because Tony has been working as a driving instructor for his brothers' driving school for the past twenty years or so, I had completely forgotten that he used to be a truck mechanic, and so when I mentioned to him that I was looking for someone with good mechanical experience to drive my back-up Land Rover for the duration of the trip, Tony immediately put his name forward. Tony knew that his brother, David, who is also my cousin, of course, would give him the time off work, and so within days I received a confirmation letter from the *Escort School of Motoring,* stating that they would be providing the back-up driver for the expedition, on the understanding that I would wear one of their lightweight summer jackets, complete with their insignia on the back, of course, during a proportion of the filming.

With things now finally coming together, I decided that it was high time I began learning all there was to know about the continent of Africa. Derek Bryant had traversed over 100,000 miles throughout the continent, and he had also visited more African countries than almost anyone on the planet, and so his book 'Africa Overland' soon became my *bible* for the journey. At the tail end of the book was a 120 page account of Derek's latest journey, in which he had covered over 17,000 miles, thus, making my journey equivalent to a walk-in-the-park? I was so engrossed with his tales on the road, and so envious of his knowledge of Africa, that I soon realised how imperative it was for me to speak to this man face to face, and so I rung his publishers, who kindly put me in touch with Mr. Bryant's agents in London. By now it was almost 5 pm in the evening, and so, as expected, I was greeted by the proverbial answer-phone message, requesting me to "Please leave a message after the tone".

Now it was time to have a look at my second book, which Stanford's had so kindly donated, and so after sitting comfortably in my armchair, I began reading 'Through Africa', which was written by Bob Swain and Paul Snyder. Unfortunately, the more I read, the greater the task of completing this mammoth undertaking seemed. The paperwork involved before one even sets foot outside the door, would give a Philadelphia lawyer a heart attack, I am sure –and the endless list of mishaps and problems which one would *more than likely* encounter throughout the journey, were enough to give Ranulph Fiennes second thoughts about setting foot on the continent! My old ten-year passport was almost out of date, and so the first thing I must do is replace that with a new one, otherwise I wouldn't be going anywhere. As soon as I receive my new passport, I must contact a dozen or more African consuls and embassies, most of which are based in London, and within a few underground stops of one-another, thank goodness, because to obtain the required visas I would need for this huge undertaking, I had decided to visit every consul, or embassy in turn, as the mere thought of my passport getting lost in the post, simply horrified me.

However, obtaining visas would turn out to be the easy part of my 'bureaucratic nightmare', as I began reading all about the *International Driving Permit*, the *International Certificate for Motorcycles*, the *Bank Guarantee and Credit Letter*, the *International Camping Carnet*, the *Proof of Ownership of Photographic Equipment* document, and last, but by no means least, the *Carnet de Passage En Douane*. This is an incredulous insurance policy that includes leaving a deposit of two and a half times the value of the motorcycle at the country of entry into Africa, which apparently would be returned by the last country one visited before finally exiting the continent. Forgive me, but this I found very hard to believe. The reason behind the policy is to make sure that people didn't take vehicles into Africa, before selling them for a massive profit, hence the reason for such a humungous deposit. With my head now spinning with confusion

I decided to *close the books* on work for the day, and do what I enjoy doing most – and that is playing football.

It is now Friday evening, and so tonight a dozen or so of my closest friends, along with *yours truly*, will be meeting at the 'Ely Leisure Centre' in Cardiff, in order to indulge ourselves in a 90-minute game of five-a-side soccer on the open-air pitch. We had been playing at this same venue for over a decade, and *come-hell-or-high-water* the majority of us would always turn up for the game. Regardless of whether it was pouring with rain, blazing hot sunshine, sleeting or hail-stoning heavily, we would play on through the blistering heat and freezing snow blizzards until our one and a half hours of non-stop running around was finally at an end. Several of the other lads, who were all a lot better at playing football than my good self, also played amateur league games on Saturday's and Sunday's, and so Friday night's were like training sessions to them, whereas for Shaun Donovan it was the highlight of my sporting week.

On Saturday morning I gave Caryl her usual time-out from the children, by taking Liam, Carl and Hayley over to the local park bright and early in the morning, because unlike their mum, the boys were both 'early risers', and couldn't wait to get the fresh air into their lungs. Since she was born, Hayley had been keeping her mum up half the night, as Caryl was bottle-feeding her in the daytime, before breast-feeding her overnight, and so by morning-time all Caryl wanted to do was to catch up on lost sleep. I had begun this morning ritual with the children a few years ago, whenever I was working on afternoon shifts, which could often be months at a time, as my fellow workers would much prefer doing the early morning shifts, and so I was quite happy swopping shifts with them. Upon our return to the house, I cooked breakfast for me and the boys, before treating Caryl to breakfast in bed, whilst I ran a bath for her. In the afternoon, we took the children out for a drive out in the country, before returning home for a quiet night snuggled-up in front of the fire, watching cartoons with the boys, before sending them to bed around seven-thirty, so that Caryl and I could enjoy the delights of some prime time television.

Sunday has always been a 'Visit grandparents day', and so in the morning I drove the children over to my mother's house for an hour or so, whilst Caryl sorted out the weekly washing, along with cleaning the house from stem to stern. Upon my return, the five of us drove around to Caryl's parent's house, where a wonderful Sunday dinner was being prepared by Caryl's mum in the kitchen. While we were waiting for our lunch to be cooked, Caryl's dad entertained us all in the living room, the children playing happily with the mountain of toys which Caryl's parents had accumulated over the years, while we chatted about anything and everything –over a chilled bottle of beer, of course. Lunch then followed, which today was of the chicken variety, and this was accompanied with one or two glasses of chilled white wine. Within an hour of demolishing a mountain of meat and vegetables, we were all being presented with huge slices of strawberry cheesecake, which filled our bellies for the rest of the afternoon, before finally saying our farewell's to two truly wonderful people at around 5.30 pm.

Our day was rounded-off with Caryl bathing the boys, while I kept an eye on Hayley, who had fallen asleep in the car on the way home, and was now safely tucked-up on the sofa. This bathing ritual was then followed by a quick bedtime story, which was read to them by yours truly, and after Caryl had given Hayley her final feed of the evening, the pair of us was finally able to relax in the peace and tranquillity of our lovely home. On Monday morning Caryl called me to the telephone, saying that a lady called 'Caroline Watts' was on the other end of the line, and that she was asking for me personally. Caroline said that she was calling me on behalf of 'London Independent Books', before following up her statement by saying that she understood that I wanted to get in touch with Derek Bryant.

After explaining the situation to Caroline, she said that Mr. Bryant currently lives in South Africa, and therefore should I require his personal services then all travelling expenses would have to be paid for in advance, along with his standard fees for arranging, and possibly participating in any future trans-African expedition. Caroline went on to say that during her last conversation with Mr. Bryant, he had informed her that he was currently on an expedition, and that he was expecting to be back at his home in South Africa by the end of April, with a view to setting off from Nairobi in May, and heading northwards, all the way to Cairo in Egypt. From Cairo he would be flying to London, where he was hoping to meet up with Caroline around the middle of July. I told Caroline that I would also like to meet up with Derek at that point in time, before explaining to her why the BBC would also want to talk to him. The latter must have gained her trust immediately, because within minutes she has not only given me Mr. Bryant's home address in Natal, but also his private telephone number as well, should I wish to try and contact him before he left South Africa to fly to Kenya.

Thanking Caroline for her call, and also for the information on Mr. Bryant, I said that I would get back in touch with her in the near future, with a view to discussing relative fees for Derek, either for information and advice alone, or possibly for his participation in our expedition across the African continent. In reality, I personally had nothing to offer Derek except friendship at this particular moment in time, and until any firm proposals had been put forward by the BBC, then I would not be in any kind of position to likewise discuss firm proposals with Mr. Bryant. However, just the thought of having this man's knowledge and expertise at our fingertips made me feel as if I was one major step closer to achieving this mammoth undertaking, which had been duly thrust upon me without any warning.

Knowing that I now had a 'friend' in Stanford's, I immediately rang Bryn, before asking him politely if he could possibly send me the third and final piece of the African map, which primarily covered the North-Western Sahara. The following morning it arrived on my doorstep, and so I immediately I set-about piecing this gargantuan puzzle together on the floor, our living room carpet slowly, but, surely disappearing from view, as I meticulously connected all three sections together, by using huge strips of sticky tape. As soon as all three maps had become one big map, I then used reams of sticky-back transparent *fablon* to give the map a strong plastic coating, as I knew that it would be taking a hammering over the coming months. To finish off my masterpiece I then attached individual batons to the top and bottom of the map, before tying a length of cord to the top baton, in order to be able to hang it like a picture-frame.

I then draped the map over the top of our huge wall unit in the living room, before standing back and gazing in awe at this seven feet high, by six-foot wide *rollable* continent. However, I wasn't finished yet; in fact the real work was about to begin. Looking closely at the major roads and minor roads, along with the virtually non-existent roads, I began dissecting them into various lengths, strategically planning each day's ride in turn, from Alexandria in Egypt, and through the heart of the mighty Sudan, before crossing into Ethiopia, and following the main route through Kenya, Tanzania, Zambia and Zimbabwe, all the way down to Cape Town in South Africa –journey's end. Whilst I was away Caryl would be able to follow my route from start to finish, although the thirty-seven segments I had duly created would not necessarily represent individual days of course as anything could happen along the way.

Apart from being a guide for Caryl, the map would also serve a far more important purpose, for I would be using it in my forthcoming presentation to Mr. Howell's and Ryan, which was due in less than a week. I had suffered two cancellations by Mr. Howell's this time, which was about the norm, going by our previous liaisons, but now the meeting had been confirmed by his secretary, and so it was 'all-systems-go'. Gareth had kindly offered to print off my *embassy* letters for me on his word processor, and so I popped into the Budget offices with my main template, along with a bunch of A4 envelopes, each one individually addressed to the relevant embassy, which Gareth could simply copy onto each headed letter in turn. In the forthcoming weeks I would have so many letters to write that I knew I would need to obtain a typewriter from somewhere, as it would be unfair to burden Gareth with such an additional workload. Unfortunately, the burning question was "Where the hell am I going to get the money from to purchase a typewriter?"

Although Jules had openly admitted to me that his personal funding for the project was currently on hold, until something more solid had been agreed with the BBC, he was always there to help me out, even if it was only by giving me some sound advice. Jules suggested that I write to Budget's head office, informing them of Budget's current involvement in Cardiff, before asking them for financial backing with my project. "With 'Budget-Rent-a-car' in Cardiff being the 'Official Sponsors', why shouldn't the biggest rent-a-car company on the planet want to get involved nationally, or even 'Globally' Shaun? Once they realise the worldwide advertising they will get when 'Steps of Jupiter -20 Years On' is broadcast in a dozen countries throughout the world, it should make sponsoring your cause a *no-brainer* for them, my friend", Jules added, his confidence and enthusiasm sending my adrenaline into overdrive, as usual.

It is no wonder that this man is such a successful businessman, because his determination and tenacity certainly knows no bounds, and he has been such a great mentor to me, let-alone my sole financier, that I truly hope that one day I will be able to repay his kindness in some way, shape, or form, as I owe so much to the man. As I drove away from Jules' office that day, a shuddering thought suddenly crossed my mind "What if Budget *are* willing to sponsor me, and they suddenly asked me all sorts of questions, such as 'Which bike do I intend using?' 'How much is it going to cost?'

215

'Who would I be buying it from?' –and 'What would be the estimated resale value of the bike once the journey had been completed?' How stupid would I look if I didn't have any positive answers to their questions? Standard road bikes would be out of the question, of course, because of the desert and mountain terrain's I would be crossing along the way, but having had very little experience with both *trail* and *trial* bikes in the past, I really didn't have a clue as to what bike would be ideal for the journey?

And so the following morning I set off on my quest to find a 'trusty steed' for the expedition. At the local Yamaha dealership I came to grips with the awesome 'Africa Twin'-a superb motorcycle, which had been specifically designed and built for such an undertaking as mine. Boasting a very powerful 750cc engine, along with relatively low petrol consumption, and a huge fuel tank capacity for those endless stretches of barren land, which Africa is so famous for, it surely was the Armageddon of overland / off-road biking machines. Unfortunately, there was one major problem with this dream bike, which I would soon discover to my peril. Sitting bolt upright on the thickly padded seat, with the bike standing perpendicular to the floor, I am left with both feet dangling approximately six inches above the ground, whilst the sales assistant I am chatting with is mustering every ounce of strength that he has, in order to stop the bike from falling over, as he continues on puffing and panting his way through the endless list of benefits and capabilities that this bike apparently possesses.

This is where being the proverbial *short-ass* has serious drawbacks in life, for I knew that this 'perfect' machine would be 'heaven-on-Earth' to a six-feet-plus individual, but completely useless to man who is only a few inches over the five feet mark. Getting off the bike with someone holding it upright for me was hard enough, and so the mere thought of trying to manipulate this great beast through miles of sodden mud trenches, or over the top of endless sand dunes, whilst being fully loaded with several gallons of fuel, a dozen litres of water –and twenty-odd pounds of luggage (including a comprehensive tool kit and a humungous padlock and chain) was quite frankly, frightening the life out of me. Standing adjacent to the Africa Twin was the Yamaha 'Dominator', a 500cc trail bike, which looked *something* like a smaller version of the Africa Twin. Unfortunately, its seat height once again left me hopping intermittently from one foot to the other, as I tried my uttermost not to keel over into one of the bikes standing on either side of me, thus creating a *domino-effect* with a dozen other perfectly positioned machines, and subsequently causing hundreds of pounds worth of damage.

With Yamaha bikes having now been eliminated from the project, I headed for 'South Wales Superbikes', a reputable Kawasaki dealership located in Malpas, Newport, in order to see what they could come up with? The motor-cycle industry was currently going through a worse recession than the car industry, and so I knew that extracting any kind of sponsorship deal was going to be a monumental task, but hey, if one doesn't try, then one will never know. As I entered the large showroom I was amazed to see a whole new breed of 'Triumph' motor-cycles standing proudly in one corner of the display area, each and every one of them looking equally as attractive as their Japanese counterparts, and according to the leaflets that were scattered all around them, they were all just as powerful too. Sitting astride a 1,200cc race-converted Triumph immediately brought the memories flooding back to me of my days racing around the highways and by-ways of Europe, subsequently giving me an adrenaline rush that I hadn't experienced in a very long time.

A young salesman was quick to note my interest, immediately pouncing on me before I had gotten off the bike, in the hope of getting a sale, of course, and thus making my *somewhat enthusiastic* friend a wedge of commission, but unfortunately this was not going to be his lucky day. When I explained to him exactly why I was here, he was somewhat disappointed at first, but then he insisted on showing me the new 'KLE 500' on / off road machine, which was standing in the middle of a row of bikes, adjacent to the main showroom window. I must admit that I was very impressed with the bikes' features, and when I was offered a quick test ride out on the forecourt, I could not resist the temptation and accepted immediately. However, this would not be as easy as it sounded, for it meant that we would have to move half a dozen other bikes out of the way, before a pathway to the door could be cleared. Nonetheless it was worth all the effort in the end...well at least for me it was.

Being on private ground meant that I did not require a crash helmet for the fifty yard *squirt* up and down the Tarmac, and so I was able to feel the wind rushing through my hair, and whiff the aroma of petrol fumes, as they belched-out through the multi-exhaust system. I felt as free as a bird, and at peace with everything around me, which was a feeling I had not experienced for many years. In fact I had become so used to driving a car that I had forgotten the immense pleasure that riding a motorcycle could give to a person -indeed, I felt as though I had been born again. Even though I had not exceeded 20 mph, because of the relatively confined space out on the courtyard, I felt right there and then that this was the motorcycle that I would be destined to take around the world.

CHAPTER 20

Africa Overland

Returning the beast from whence it came, I knew that I was in love with it, and *by hook-or-by-crook* I was going to have it. The seat height was much lower than the last two bikes I tried out, and I felt comfortable, so long as the balls of my feet were able to come into contact with the ground on both sides of the bike —which they were. I was just about to take my leave of the place, when suddenly I spotted a familiar face at the other end of the showroom; it was my old friend, Raymond Hill. I had enjoyed many hair-raising conversations about professional track racing with Raymond, when he worked in the main Kawasaki dealership in Cardiff, back in the seventies, and so knowing that he would be extremely interested in my venture, I went over to have a chat with him. Raymond said that the whole adventure was "Epic", before adding that he would be able to lower the bikes' seat height a further one-and-a-half inches for me, without having to cut away any of the padding, thus reducing the comfort.

Raymond also showed me parts of the bike that could be removed, in order to lighten the load, before stating that a larger tank would have to be fitted, in order to be able to cover the long distances between fuel stations in Africa. Raymond then moved on to the *luggage* department, saying that the German company, 'Krauser', already made 35litre and 45litre top boxes for the bike, before adding that extra welding would be necessary, in order to compensate for the additional vibrations I would encounter on the rough and uneven African roads. Other adjustments to the bike would include strengthened cable leads, protective metal plates for the engine and gearbox, and a set of chunky, off-road tyres. "Once that lot has been sorted, all you will need then is a set of rear panniers, and a decent sized tank bag, and you'll be ready for the off," Raymond said enthusiastically, before adding that when he retires, he has plans on doing a similar road trip across Africa.

Raymond then invited me into the office, where we negotiated a sales deal, which would include a Krauser top box, a set of rear panniers, and a tank bag. Also, in return for a few 'complimentary' modifications, which we had discussed earlier, I told Raymond that we would advertise his company on the new *enlarged* petrol tank. With the figures having been set out in black and white, my next major objective would be finding someone who would pay for it all, but I decided to worry about that at a later date, for Raymond had long since filed his copy of the quote, and he was now far more interested in talking about the amazing continent of Africa. Raymond told me about an *overland* company called 'Safari's in Africa' that was situated only a few miles from his dealership, adding that the owner's name was 'Joshua Ball', and that he was an expert on African travel. Raymond then wrote down the address on a piece of paper for me, before hand drawing a map, and verbally outlining a set of directions on how to get there.

I thanked Raymond for all his help and advice, before setting off to find this intrepid explorer, my eyes peeled for two large *overland* trucks, one of which Raymond said would probably be parked on the driveway outside his home, and the other one situated on a hard stand directly opposite his front gate. The trucks had been painted in typical khaki and green camouflage colours, making them look more like army trucks, and so they should be very easy to distinguish from other trucks on the road. However, life is never that simple, and so despite following Raymond's instructions to the letter (or so I believed), and apart from driving up and down every street in the vicinity, I was unable to find Joshua's house anywhere? Feeling rather frustrated with my negative result, I decided to do what I should have done at the start, and that was to ask somebody local for their assistance. However, trying to find someone who actually lives in the vicinity of where you are searching can often be equated to looking for 'a needle in a haystack', unless one strikes it lucky.

"Today must be my lucky day", I say to myself, as I spot a man working underneath his car on a nearby driveway. Pointing to the far end of the street, the gentleman kindly informs me that Joshua Ball lives in the house on the corner, before adding that I will find all of the windows to the house covered in whitewash, and as there are no trucks parked-up, it probably means that Joshua is away on one of his overland adventures. "Perhaps not so-lucky", I murmur under my breath. "However, Joshua's dad –'Joshua Ball Senior', only lives in the next street; just turn right at the end of this road -only lord knows what number he lives in?" my friend added, after seeing the look of disappointment on my face, and trying to be as positive as possible. Thanking him for his assistance, I duly followed his instructions, before ending up in a small cul-de-sac of houses, some with their front doors, and others with their rear entrances facing towards the roadside, which I thought was rather odd?

It was the kind of estate that not only looked as though it had been built to a cat-burglars' standard requirements, but that it had also been designed to confuse even the most dedicated of postmen? As I exit my car, tentatively pausing to survey my surroundings, I suddenly notice an elderly lady looking down at me from the confines of her bedroom, the main curtains in her bay-fronted window, having been pulled back at an acute angle, thus revealing only her head and shoulders –and a pair of piercing eyes that were currently focused on this 'Stranger from out of town'. However, instead of feeling intimidated by the local 'neighbourhood watch woman', I was just happy to see signs of human inhabitation in this mini ghost-town, and so I duly called up to the lady:

"Could you tell me where Joshua Ball Senior lives please?"

"Never heard of him" was her swift reply, before promptly disappearing behind the veil of her flimsy net curtain.

Figuring that it was finally time to admit defeat, which is something I rarely do in life, I swung open the drivers' door of my car, with a view to heading back home. However, before I had a chance to get back inside my vehicle, that same lady was now bellowing at me again, saying that Joshua Ball Senior lived in the house that was directly opposite hers. Why this woman suddenly had a change of heart will remain a mystery to me for the rest of my days, but I am so glad she did, for within minutes I am standing face to face with Mr. Ball Senior. After introducing myself to Mr.

Ball, we chatted freely for a few minutes, and then I asked him if his son is on one of his amazing expeditions, traversing the great plains of Africa as we speak, to whit he nonchalantly replied "Oh no Shaun' he's on a stag weekend in Liverpool". At this point I decided to leave my telephone number with Mr. Ball Senior and head off home.

Caryl went shopping for our weekly supply of groceries early the next morning, leaving me to look after our three young children. Liam, who has just turned four, was happily watching a cartoon on the television, whilst Carl, who is two and a half, was sitting aloft in his high chair, contentedly munching-away at his breakfast. Hayley, the baby, was peacefully sleeping in her bouncy chair, and so everything was serenely quiet in our household. This seemed like an excellent opportunity for me to give the local stationary store a call, in the hope of getting them to donate that all-important typewriter to my cause. Sitting comfortably in my armchair, I rattled off the introduction routine to the manager, without stopping once for a breath, and feeling very confident with my efforts, I prepared myself for the ultimate question: "Would your company be interested in being a part of my exciting business venture?"

However, I had only managed to get halfway through my spiel, when Liam decided that he preferred a more comfortable seat than the floor, in order to watch the television, and so he promptly climbed aboard his dads lap for a better view of the screen. Entangling his one foot in the telephone wire, meant that I had to change hands with the receiver as swiftly as possible, in order to release him, whilst at the same time keep my conversation going as best as I could. Within seconds Carl decided that he had had enough of sitting in his high chair, and so he stood up on his seat, with the intention of climbing out of the chair, his young mind blissfully unaware that he was around three feet off the ground! Asking the gentleman to "Hold the line for one second", I immediately jump out of my armchair, still clutching the phone to my right ear, but unintentionally throwing Liam to the ground with my reflective actions. As I lunged forward, in order to catch Carl before he falls, head-first over his eating table, I neglected to see Hayley's bouncy chair directly at my feet, inevitably tripping over the blessed thing, which in turn woke Hayley up, causing her to cry out loud with the sudden shock of being *booted* by her dad.

Carl has seen me coming, of course, and so he assumes that it is a game of 'catch-me-dad', and so he lunges out of the high chair, hitting me square on the chest, his two stone body weight knocking me backwards off my feet, over the top of Hayley's chair (again), and back into the armchair I had just exited. Luckily, I have managed to keep hold of Carl with my one free arm, but during his death-defying leap, the top of Carl's head has inadvertently made serious contact with my chin, thus causing my beloved son to join in the crying chorus with his baby sister, Hayley, who must surely be wondering what the heck, is going-on? Liam is also whinging at me for frightening him, even though he is unhurt, and I am still trying my uttermost to talk down the phone in a very placid tone, whilst lying virtually upside-down in my armchair...what a palaver.

I cannot imagine what the gentleman on the other end of the line must be thinking right now, having undoubtedly heard all the commotion going on in the background, but so long as I still have him on the line, I am going to carry on with my sales pitch, just as if nothing had happened. If all this wasn't enough to drive me crazy, there is now someone pounding at my front door. Looking over at the front window, I can see my neighbours' head pressed against one of the panes of glass, as he tries his best to peer through the net curtain, to see if anyone is at home, his face contorted with anguish and frustration. Whatever next? Scrambling to my feet once again, I walk over to the window, before pulling back the curtain, in order to show him that I am busy on the telephone, to whit he simply responds by ringing the doorbell, before pointing to his wrist watch, indicating that time was of the essence.

If only he knew how important this phone call was to me, he just might appreciate my reluctance in helping him with his plight...whatever that may be? However, like the sucker I have always been, I ask my potential sponsor to kindly hold the line once more, while I find out what *earth shattering* problem my neighbour is trying to rectify? Placing the receiver next to the telephone consul, I immediately opened the front door. "Can I borrow your set of ladders from under your car port please Shaun; you see I desperately need to reach my bedroom window, which thankfully my missus has left open" my neighbour grovels, as if it was a matter of life and death. "Sure, just take them," I say, before attempting to shut the door, but my neighbour is too quick, slipping his foot in the doorway before I have chance to close it properly. "Sorry Shaun, but I'll also need to borrow a screwdriver, as the safety hooks they rest on are screwed shut. Sorry for being such a pain buddy, but I have locked myself out, and the keys to my car, and also my garage door, are both on the door keys, which I have stupidly left in the kitchen. I'm on my way to work, and if I don't get there on time my boss is going to hang me!"

I empathised with the predicament he was in, of course, but unfortunately I currently had more important issues of my own to deal with, and so I told him that he would have to give me five minutes, in order for me to finish my "Very important" phone call. Returning to the telephone, where, unbeknown to me, Carl has been continuing my phone call in his own inimitable style...in other words talking lots of gobbeldy-gook with the occasional legible word thrown in for good measure, I received the answer that I had half-expected. The manager politely refused my request, saying that all of their sponsorship funds for this year had been exhausted, before welcoming me to try again next year. Because of the crazy set of circumstances that had just occurred, I had given him the time he needed to confer with his colleagues, thus stacking the odds of me winning the deal firmly against me, but it was unavoidable, I suppose, and so I graciously accepted defeat.

After helping my neighbour with the ladders, I return to the house, before slumping into my armchair —and then bursting into tumultuous laughter, as I realise how hysterically funny that little episode would have looked as a prearranged comedy sketch. None of my three beloved children can understand why their daddy is giggling like a schoolboy, but they all decide to join in my joviality, their wonderful smiles, and innocent laughter reminding me that the world is not such a bad place after all. Shortly afterwards, Caryl arrived home, both of her hands and arms heavily laden with huge bags of groceries and toiletries, and so immediately I jump up out of my chair, in order to assist her in carrying our weekly shopping load into the kitchen. As I relieve her of the two heaviest bags, Caryl says nonchalantly: "Nice to see that the place is just as calm and as tranquil as when I left it Shaun; I hope you have enjoyed quietly relaxing with your children, while I have spent the last couple of hours tramping up and down the supermarket, before loading the boot of the car on my own... and then, to top it all, I ended-up getting stuck in miles of congested traffic".

"Sometimes there is just no justice in this world", I think to myself "It seems that I am 'damned if I do' -and 'severely damned' if f I 'don't!" After making my beloved wife a nice hot cup of tea, in the hope of *getting into her good books*, I duly rang the shop once again, having already resigned myself to the fact that I was going to have to spend money in order to get what I wanted, but at least I managed to negotiate the price of a new typewriter down from £50 to only to £35, £15 being a lot of money to me right now. More importantly, it meant that I could get on with all of my important letters, rather than having to contact Gareth every time I needed to ask companies for any kind of sponsorship. For the next few days I practiced my typing skills on my new machine, the memories of owning my very first typewriter as a young lad, now sitting at the forefront of my mind, as I plink-plonked away at the keyboard, using the *two-finger* method -a ritual that has shamefully never changed in almost fifty years of typing, albeit on a computer keyboard now, instead of a typewriter, of course.

Joshua Ball had rung me earlier on in the week, and we had arranged to meet on the Wednesday morning at his home in Newport. I had already decided to go to the passport office as soon as it opened at nine o'clock on that same day, and so, knowing that I would probably be there for a couple of hours, at least, I told Joshua to expect me sometime between eleven and twelve o'clock. This was not a problem for Joshua, as he always worked from home -whenever he was not crossing continents, of course, and Joshua had already told me that he was very much looking forward to having a long chat with me. The day soon dawned, and after parking my car in one of the multi-storey car parks, I made my way to the passport office, which was only a few blocks away. Apart from simply applying for a new ten year passport, as my previous passport had expired over two months ago, I would also be adding Hayley's name to Liam and Carl's names, as 'dependent children' on my new travel document. I was also doing likewise with a new passport for Caryl, as hers had expired over a year ago, and she had since lost the old one altogether.

I was pleased to see that the offices had been restructured to cope with the onslaught of spring travellers, who had made last minute plans to fly to the sun, but had obviously refused to purchase a simple one year passport from their local post office, choosing instead to buy a ten year passport, and thus pack out the upstairs waiting area to capacity. Within an hour my application had been duly processed, and I was so pleased to hear that my new passport would be posted to my home within the next ten to fourteen days, as opposed to having to collect it from this very same office, in a few hours' time, as I had done back in 1982. As for Caryl's new passport, well unfortunately she had forgotten to sign the application form, and so she would either have to come back to these offices on another day, or simply pop the form in the post. By ten thirty I was on my way to Joshua's house... or so I thought? Pulling up outside the house, I could see that the windows were still white-washed, just as they had been last week, which I thought was rather unusual?

However, upon closer inspection I could see that the house was empty, with not an item of furniture inside, and no signs of human inhabitation whatsoever? Looking through the letter box I could see a handful of letters, along with a small selection of postcards lying on the floor in the hallway. Regressing myself for not checking-out the house more thoroughly on my first visit, I knew that my only recourse would be to call upon Mr. Ball Senior again, to see what on earth was going on? As I exit the car, after reversing it into one of the communal parking areas, just up the road from Mr. Ball Senior's house, I am suddenly confronted by an elderly man, who enquires the name of the person I have come to visit? Assuming that he must be a lifetime member of the same *neighbourhood watch* group as that lady in the window, I tell him my predicament, in the hope that he will let me pass. Surprisingly, the gentleman assures me that Joshua lives with his dad, but if that is the case then who's house is...or *was* that one at the bottom of the hill?

The plot thickens! This was becoming more like an Agatha Christie novel by the minute. Confused as ever, I knocked on Mr. Ball Senior's door (which is around the back, of course) but there is no answer, and so I decide to retire to the comfort of my car, and simply wait for someone to turn up, in the hope that they can explain this puzzle, and thus unravel the mystery that was starting to drive me insane. "Could there be only one 'Mr. Ball'?" I said hypothetically -and "Could he be some kind of psycho, living out the lives of two different people?"After all, I had only ever spoken to the son over the telephone, and so he could have easily disguised his voice. As my imagination begins to put the *fear-of-god* into me, and I consider getting the hell out of here before it is too late, my thoughts are dispersed by a thunderous 'bang' on the roof of my car, causing me to jump clean out of my skin.

It was Mr. Ball Senior, who said that he had recognised me sitting in my car, and so he had simply tapped on the roof in order to get my attention, as he could see that my mind was obviously elsewhere. However, in my current state of mind, it had sounded like an explosion. Looking closely at this frail old pensioner, I cannot conceive that he is a mass-murderer, who has a torture chamber hidden in the cellar of his lovely home, concealing the bodies of a dozen unsuspecting victims, all hanging from giant meat-hooks on the walls of his rat-infested dungeon...or that he possesses a huge kitchen table, hidden away in his attic, which is splattered with a plethora of blood-soaked limbs, in readiness to be carved into tiny pieces for tonight's supper menu! Putting all of those ludicrous thoughts to one side, I decide to follow Mr Ball into his back room (which is just where I expected it to be -around the front) and listen very carefully as all is revealed to me by *pops*.

"Joshua lives on the other side of Newport, although I am at a total loss as to why he didn't give you his address over the phone? (So much for the old man's statement that Joshua lived with his dad!) That house at the bottom of the hill belongs to my daughter, who has already moved into her new home, and so she is currently in the process of selling that one. I live on my own... is that all clear to you?" I simply nodded, as Mr. Ball Senior continued on with his explanation: "Joshua used to leave his trucks outside my daughter's house, simply because he lives on a main road, and so there was nowhere to park these rather cumbersome vehicles during the daytime...it's as simple as that Shaun." Mr. Ball Senior then phoned Joshua, to inform him why my arrival at his home had been delayed, before giving me directions on how to get to Joshua's house. Within ten minutes I am finally knocking on Joshua's door.

To the left of where I am standing is a stone staircase, leading down to the basement area, and in front of a large window, at the bottom of the cellar, stands a battered trails bike, leaning precariously against the bottom few steps. High railings surround the front garden, complete with a wrought iron gate, which has been propped-open with a single house brick. Joshua answered the door, and greeted me kindly, before leading me upstairs into his living room-cum-working quarters on the second floor.

As we enter the room, I can see Joshua's wife, who is sitting on a sofa, with her back turned against me, which I thought was a little peculiar at first? However, after placing my butt in an armchair directly opposite her, I can see that she is currently in the process of changing their son's dirty nappy, which was a sight that was certainly not for the squeamish. Joshua left the room for a couple of minutes, before reappearing with a stack of photograph albums and a pile of travel magazines for me to have a browse through at my leisure.

As I begin sifting through the deluge of literature, which has been thrust upon my lap, Joshua spends a few minutes sorting out his favourite pictures, from his numerous expeditions throughout Africa, before handing a pile of them over to me. I am amazed at the beautiful colours in every single photo, from the native's colourful clothing, to the multi-coloured flowers, and the incredible animals to the unbelievable landscapes. It is no wonder that Africa is renowned for being one of the most beautiful continents on the planet. The more I looked at the photos, the more I wanted to be there, to actually see these wondrous sights in the flesh, and to be able to taste the bitter-sweet truth about living off the land. I also needed to physically understand the harsh reality of life in this Dark Continent, in the hope that it would make me aware of just how lucky I really am —and also to appreciate the gift of life itself. As I sat there, contentedly daydreaming of my forthcoming journey through Africa, my thoughts are dispersed by an earth-shattering statement from my host:

"As far as I know Shaun, there is currently no available route where one can travel from the northern coast of Africa all the way down to the southern tip, due to political warfare and border closures. Mind you, borders are opening and closing all the time, and so no-one really knows what is going on from one day to the next?"

Joshua then vacated the room for a few minutes, leaving me to chat to his wife about the times that she had accompanied Joshua on his many adventures across Africa. She said that the journey's were very exciting, adding that they always had lots of *fun* along the way, which somewhat surprised me, for here was I, gathering as much information as possible, in order to enable me to get through this daunting journey with all my faculties fully intact, and this lady describes it as "Fun!" By now Joshua has returned to the room, saying that he had just contacted the Foreign Office, who is still strongly advising people not to attempt travelling through the southern regions of Sudan, due to the current war situation, which is still in progress. Joshua then handed me an article on Southern Sudan from one of his travel magazines, which duly enlightened me on some of the atrocities that were being perpetrated in that part of the country, before advising me to choose a different route for my own safety.

"Why not start your crossing from Algeria, instead of Egypt?" Joshua says, before opening-out a map of Africa that simply engulfed his large coffee table. "Then go south, through the Sahara Desert into Nigeria, before dropping-down into Zaire, thus avoiding all of the dangerous zones", Joshua added, his right index finger sliding across the map with the greatest of ease. However, as easy as Joshua had made it look on paper, I had some serious reservations about traversing...and possibly getting totally lost, in the middle of nine million square kilometres of desert. Joshua admits that it will certainly not be an easy trip to complete, both stamina-wise, and bureaucratically, as border crossings into the aforementioned countries can be a nightmare. "In fact, one of my trucks is currently stuck at the Nigerian border, as we speak, while the driver is doing his uttermost to obtain permits and entry visas for him and his passengers, without having to pay too much of a bribe to the officials at the border post, of course".

It seems that there is no way that I am going to be able to follow Ted's route identically, but without being able to do this, how am I going to call my documentary "Steps of Jupiter -20 years on?" Joshua keeps stressing the fact to me that an eastern crossing of Africa would simply be inviting trouble, calling it "Potential suicide" at one point, and so I am now seriously considering taking my only other alternative, if I want to do an eastern route, and that is to board a ship at Port Sudan, and sail around Ethiopia, into Mombasa, in Kenya. I decided to worry about which route I will eventually take at a later date, because now I had more important issues to sort out, such as asking Joshua to loan us one of his trucks for the expedition. Unfortunately, my request was side-stepped by Joshua, as he went into a spiel about the eligibility of driving, along with the feasibility of handling one of these *monster-trucks* of his.

"The problem with me lending these large vehicles to non-experienced drivers is that nowadays a person has to pass a P.C.V., or a P.S.V. test, before qualifying to drive one of my trucks, whereas beforehand a standard car licence would have sufficed. Also, these trucks are huge, as they can accommodate up to twenty people, and so they would probably be too big for a small crew to handle. Mind you, if you did get stuck in the mud, you would not have any worries of being trapped there for days on end, because the people in the nearest village to where you have come to a halt, will soon be flocking in their droves to rescue you from your dilemma, as there is nothing they like more than digging out a sunken truck".

Joshua had obviously decided that he could not risk someone else driving his trucks across such a dubious terrain as the Sahara Desert, and who could really blame him, after-all, these trucks were his livelihood. I also accepted the fact that Joshua could probably ill-afford to 'donate' a truck for three months or more anyway, and there was certainly no way that I could afford to pay him for the loan of one. Taking everything into consideration, I agreed with Joshua that a simple Land Rover vehicle would be much more suitable for the journey in question, and on that note, I duly thanked Joshua for his time, and also for his sound advice, before wishing him well with his future expeditions. If nothing else, the meeting had certainly been an informative one, even though I was no closer in obtaining a back-up vehicle for my African crossing, but that is the way the cookie crumbles, I suppose.

CHAPTER 21

The Penultimate Meeting

In the afternoon I concentrated solely on preparing a detailed itinerary for tomorrow's all-important meeting with Jim Howells and Ryan Willis. Having attended numerous management courses, along with countless business meetings in the past, I did not find the task too difficult, and by the time I had completed the document I was very proud of my finished product. As a matter of course I rang Robert Staines to inform him about tomorrow's meeting, as I have always considered him to be a major part of my project, and therefore entitled to be kept abreast of its progression. Upon mentioning my growing sponsors list, and saying how I had failed to obtain a vehicle from Joshua Ball, Robert once again stretched out his forever helping hand to me. He said that he would be travelling to London in a few days time, to sort out some business plans for his own company, but if I could type him a list of all the items I required for the African journey, from a cutlery set for myself, to a Land Rover for the film crew, then he would take time out to call in at the offices of a film company called *Grey-stoke,* as they specialise in obtaining sponsored merchandise for television and film productions.

Robert warned me that he could not give me any guarantees of success, as he could only 'request' their assistance, although he was confident that they would be willing to help him as much as they could. Once again I was indebted to Robert for giving me his valuable time, along with his endless efforts in assisting me with my cause. Needless to say that I spent the whole evening detailing a complete inventory of items I would require for the journey. As soon as I had completed the list, I compared it to the catalogue of equipment I had taken to Europe with me ten years earlier, and it was over double the amount of items. This was primarily because I no longer possessed any standard motor-cycle clothing, such as boots, socks, gloves, over-mittens, or a scarf. I had even sold my last crash helmet and my prized racing leathers to a close friend of mine a few years ago.

The meeting was scheduled for 9.30 am in the morning, and so I rang Gareth, and arranged to meet him in Budget's reception area at nine o'clock sharp, in order to collect the letters for the foreign embassies, which I intended showing to Jim and Ryan. Everything had been meticulously prepared, and the stage had been set to perfection, and so all that I had to do now, was to try and get some sleep during the next ten hours, or so. And so the day dawned. I woke just after 6am, having only slept soundly for a few hours. It would be a very long while before anybody else in the house would even consider surfacing, and so I had plenty of time to go through my *practice presentation* for the final time. Today was not a special day for Mr. Howell's, and Ryan would only be attending the meeting, in order to ask for Jim's advice about which way to play the game of programme selling, but for me it was my whole future that was at stake.

I have worked so hard for some kind of recognition these past six months, and yet if all does go to plan, and I am given the go-ahead for the journey, it will simply be a case of 'out of the frying pan, and into the hottest inferno on Earth'. Realising how tense I had already made myself, I sit in silence at the bottom of the staircase, before telling my mind to forget about the past, not to even contemplate the future, and to just to concentrate on 'TODAY'. Looking up at the top of the staircase, I can see my beloved son, Liam, standing still on the edge of the landing, his gorgeous brown eyes staring longingly at me, but his lips are sealed, for he is not uttering a single word. Lord knows how long he has been standing there, but it is as if he knows the pain that I am going through right now, and therefore he does not want to disturb me in my hour of need.

Perhaps he is just afraid that I will send him back to bed, which is so often the case on a normal day...but not this morning. Beckoning Liam to come to me is all the encouragement he needs, and within seconds my beloved son is cuddling-up to me on my lap, and I can feel the tentativeness of his embrace, reassuring me that if ever I lost everything else in my life,

I would never lose the love that my wonderful son has for me –or that I have for him, of course. Huddled-up together on that bottom step, I can feel a strange energy slowly emanating from his little body, and it is as if my sons' lack of fear and knowledge of the outside world was being transmitted over to me, thus giving me the strength to overcome my own morbid fears, along with the power I would need to overwhelm my peers. In that short space of time I felt as if all of my confidence had miraculously been restored to me, and I was now ready to face anyone and to overcome anything that might dare to stand in my way.

I felt more than good –in fact I was 'on top of the world', and all because of a 'Cwtch'. At 7.30 Caryl is sharing breakfast with us, and by eight o'clock I have loaded the car both with my brief case, and also that humungous map of Africa, which I have managed to squeeze into the space above the front passenger seat, with only a few inches to spare. The caterpillars that must have been lying dormant in my stomach for some time have metamorphosed into the proverbial butterflies, and by eight-thirty they feel like they are the size of golden eagles! Reversing the car out of the driveway, I say my fond farewells to Caryl and Liam, who are both standing at the garden gate, the pair of them waving frantically at me. "Good luck" I hear Caryl shout as I shift the gearstick from reverse back into neutral, followed by Liam bellowing "Buy me pop and sweets daddy". Having brought the car to a complete standstill, I look up at our bedroom window, before blowing a few poignant kisses over the palm of my right hand, one to Hayley, who is finally *out-for-the-count* in her cot, having been awake on the hour, every hour, this past night –and the second one to my beloved son, Carl, who is still sleeping soundly in his bed.

None of my children have any idea of how significant this day is in their papas' life, and yet I am patently aware of how vital it is that we keep moving forward with this project, otherwise the whole future of us as a family could be in serious jeopardy, and the mere thought of the possible repercussions for my failure is simply eating me up inside. It had been so long since I had driven my car during the rush-hour that I had forgotten how horrendous the traffic can be. During weekdays getting through the nearby village of Llanishen is a complete and utter nightmare, simply because the high school sits right in the centre of the main road that runs through the village, and the parents will insist upon driving their cars as close to the front gate as possible, before pulling over at any point in the road (or even on the pavement, in extreme cases) before letting their children out of the car, thus causing traffic jams in both directions.

By now I know only too well that if things don't improve really soon, I am not going to have time to collect those all-important letters from the Budget offices. Suddenly there is a break in the traffic, and so I manage to slipstream my way through a small gap between a row of cars, before taking a sharp right off the main road, and from then on it is plain sailing all the way –well at least until I reach the centre of Cardiff.

It is now nine o'clock, and I have just arrived at the Budget-rent-a-car offices, but there is no sign of Gareth? To make matters worse, the receptionist has just informed me that he is supposed to be working in their Newport office today, which is over ten miles away. At 9.15 I decide that I cannot wait any longer, and so I get back in my car, but just as I am about to reverse out of my parking space, Gareth pulls up right behind me. He is full of apologies, of course, blaming the horrendous traffic for his lateness (as we all do) but I have no time to listen to his excuses, and so I roar off, heading in the direction of the nearest multi-storey car park. Running all the way to the Castle Arcade, I can see that the entrance to the balcony is padlocked, and so I am unable to take my *shopping list* to Mr. Staines, even if I had the time to do so. I am down to a jogging pace as I reach the centre stairway, which leads up to the Ffilm Cymraeg offices, and I am very pleased to find that this doorway is open. Skipping alternate steps to the top, I dive around onto the balcony, before ringing the office doorbell –it is now nine-thirty precisely.

Mr Howells' secretary asks me to take a seat in the foyer, before informing me that Jim is upstairs, doing his usual morning *phone-around*, in order to keep up to date with what is happening in the filming world. Ryan is the next person to enter the building, and so I hand him his copy of the itinerary, for him to read through before the meeting starts. I have just finished checking over my embassy letters, which all look perfect, and so this is certainly a good start to the day. As Ryan and I begin chatting, our conversation is interrupted by the sound of a buzzing noise, which is emanating from the switchboard on the secretary's desk, thus signifying that it is time for us to ascend the wooden staircase and meet our host. Ryan is now carrying my briefcase for me, leaving me to do battle with my *seven feet tall* African continent, which is causing me a fair amount of grief, as I attempt to negotiate the swirling staircase.

Ryan, along with the front end of my map, enter Jim's study a few seconds ahead of me and the tail end of my oversized scroll –much to Jim's amusement, I might add. We all shake hands and the usual greetings are exchanged between us, before Jim and Ryan break into a conversation in Welsh, which, I shamefully admit, somewhat loses me after the second sentence, and so I decide to concentrate my efforts on more important issues, such as where to hang my two-metre square sheet of canvas. Thankfully, there is a tall cabinet standing against the back wall, and so, having already attached a long piece of string (using half a dozen pieces of sticky tape) across the top of the map, I managed to hang my presentation map from the top of the unit. I have already given Jim his copy of the itinerary, which he is duly studying with great enthusiasm, it seems, but now it is time for me to take full charge of the meeting. Asking them both to make themselves comfortable, I begin one of the most important lectures of my life.

Reading 'Section A' out aloud, I proceed to show my captive audience the proposed route. Standing proud, and using a short bamboo stick as a pointer, I begin bellowing-out my instructions, like a general issuing his battle plans to the troops on the eve of a mighty confrontation with the enemy. By the time I reach Cape Town on the map, I have covered everything from road conditions, to border problems, inevitable train journeys to river crossings –and a host of other unavoidable situations which might occur on the journey south. I have also given my *students* an itemised breakdown of the mileages and estimated travel times between each village, along with a daily rota of where I expect to be at the end of each individual day's riding. Like a seasoned (and a true professional) lecturer -which I am not, I ask if there are any questions which Jim and Ryan would like to ask me, in the hope that I will be able to answer them, of course, but I am met with a resounding "No thanks, you're doing just fine" from both of them, which is exactly what I wanted to hear, of course.

Moving swiftly on, I ask my learned colleagues to pay particular attention to "Section B"–'The Sponsors'. Assuring my peers that Budget would be willing to sponsor the bike, and relying on Jules' confidence in being able to sort something out personally, should their head office refuse to assist us, I then moved on to the current list of 'additional' sponsors. Ryan congratulated me on getting Stanford's of London involved in the project, before asking me how on earth I would be able to carry two spare tyres on a motorcycle, after reading about Windway Motor Services donation. I told him that the tyres would either have to be strapped to the rear panniers, or the top box, or failing this I would just have to loop them around my waist, or over my shoulder for the first half of the journey. That was a relatively easy question to answer, but now I was about to come under fire from Jim.

He wanted to know why I had applied for a driver for a Land Rover, after reading the letter confirming the offer of one from the Escort School of Motoring. "Surely we agreed at our last meeting, that you would be using a hand-held video camera, didn't we Shaun?" Jim exclaimed. This was where I expected Ryan to step in, as it was he who had dismissed the idea of me doing my own filming in the first place, but now he was agreeing with Jim, about how 'practical' it would be for me to use a hand-held camera, and more-importantly, how much less the budget would be without having to fork-out salaries for a two man crew. Continuing-on with other lines of sponsorship, I said that I had not yet received a reply from Virgin Airlines, about my flight seat from Cape-Town to the UK, before adding that I would not be applying for any cross-channel, or Mediterranean ferry tickets, until I had spoken further with the Automobile Association, regarding timetables, both home and abroad.

Rounding-off section 'B' of my itinerary for the meeting, I said that I would only apply for any further donations as and when a letter of commitment had been received from the broadcaster, thus giving me strong credibility when pitching my prospective sponsors. Jim agreed with me, adding that we had better "Hang-fire" with any sponsorship deals, until we had been given the green light to go ahead from the respective broadcasting company. Moving on to Section 'C', I asked Jim and Ryan to read their copies of the letter I had received from Ted Simon, in relation to his participation in the venture. It clearly stated that Mr. Simon would be happy to become involved in the production, on the proviso that his son, William, would be able to join him, as and when we meet up again at selected places in Africa. Both Jim and Ryan were happy to agree on Ted's condition, as was I...without question, of course.

Jim also liked the idea of Ted meeting up with me in Cape-Town, at journey's end, as well as chatting with me, and waving me off at the ferry port at the beginning of my expedition. In fact, he even suggested flying a crew out to California, to interview Ted about the proposed journey, several weeks prior to take off. This was undoubtedly the best meeting I had had to date, and so without further ado, I swiftly moved on to the next item on the agenda, which was the 'Matt Taylor Appeal Fund'. Unfortunately, this met with complete disapproval from Jim, because, even though he was all in favour of doing the journey in aid of charity -and completely sympathised with Matt's plight, he felt that the substantial amount of money that a project of this magnitude could raise, would be better off donated to an organisation, rather than an individual. This meant that I would now have to readjust my official sponsorship letters, which currently included raising money for Matt.

Carrying straight on to the final section, which I had entitled 'Experienced African Overlander's', I enlightened my peers about my meeting with Mr. Ball, along with my forthcoming rendezvous with Mr. Bryant, in July, although neither of these would be crucial factors any more, owing to the fact that their guidance on vehicle strengthening and loading preparation of a Land Rover, would no longer be necessary without a back-up crew. A chat with Mr. Bryant would still be very beneficial, of course, even if only to ascertain how he managed to get through the Southern Sudan in one piece, and so I fully intended keeping my appointment with him. Unfortunately, what I originally thought was going to be the major talking point of the meeting, had soon fizzled-out into no more than a mention of names, soon to be forgotten by the men who mattered the most.

All that was left to discuss now was whether or not we would still require Robert Staines to proceed with the involvement of African film organisations? When Robert had approached them in Southern France at the film festival, he said that they had shown tremendous enthusiasm with the idea, only requiring a firm commitment from a British Broadcasting Company, and they would be more than happy to begin negotiations on a partnership production. Both Jim and Ryan agreed that this could be very beneficial to our cause, and so Jim said that he would write to Robert, confirming his approval for his company to become involved in the production. Secretly, I hoped that this would also restore my chances of having a back-up team, along with the possibility of bringing in a few light aircrafts, so that the pilot's could survey my every move across the wide open plains of Africa, just as Robert had implied during one of our earlier conversations. The thought of being a 'live overlander', has always appealed to me far more than being a 'dead hero'.

Handing out the letters that I would be posting off to various African consul's and embassies, I rounded off my presentation, by saying that I also had in my possession, documents relating to each country's visa, permit, and vaccination requirements, together with individual lists of African Motoring organisations, various rainfall and temperature charts -and also an updated fuel pricelist, should they wish to see them at a later date. Feeling very confident that I had *spread-my-wampum* with the utmost professionalism, and a mass of enthusiasm, I tentatively gathered-up my various piles of paperwork, before slotting them into my briefcase, and announcing quite calmly "Well, gentlemen, that concludes my presentation of our forthcoming production: 'Cardiff to Cape-Town' -I trust that it meets with your approval."

Without saying another word, I simply sat back in my chair, and waited for either of my compatriots to speak first. It was the first *silent close* that I had done in a very long time. Jim broke the silence by asking me if I had managed to speak to James Gerwyn since our last meeting, to whit I replied that I had tried several times, but had been unable to contact him. "Well it just so happens that I have a meeting this afternoon with James, and as he is the official 'Head of Drama's and Documentaries' for the BBC, I will be happy to present your project to him Shaun", Jim stated with all sincerity. It was wonderful knowing that Jim was backing me all the way, and I couldn't have been happier. "How much do you think I should ask him for?" said Jim, before turning to look at Ryan, and thus putting him right on the spot. Ryan was speechless; he had not envisaged Jim asking him this kind of pertinent question at such an early stage.

"Five, ten, twenty thousand pounds", Jim prompted Ryan "You're the producer Ryan, tell me what you have in mind, my friend" Jim added, the sheer look of hopelessness on Ryan's face having already told Jim that Ryan didn't have a clue. "Um, as much as we can get, I suppose -twenty-five thousand pounds, at least, I would say Jim" was Ryan's somewhat hesitant reply. At this point I interjected the conversation, by saying that I would draw up a comprehensive budget plan for the entire trip, on the understanding that Ryan would work on a separate budget for the filming costs. "We could then amalgamate the two, in order to produce a final budget", I added, which received an instant approval from Ryan, who had obviously welcomed my intervention.

However, now it was time for me to confess my inadequacies, as I admitted to both Jim and Ryan how pathetic my camera skills are. I had never even changed a film in a 35mm camera, let alone worked on movie filming, and so the sheer thought of doing this side of things was frightening me to death. To pacify my trepidations, Jim said that I would go on a crash course on filming techniques, which would teach me all I needed to know about producing top quality footage. "Otherwise everything would be a complete and utter waste of time, effort and money", he added. Ryan then suggested that I come along to their television studios, where I would be able to watch how video tapes, which had been sent to the studio from three other world traveller's, who were traversing the planet by various means of transport, were being cut, in readiness for the voice-overs to be added, and the music to be incorporated into the documentary. Posting video tapes back to the UK would be yet another one of my ever-increasing list of jobs that I would have to find the time to do -and also the money to pay for postage, during this epic journey of mine.

The next suggestion was to involve a local radio station, such as Red Dragon, by using a kind of *phone-in* situation, with me reporting back to their office at every opportunity throughout the trip, which could either be broadcast live, or within a small space of time after I contacted them. Either way, it would be a great way of keeping the public informed of my progress, and so I said that I would be very happy to do it "Providing that I can find a payphone in the middle of the desert, of course", I added sarcastically. The meeting had lasted for over two hours, and we had covered virtually everything under the sun, and so Jim decided to call a halt to the proceedings, before offering Ryan and I refreshments, in the form of tea or coffee -or even *a drop of the hard stuff,* if we preferred? However, as it was still relatively early in the morning, both Ryan and I politely declined Jim's generous offer, the pair of us settling for a caffeine fix instead. I had grown accustomed to this *concluding stage* of my meetings with Mr. Howells', where we would discuss every topic conceivable in a matter of minutes, before being politely asked to vacate the premises, in order to allow Jim's next client a chance to *spread his wares*.

The meeting was deemed as a success, even though there was still a very long way to go. Each of us would have our own tasks to complete in the weeks ahead, and no one was going to say that *Shaun Donovan* let the side down by failing in his part, and so I immediately began working on the budget. This was going to be an arduuss task, and it would also take several days to prepare, as it would have to include taking every possible scenario into account, before coupling the financial allowances, which we had already discussed in part, with the average day to day expenditure, in order to come out with a final 'estimated' figure. Jim will be responsible for obtaining a firm commitment from the BBC, primarily in the form of financial backing, and Ryan had taken it upon himself to approach Sony for a range of photographic / filming equipment for the trip. Ryan said that he was quite confident that Sony would be willing to loan us a selection of camera's and video cameras, as a *gesture of good will*, primarily because Ryan's film company had recently purchased an enormous amount of equipment from them.

Once these preliminary goals had been achieved, it would then be up to Robert Staines to follow through with his side of the arrangement.

CHAPTER 22

The Budget

I couldn't have been happier with the way that my 'non-paying' business was running, but at home things could not have been worse. Minor mishaps, soon turned into major catastrophes, as general *wear-and-tear* components on household appliances, began to fail, one after the other. First to go was the motor on the washing machine, which could not be replaced, simply because the machine was so old that the part had long-since become obsolete. Apart from a daily change of clothes for Caryl and me, we also had three very young children to clothe every day, which included a stack of baby-grows for Hayley, who would inevitably have to be changed every three to four hours. I did not have enough ready cash, in order to place a deposit on a new washing machine, let-alone buying one outright, and so things were looking pretty hopeless right now. Caryl and I were in a desperate situation, which required desperate measures, and so, as much as I despise begging, we went cap-in-hand to the Department of Health and Social Security (DHSS) who kindly offered us a loan, in order that we could purchase a second-hand washing machine.

A small proportion of the loan would be deducted each month from my dole cheque, in order to pay back the loan in its entirety. With that major hurdle now thankfully behind us, our second most important appliance decided that it was now its turn to play us up, thus increasing our frustrations to an all-time high. Our *ever-so-reliable* vacuum cleaner, which had been given to us as a wedding present by my parents over eight years ago, began a series of problems, which started with the drive belt snapping during the middle of one of Caryl's major clean-up campaigns. However, a change of belts soon sorted that problem out, but then the wheels suddenly fell off, after the plastic retainer that housed them duly snapped, thus temporarily immobilising the cleaner until I could buy a tube of superglue, in order to stick the piece back on. Unfortunately, by the time I got around to doing this, the wheels had disappeared altogether, having probably been deposited into the pedal bin under the kitchen sink, which was a favourite storing place of my beloved son, Carl, before ending up in the main dustbin a few days later, never to be seen again.

Having to use my full bodyweight to force the cleaner across the living room carpet on its plastic hinges turned out to be a total disaster, as the sharp edges only dug grooves into the carpet, before coming to an immediate and unexpected halt every few seconds. One day I was caught totally unawares by one of these abrupt endings, which ended up with me falling over the top of the cleaner, whilst bringing it crashing down to the floor with me. This resulted in the pair of us rolling over on the carpet, like a pair of drunken revellers, who had let alcohol get the better of them. Unfortunately, my acrobatic antics had done more damage than I expected, which I only realised when I went to pick up the cleaner, only to find that the main handle had snapped off completely above the dust bag, thus rendering it useless...and damn near impossible to repair. To our great dismay, the social security remained unsympathetic to our cause. "Unfortunately, a vacuum cleaner is NOT a household necessity, and therefore we are unable to even consider giving you a loan for a new one...I'm sorry", said the young lady behind the counter, who seemed genuinely saddened about being unable to help us with our plight.

From that day forward Caryl and I had no choice but to use one of those old-fashioned *push-sweepers,* in order to collect up the majority of the food crumbs which my children had unwittingly deposited around the living room, and twice a week Caryl would drive the seven miles to and from her parents house, in order to borrow their carpet cleaner for a few hours, so that it could churn up the remaining bits and pieces that our dust pan and brush had failed to collect. If that wasn't enough to cope with, there were more surprises in store for us as the weeks progressed. The electric kettle burned-out its element, forcing us to revert to saucepans for boiling water, until such time as a *kettle-donor* could be found, and the central heating system also packed in, but that would now have to wait until September, because at least we still had the gas fire to keep us warm during the somewhat *nippy* mornings and evenings.

Between Liam and Carl they had not only managed to break the second hand television set, which Caryl's parents had given us about a year ago, but they had also damaged a replacement television, which our neighbour had kindly loaned us only a few weeks ago. Besides all of this, my two beloved sons had also managed to put out of action no less than 'three' cassette players, all of which were now sitting on a bench in my garden workshop, in the vain hope that one day I might get around to fixing them? Our attic was also beginning to look like the proverbial 'Aladdin's Cave', after the amount of damaged pictures and fractured ornaments I had stashed away up there, for fear of them being totally destroyed by my trio of mini-vandals. To top all of the above, our living-room carpet currently boasted an array of coffee and tea stains, along with a selection of fruit juice and Ribena patches dotted here, there and everywhere. Also, the occasional trace of *undesirable substances*, which had been kindly deposited by my beloved daughter, Hayley, during one of her *nappy-free* times, could be found lurking in various corners of the room.

The interior walls of the house that were once adorned with several gallons of *brilliant white* matt emulsion, were now of the *psychedelic* variety, thanks to a vast number of glorious multi-coloured etchings, all of which were perpetrated by my entourage of budding Van Gogh's, each one of them using only the finest felt-tip pens, and thick waxed crayons, in order to create their individual masterpieces. Also, the wallpaper in Liam and Carl's bedroom, which used to rest neatly above skirting board, alongside the doorway, had *mysteriously* peeled away from the plasterboard walls, even though neither of my beloved sons would admit to administering a helping hand, of course. Damp and condensation had also played a major role in the dilapidation of our wonderful home, with both the door frames and the window frames being in serious need of repair. In fact, on one occasion, I even found a small mushroom growing in one corner of our hardwood conservatory.

Our previously immaculate lawn was now flooded, after a volley of torrential downpours had turned the neatly cut grass into a quagmire of muddy pools, thus making the garden unsafe and unhealthy for the children to play in. Also, the once colourful, swing, slide, see-saw and climbing frame combination unit, which I had built for the children a few summers ago, had now rusted so badly, that the whole thing was not only unsightly to look at, but it was also too dangerous to use. Looking all around me, I could easily understand what was making Caryl's life so miserable, but what could I do about it; I had all the time in the world to put things right...but no money to do it? Jim Howell's had not managed to meet up with Mick Chappell, who is the current head of English Language Programmes for BBC Wales, as his original appointment had been cancelled via Mick's secretary.

By this time I had prepared a ten page budget, which included a complete breakdown of fuel costs, both in Europe, and also in Africa where prices ranged from eighty pence a gallon to over four pounds, if bought on the black market in Tanzania. Tolls for European roads had also been taken into account, especially in France and Italy, where they can become quite expensive, depending on which route one takes. Bridge crossings can also set you back a few pounds, as I discovered on my previous journeys throughout Europe. As far as I know there are no tolls on any African roads, or charges for any bridge crossings on the continent, but things can change overnight in Africa, and so I wasn't ruling this out entirely. Although my cross-channel and trans-Mediterranean ferries would hopefully be sponsored by the Automobile Association, I knew that at some stage I would have to board the ferry that sails down Lake Nasser from Egypt to the borders of Wadi Halfa, which is the gateway to the Sudan, and so I had allowed a percentage of my expenditure costs to cover the price of this ferry.

I will also be taking the train from Wadi Halfa to Atbara, as this is apparently one of the most difficult stretches of the Nubian Desert, which covers around 400,000 square kilometers along the eastern side of Northern Sudan. The carriages can easily accommodate the bike, and after having ridden the length of Egypt, I am sure that I would be extremely happy to cadge a lift for a few hundred miles. Overnight accommodation would be spent mostly under canvas, wherever and whenever it is safe –and hopefully 'free' to pitch my tent. However, whenever I reach one of the major cities, I will be treating myself to a hotel room for the night, so that I can enjoy the luxuries of a hot bath, whilst washing out the multitudes of gritty sand and grimy dust particles from every sweaty orifice in my body. Working on an average of six nights under canvas for every one night spent in a hotel or a guest house, I put together an overall budget for the anticipated seventy day crossing of continents.

Food and drink would turn out to be the most difficult commodity to assess for the entire duration of the journey, simply because prices would vary considerably from one country to the next, and so I decided to take an educated guess on this one -and hope for the best. I would be taking a fair amount of dried food with me, which would have to be mixed with clean water, of course, and so in certain areas of Africa it could be cheaper to purchase fruit, or other sources of sustenance, rather than having to pay exorbitant amounts of money for that all-important commodity...water. As for spending money for the journey, well this would be quite nominal actually, as I would certainly *not* be buying any souvenirs along the way, not only because of my *ever so tight* budget, but also because there would simply be no room on the bike to store any additional items. Knowing that absolutely anything could happen on the journey, I insisted on having an emergency back-up of funds, which would be made up of various currencies, along with a handful of traveller's cheques. This money would cover me in the event of any of the following misfortunes:

THEFT: *To be able to replace any items, such as photographic equipment, clothing, tools, spare parts, or even medication, that is stolen from the bike, or from me.*

ACCIDENTAL DAMAGE: *Should a moving part on the bike be damaged beyond repair, such as a buckled wheel, or a twisted fork leg, then these will have to be replaced at the first opportunity.*

MECHANICAL BREAKDOWN: *Although numerous spares will be carried on the bike, I may need professional help if the fault cannot be rectified by me.*

ALTERNATIVE TRANSPORT: *If I run out of fuel, and I am unable to purchase any from passing vehicles, I may have to load both the bike and myself onto other forms of transport, such as a pick-up truck, or a recovery vehicle.*

<u>BRIBES:</u> *Used mainly at border check-points, these can be in the form of small gifts, money, food, or even photographs, and so it is always advisable to carry an instamatic camera, as apparently border guards often enjoy having their photo taken with foreign travellers, before showing off the photo to their friends, family, and fellow workers.*

<u>DOCUMENT LOSS:</u> *Replacing lost visas, entry permits, vaccination certificates or even my passport, heaven-forbid, can be both a costly and lengthy process, let-alone the inconvenience of it all.*

<u>ROUTE CHANGES:</u> *Anything can happen overnight in Africa, possibly causing me to have to change my pre-planned route, such as roads flooding, border closures, or even military uprisings, and so additional fuel costs will inevitably be incurred, along with possible air or shipping costs, should I have to avoid these areas completely.*

<u>MEDICAL EXPENSES:</u> *Although I will be covered by personal insurance for any major illness, I am sure that the local 'witch-doctor' will not accept a credit card, or a medical document for services rendered.*

<u>FINES:</u> *Who knows how many 'unwritten laws', or 'fabricated regulations' I may break during the journey, in order to service the pocket of a uniformed official, and so, once again, I must be prepared for this eventuality.*

I also told my peers that I would require access to a reserve account of available money, which I could withdraw from a bank in Nairobi, or Mombasa, in Kenya, as this would be the halfway stage of the journey, and by then I may have had to use some –or possibly even 'all' of my backup funds. The only other time I may need cash, is to post my used video films back to the UK, or if I need to use a telephone, in order to call the radio station. Apart from all the practical expenses, there would be numerous other costs involved on the paperwork side, and vaccinations would certainly not be a cheap item either. However, the most costly expense of the whole operation would be in obtaining the 'Carnet de passage En Duane'.

This is a pretty complex, and also a very costly document, that has been instigated primarily to ensure that the owner of the vehicle that is traversing Africa, does not sell the vehicle during their travels through the many African countries, otherwise they will forfeit their deposit, which is normally around two-and-a-half times the value of the vehicle in question. Upon exiting Africa the deposit is reputedly refunded in full, although I must admit that at present I am very apprehensive about 'if', 'how' and 'when' that money would be returned into one's account? With all of the above totalling around £ 5,000, I next had to add my own personal fee for actually doing the trip. Knowing that my social security benefits would be curtailed the moment that I left British shores, I would obviously require enough money to cover all of my monthly outgoings, including my mortgage payment, which is something that I had not had to pay since being made redundant all those months ago.

I then multiplied this figure by three –three months being the estimated time that I would be away from the UK, before throwing in a couple of hundred pounds for any miscellaneous expenses that Caryl might incur whilst I was away. Ryan had also told me beforehand, that I should also include any expenses that I had incurred to date, such as the cost of the trip to California, to visit Ted Simon, my two jaunts to London, the first time when I visited the newspaper library, in order to research Ted's articles, and the second time when I went to pitch Stanford's for sponsorship. My day out to the Museum of British Transport in Coventry, when Jules and I did a 250 mile round trip, in order to pay homage to Ted's *Round-the-world* motor-cycle, should also be included, as all of these jaunts were significant ventures in bringing this programme together.

As soon as I receive any kind of lump sum from the BBC, my first intention is to reimburse Jules the money he had invested in me, by paying for my trip to America, even though he had never asked me for a single penny since the day I returned to the UK. Also, as Jules was officially my partner in this venture, he should be entitled to a percentage of my earnings, even though Jules had also never mentioned any kind of payment for him in the future. However, Jules could rest assured that whatever profit I made from this production, he would receive a healthy chunk of money from me. This was not only because he had forked-out his own money to sponsor my trip, but because he had also kept his faith and belief in me since day '1', which was something that everyone else around me had doubted from the off. Apart from spending around £100 on stationary, which included purchasing the typewriter, and also the three 'Diary' photo albums, my biggest outlay to date was the phone bill, which had amounted to over £300 during the last six months, and so needless to say that this was all included in the expenses sheet.

After stapling all the sheets together that evening, I popped over to Jim's house, where I presented him with the comprehensive invoice. I also took Derek Bryant's biography on Africa with me, as Jim had asked me if he could borrow the book, in order to familiarize himself a bit more with the *Dark Continent*. In the manuscript Derek Bryant states that travelling alone in Africa is NOT recommended, and so I kind of hoped that upon reading this, Jim might change his mind about giving me a back-up crew. Jim immediately invited me into his study, where we discussed the project in detail over the course of an hour or so. Jim then graciously complimented me on how I had handled things to date, adding that he would push my idea to the limit, primarily because of his own personal belief in me, which was something that I desperately needed to hear right now. Jim finished off our meeting by saying that he would be meeting up with Mick Chappell tomorrow morning, and the mere mention of this man's name caused my goose bumps to reappear in a flash. I was ecstatic.

However, two days later, Jim's secretary called me to say that the meeting with Mr. Chappell had once again been cancelled, and I knew that there would be no chance of another meeting for the next two weeks, or so, simply because Jim was going away on holiday for a fortnight. Replacing the receiver with a somewhat heavy heart, I felt as if another half a dozen rungs had been added to my proverbial ladder of success...I was not a happy chap. The following morning Caroline Watts rang me to say that she had received a letter from Mr. Bryant, who apparently wanted to know why he had not heard from *Shaun Donovan*? This took me by total surprise, as I never knew that Caroline had even contacted him after our last conversation.

Explaining to her that I was still waiting for a definite proposal from the BBC seemed to be no excuse as to why I had not written or telephoned Mr. Bryant, as Caroline duly informed me that it would probably be too late by now anyway, because Derek was already on his way to Nairobi. All I could do was sit back and take this one on the chin, as I humbly apologised to Caroline for not getting in contact with Mr. Bryant sooner.

However, when I then told her that it looked like I would be riding solo through Africa, she openly exclaimed "Utterly preposterous" before adding "Whoever made that decision must be stark raving mad!" Her statement did nothing for my confidence with regards to going it alone across Africa, but at least I could rest easy now, knowing that Caroline –and very soon Mr. Bryant, would be fully aware of the current, and somewhat volatile, situation. Caroline said that she would now wait to hear from me, before taking matters any further, and so I thanked her for her patience, before replacing the handset, and moving on with my life.

Little did I know then that a few days later I would be leaving Caroline a message on her answer-phone, saying how imperative it was for me to have a meeting with Mr. Bryant when he arrives in London. Jim's meeting with Mick Chappell had finally been confirmed, and so I could only hope that it would be a case of 'Third time lucky'. On the night prior to the meeting, I called in at Jim's house, carrying with me a small token of my appreciation for all that he had done for me, but unfortunately no-one was at home, and so all I could do was leave the present with a neighbour, before dropping a note through Jim's letter box. The following evening I returned home after visiting my brother, Gary, at his home in Canton, only to be informed by my wife that Jim Howell's had rung the house during my absence. Jim had given Caryl his home number, before asking her to make sure that I phone him as soon as possible please. Jim's lovely wife answered the phone, saying that Jim was outside, working on the garden, and so she would go and get him straight away.

"Hello Shaun", Jim said quite excitedly, his normally calm, cool and collected voice, now bursting with excitement. "Thank you so much for the lovely present –it now has pride of place in our home. However, I am sure that you would rather talk business, and so here is the latest news. I finally spoke at length with Mick Chappell today, outlining your project to him as best as I could, but what I want you to do, is to call Mick's office in the morning, and speak with Mick's secretary, whose name is Anna-Marie –have you got that so far Shaun?" Like a doddering old fool, I simply nodded to Jim's request, before suddenly realising that he was on the other end of a telephone, and so I duly followed-up my nod with a definite "Yes Jim". My mind was racing, and my heart was thumping even faster, as Jim read out the telephone number for me to jot it down on a piece of paper, and it took me several seconds before I realised that it was exactly the same number as James Gerwyn's, only with a different extension line.

By this time Jim was halfway through explaining his all-important message to me. "I have booked you in for a meeting with Mick Chappell next Tuesday, and so all you have to do is arrange an appropriate time with his secretary that will suit both you and Mick. Once you have done this, I want you to ring my secretary, and ask her to book you in for a meeting with me on the following Monday morning ...now is that okay with you Shaun?" Immediately I gave my affirmative answer, which Jim obviously knew I would, as he continued on with his speech. "Now what we will have to work out Shaun is an exact figure for you to ask Mr. Chappell for on the day in question, and so when you get some time, I want you to go over your budget with a fine toothcomb, before coming up with a total amount that you can openly ask him for".

Jim then told me to arrange a preliminary meeting with both him and Ryan prior to my appointment with Mr. Chappell, in order to prepare everything down to the very last detail, before stressing that it would be better for me to pitch Mr. Chappell on my own. Jim concluded our conversation by saying that he would have a word with James Gerwyn before my meeting, to see if he could help things along in any way, such as having a preliminary chat with Mr. Chappell about the proposed production. By doing this, Mick would not only have a clear understanding of the merits of the project, but it would also give him time to assess the benefits that creating a series like this would do for the television ratings, as this was clearly an adventure that had never been captured on camera before, and therefore it should generate considerable interest from a great proportion of the general public.

The following day I told Robert Staines all about my meeting with Jim, and Robert advised me to ask Mr. Chappell about the BBC's policies on South Africa. Robert told me that he had already received Jim's full approval of his company's input into the project, but without the approval of everything, from start to finish by Mr. Chappell this whole project could collapse overnight. Within minutes of returning home that afternoon, everything was fixed. I would meet Jim at 9.30 am in his office on the first of June, before meeting Mr. Chappell at the BBC's headquarters in Llandaff, the following afternoon at 3 pm. Unfortunately, Ryan would be on holiday during this week, and so he would be unable to join Jim and me for our preliminary meeting.

On the weekend I happened to cross paths with my cousin, Steve, who has been working as a VT (video tape) editor for BBC Wales for many years. When I told him that I had a personal appointment with Mr. Chappell he was simply flabbergasted, insisting that the majority of people who work in the Llandaff Television Studios are not even allowed to take the lift up to the floor where his office is situated. My brother, Gary, who had also worked for the BBC, in the scenery department for over a decade, echoed Steve's sentiments, saying that he had rarely seen Mr. Chappell in person, and that he had never actually spoken to him face to face. Suddenly I felt like I was seeing 'God' long before my time was due, but in all seriousness, I also knew that I dare not screw-up this incredible opportunity, for there was no way that I would ever get a second chance.

CHAPTER 23

The Build Up

The summer was now well and truly upon us, the long, cold winter having dissipated into nothing more than a deep, dark memory of days gone by and it looks likes the warm summer sunshine is here to stay. For hour after hour Caryl and I played with our children in the garden, before taking them down to Victoria Park in Canton, where we frolicked about in the cool, shallow waters of the outdoor swimming pool to our hearts content. The lawns which surrounded our house had long since dried out under the sun's blistering rays, thus enabling me to cut the grass back to a decent length, and my brother, Gary, had kindly given me a few gallons of brilliant white matt emulsion, so that I could give the interior of our house a new lease of life, as well. It was now a pleasure getting up in the morning to the intoxicating smell of newly painted walls, and to open out our patio doors, before breathing in the unmistakable aroma of freshly cut grass.

For once I was the envy of all of my friends, who were either stuck in their stiflingly hot offices in work, or sweating their pants off in the dusty heat of a building site, but if truth be known I would have swopped places with any one of them at the drop of a hat. However, the warm sunshine of any summer's day in Britain could not compare to the heat that I was about to face in Africa. Whenever I read or hear about great journeys across scorching deserts, the person who undertook the journey almost always remarks on how "Unbearable" the heat was, especially around mid-afternoon / early evening. However, having endured temperatures in excess of one hundred degrees, which were often accompanied with stifling humidity levels, without feeling too uncomfortable in the past, I hoped that this factor would not present too much of a problem for me, and that my body would soon acclimatise to the adverse weather conditions. Upon saying this, I must also take into account the fact that on all of those previous occasions there had always been an ample supply of water nearby, either to drink, or to bathe in, whereas this time there is likely to be no signs of water for hundreds of miles in any direction.

Being unable to bring my body temperature down to a respectable level, by having a cool dip in a swimming pool, or a short swim in the sea, is not something that I have been used to beforehand, and so being out in the blazing hot sun all day, without any respite from the sweltering heat, would be an entirely new experience for me. I might also consider using suntan oil for the very first time in my life, in order to protect my body from the sun's immensely powerful ultra-violet rays, as skin cancer is something that I could well do without. I will also be taking a selection of prescribed medication, such as salt and malaria tablets, to be administered as and when required, as the mere thought of catching malaria from a mosquito bite scares me half to death. The days passed quickly, and now it was time for what will undoubtedly be my most important meeting with Jim Howells. In total contrast to our last little *get-together* my ever-present bundle of nerves were conspicuous by their absence, for I knew that today we would simply be determining how much money we would be asking Mr. Chappell for, in order to begin the production in earnest.

I had driven almost halfway to Jim's office when I suddenly realised that I had left a handful of important documents back at the office I had been working from (Caryl's grandmothers old room, in Caryl's parent's house) and so I had to make a quick diversion, which inevitably made me a quarter of an hour late for my meeting with Jim. However, my feeble excuse, along with my profuse apologies, was soon accepted by my forgiving mentor, and within minutes our meeting was underway. Jim expressed how imperative it was that I should make the budget as minimal as possible, albeit large enough to cover all of my outgoing expenses before, during and after the journey had been completed. Jim then asked his secretary, Erin, to ring Ryan's office, to see if he had received any news from Sony regarding the photographic and video equipment, but Ryan's secretary told her that as far as she knew, Ryan was still waiting for a reply from his letter.

However, the good news is that Jim had spoken to James Gerwyn a few days ago, and he had accepted Jim's offer of being the Executive Producer of the programme, overseeing Ryan, who would initially be putting the six part series together. Jim then said that it was now time for me to learn how to speak to a man of Mr. Chappell's stature, adding that one wrong statement from me could be the difference between getting a "Yes" or a "No" answer for the production –the latter being something that I could ill-afford at this crucial stage of the game.

"Remember at our last meeting, when Ryan asked you if you would still go ahead with the journey if nobody was willing to sponsor the trip, and you quite honestly answered that without someone being able to cover your personal expenses whilst you were away, you could not even consider to doing the journey. Well, Mr. Chappell is going to ask you that exact, same question, only this time, without any hesitation whatsoever, you must look him straight in the eyes and state quite categorically 'Yes, of course I would'. If you hesitate, or worse still, give a negative response to the question, Mr. Chappell could easily believe that you are just trying to get someone else to pay for your 'freebie' trip across Africa, and believe me when I say that this man does not suffer fools lightly".

Only a handful of people know that it was Jim's idea for me to do this trip in the first place. It was also his idea for me to go it alone, in order to express my innermost feelings along the way, just as Ted Simon had done in his book, the only difference being that I would be talking directly into a video camera. My idea was to make a movie; to recreate Ted's incredible journey, and to make it as exciting as possible to millions of biking enthusiasts all over the world. I never had any intentions of doing a trek across Africa, let-alone at someone else's expense, and so I would be truly mortified if Mr. Chappell thought that for a single moment. However, although I have never been a blatant liar, I am usually pretty good at bluffing my way into things, in order to get what I want, and so a little *white-lie* might be in order. Telling Mr. Chappell that I have other parties that are interested in my project, as a kind of *urgency pitch* might be worth a shot I suppose, but then did I really want to take the risk of him calling my bluff, before sending me packing?

And what if he does say "No" anyway, what would I do then? Would that be the end of everything as we know it, or would I just go back to the drawing board, and start all over again? My mind boggled with a multitude of questions that I may never have to answer, and so I decided to just do what I had to do, and be honest about everything, just like I had been from the start. After a very lengthy discussion with Jim, which included going through my budget sheets with a fine toothcomb, we came to the final conclusions:

£ 5,000 would be the bare minimum that it would cost for me to complete the journey in question, which would include the reserve cash fund and the back-up money, of course.

Another £ 5,000 would cover all of my expenses back in the UK whilst I am out of the country.

£1,000 would also be required, in order to pay back Jules Gilmore for the money that he had given me to go and pitch Ted in America, and a final £1,000 would be needed, in order to pay for the various visas, vaccinations, and a mass of legal documentation that would be required for me to complete the journey.

This minuscule figure of twelve thousand pounds in total would only be acceptable on the understanding that I am given a letter of commitment from the BBC, as proof of their involvement in the project, in order that I could show it to other prospective sponsors as good credibility, before asking for a financial commitment from them. If the BBC also agreed to involve a South African broadcasting company in the project, then the costs could be split between both parties concerned. In return I would guarantee the BBC three 30-minute programmes in total for *Part I* of 'Steps of Jupiter-20 years on' which combined will show the journey from start to finish. Each episode would include various extracts from my liaisons with Ted Simon along the way, where we would chat about the changing faces of Africa over the last twenty years. Jim said that this would be better than our initial idea of doing two 45-minute programmes, because the more 'slots' of air-space that a series takes up, the greater the revenue when selling the series to another broadcasting network abroad.

Jim also said that if I am able to make a one hour programme out of twelve hours filming, I am doing exceedingly well indeed. By now I was getting quite excited about my meeting with Mr. Chappell, and so I asked

Jim how long it would take before getting an answer –and hopefully the 'go-ahead' on my journey? Jim shook his head gently from side to side for a few seconds, before shrugging his shoulders and making his final speech of the day:

"Let's put it this way Shaun, at the end of your presentation, Mick is not going to say to you 'Fine; here is a cheque for twelve thousand pounds; now just go away and do it –and don't forget to send me a postcard along the way'. You will probably have to wait two to three weeks before you get your answer, either good or bad. However, what you must express on the day is the urgency for an answer from Mick, sooner, rather than later, simply because of the minimal amount of time you have left in order to obtain all of the papers you will require. You will also need to get as many of your sponsorship deals confirmed long before September arrives otherwise you could find yourself way out of pocket with all your additional expenses."

Jim was running late with his schedule, and so we never had our usual informal natter at the end of our *official* meeting, but as I made my way down the stairs from his office on the first floor, Jim leaned over the banister rail and shouted "Don't forget to take that amazing map with you Shaun –the sheer enormity of it overwhelmed me, just as I am sure it will him". Smiling like a Cheshire cat, I gave Jim the 'thumbs-up' sign, letting him know that I fully intended taking the map with me anyway. Whistling out loud as I walked –and even skipped once or twice, down the arcade, I felt as if another great step had been climbed towards the top of my never-ending ladder. Jim had instilled in me every ounce of confidence that I needed, and he had also paid me a great compliment, that I will never forget. Halfway through his speech on how I could best sell myself to Mr. Chappell, Jim simply threw his hands in the air before saying: "Oh what the heck –I am trying to teach my grandmother to suck eggs here –just be yourself Shaun, and you will do fine".

I drove across town, purely to see if I could catch Jules at the Budget offices, but once again I had managed to miss him by only a few minutes. The new offices, which were originally due to open in Swansea at the end of March, now looked as though they would not be opening until the autumn, due to financial restraints, which in turn was due to the current economical climate –I.E. the recession. Ever since the inception of this additional franchise, Jules had intimated to me that *if and hopefully when* this new depot opens up, then he would love me to run the entire operation. However, as he also knew that I would only take him up on his offer if my current project fell through at the eleventh hour, he openly admitted to me that he wasn't *holding his breath* on that one? As the job with Budget was the only back-up plan I had, should everything else go belly-up, this latest delay was a real set-back for me, as our current financial situation was dire, to say the very least.

In fact I think it is fair to say that Caryl and I were now in a critical situation, and very close to *breaking point*. Jules was now working a lot more hours, primarily because of various problems that had arisen in the Newport depot, and consequently we had been seeing a lot less of each other. Over the last few weeks I had left several voicemail messages on his home telephone, the majority of my calls having been finally returned a number of days later. Unfortunately, Jules had been forced to lay-off staff, in order to sustain the business, thus forcing him to work longer hours, in order to keep the never-ending paperwork up to date in the office, and so he was always too busy to meet me for lunch, like he used to, and his evenings were invariably taken up with having *emergency meetings* with his business partner, or by entertaining his lady-friend.

Not wanting to waste a single moment of my time, I drove straight over to the BBC studios in Landaff, in the hope that James Gerwyn would be able to spare me a few minutes of his valuable time, as I desperately wanted to know exactly what his role in the project would encompass, and I also needed to know what he could offer me, help-wise? The receptionist kindly rang James' office for me, but his secretary, Elisabeth, told the young lady that Mr. Gerwyn was currently in a meeting with Mr. Chappell. However, Elisabeth then told the receptionist that she would ring Mr. Chappell's secretary, asking her to ask Mr. Gerwyn if he could spare me a moment of his time, before ringing her back with his answer. Whilst waiting patiently for the outcome of all this toing and froing, I began reading a pamphlet about a selection of books that had been published through the BBC, taking particular note of Michael Palin's book 'Around the world in 80 days'.

I was astounded to read that it had already sold nearly half a million copies, and so considering that I intended writing a follow-up book based on the 'Steps of Jupiter' series, I began looking forward to a very prosperous future indeed. My dreamy thoughts are interrupted by the receptionist, who is now calling out my name, whilst waving a receiver in her right hand, before stating that she now has Mr. Chappell's secretary, Anna-Marie, waiting on the line to talk to me. After cordially introducing herself, Anna-Marie said what a coincidence it was that I had called into the BBC studio's just as she was about to give me a call for an informal chat. Unfortunately, her opening statement immediately set all the alarm bells ringing in my head, and so I knew exactly what was coming next:

"Mr. Chappell has had to reappoint your meeting to next Thursday morning, due to his hectic schedule, I am afraid Shaun –oh and Mr Gerwyn has also asked me to contact his secretary, and book you in for an appointment with him next Tuesday morning...is that okay with you?"

I had to say that it was fine with me, of course, but I was feeling really disappointed about the reappointment. Mr. Gerwyn's secretary then called me in the afternoon, in order to reschedule my meeting with James for another two weeks, due to the fact that his wife had just had a baby, and so my new dates were as follows: Thursday, 11th June for my meeting with Mr. Chappell, and Tuesday, 16th June for my meeting with Mr. Gerwyn. Although I would have much preferred seeing James Gerwyn before Mick Chappell, I was not in a position to be choosey, and so I duly accepted the proposed dates.

Although no-one had ever asked me for anything in return for their help with my project, I consider myself to be a man of honour, and therefore I always like to repay what I personally deem to be a debt of gratitude, if you like, and so I decided to give Robert Staines a small token of our friendship, not only for his continued assistance, but also for his belief in me...something that had never waned since the day we first spoke about my cause. And so the following day I drove into the centre of Cardiff, where I was lucky enough (or so it seemed at the time) to spot an empty parking space in one of the *Loading Bay* areas, in front of the central market, on the main road. Having parked-up the car in a neat and orderly fashion, I proceeded to read the inscription on the little blue plate that was perched about halfway up the lamppost, which stood adjacent to the parking spot. It stipulated that the 'maximum loading time' was half an hour, and so as I only intended dropping-off Robert's present at his office, before returning to my car, I duly locked my drivers' door and headed for the side entrance to the Castle Arcade, which was literally across the road.

Unfortunately, the main entrance to Robert's office was locked, and so I peered in through the side window, but there was no-one in sight? As it was almost 1 pm, I had a hunch that Robert might well have locked himself away in his back office, in order to enjoy a peaceful lunch, and so I duly rapped on the main door, pretending that I was a delivery boy –just in case he had ordered a snack from one of the many restaurants that were situated on the ground floor of the arcade, to be delivered to his office. My theory turned out to be half right, as Robert was indeed in the back office, but as he came to my aid, whilst gnawing on a sandwich with his right hand, and opening the door with his left, it was patently obvious to me that he was in the middle of enjoying a packed lunch when I disturbed him. Having made the humblest of apologies for my intrusion, I duly handed Robert the gift, before turning to leave him in peace.

However, Robert would not hear of this, beckoning me to make myself comfortable in his secretary's chair, whilst he pulled up another one on the opposite side of the desk. "So how is everything going, my friend?", Robert asked keenly, his sincerity in wanting to know all *the gory details*, as strong, and as passionate, as it ever was. As I knew that I could trust Robert as a friend, as well as a partner in my proposed production, I poured my heart out to him, telling him all that had transpired since we last chatted together, including my serious doubts about the future of my project, as time was running seriously low by now, and how I was beginning to get really scared of what the final outcome would be? As always, Robert was a tower of strength, cordially congratulating me on all that I had achieved so far, before openly admitting that I still had "A long way to go". He also stressed the fact that I must get Mr. Chappell on board with the production, as he was the 'One Man' who could make everything possible for me. I told Robert that I intended preparing a twenty page itinerary for our meeting, so that I could get him fully up to speed with the story to date, but Robert advised against it, saying that I would probably only have an audience with him for around half an hour, and a lot of valuable time could be wasted trying to explain everything.

"All he will want to know are the main points of the storyline, including 'The bottom-line', of course...in other words 'How much is it going to cost him?'" Robert added, in a commanding tone, causing me, once again, to bow down to his professionalism. Sitting back in my chair, I let Robert continue-on with his lecture: "If you really want Mr. Chappell to have an avid interest in your project, along with his undivided attention throughout your meeting with him, then it is essential that you send him an analysis of your project, in the form of an elongated synopsis, rather than a full-length dossier, otherwise he will not take the time out to read it. Deliver it to the BBC studios around 24 hours prior to your meeting with him, thus giving him enough time to read it through thoroughly and to hopefully conjure-up a picture in his own mind of exactly what you are trying to create on screen". Without further ado, Robert then picked up the phone nearest to me, before thrusting it in my face, and ordering me to ring the BBC studios "Right now".

The receptionist immediately put me through to Anna-Marie, Mr. Chappell's secretary, and so I duly informed her that I would be dropping-off the aforementioned document for Mr. Chappell's personal attention on the Wednesday morning –the day before our meeting. By the time I had replaced the receiver, Robert had already lit up a cigarette, and he was now pacing up and down his office like an army sergeant on a military parade ground, his rigid expression disappearing through a cloud of grey smoke, before reappearing once again a few seconds later, his eyes wide open, and a beaming smile now adorning his face as if he had just been enlightened with an incredible idea...and then it disappeared again. "What about South Africa, Robert?" I asked tentatively, but my question obviously fell on deaf ears, as Robert continued marching from one end of his office to the other, his brain otherwise engaged with far more important matters than simply answering my question –or so it seemed...and so I tried again: "Jim Howell's said that the BBC's policies with South Africa are quite relaxed at the moment, and so he believes that bringing a South African film company on board should not present any problems".

At this point Robert's transcendental state of mind finally returned to the land of the living, as he suddenly stopped dead in his tracks, before turning towards me, and glaring deep into my eyes. "Eureka" I say to myself, in the hope that Robert is about to solve all of my problems, but then life is never like that, and so I just sit back in my chair, calmly waiting to hear what my great friend has to say next? Robert opens his mouth with the full intention of speaking, but before a word can be uttered, his right index finger suddenly crosses his lips, as if telling himself to be silent, as his eyes roll to the heavens in deep contemplation once more. Within seconds all hell breaks loose, as Robert immediately stubs out his cigarette in the ashtray on his desk, before bursting into his command performance of how to 'close' Mr. Chappell.

"Tell Mr. Chappell that my company is quite willing to act on the BBC's behalf, in order to bring a South African broadcaster on board with the project –and make sure you tell him that it looks like you have a Land Rover available, should he decide to include a back-up crew in the project, as this will show him that you have done your homework right down to the very last detail. To finish off your presentation you must give Mr. Chappell the following three choices on how to do the filming:

Number One: *Say that you will be happy to take a camera along with you on the motor-bike, and do all the filming yourself, which will be the most economical way of doing things, albeit the least professional way, of course, due to your lack of knowledge in the filming industry, which he will already be aware of, I am sure.*

Number Two: *Tell Mr. Chappell that if he is willing to include a back-up crew in the project, then not only will you provide the vehicle for the job in question, but you will also provide a driver, who is a fully qualified Land-Rover mechanic, for the entire Journey from Cardiff to Cape-Town, at no extra cost, other than his travelling expenses and his food allowance.*

Number Three: *If Mr. Chappell wants to reduce the filming costs to a bare minimum then tell him that I will do my uttermost to bring a South African broadcaster on board, thus halving the cost of the production, give or take a few pennies, which will be like music to his ears –have no doubt about that.*

What more can he ask for Shaun; he will have three golden opportunities of getting this exciting production of yours on the road at minimal cost to the BBC, and this 'choice close', in other words 'Which option(s) would you like to choose Mr. Chappell?' should eliminate the chances of him saying "No" at the first hurdle, although you will still have a long way to go, of course".

I could not thank Robert enough for his brilliant performance, and I only wished that he could be standing by my side when I confront Mr. Chappell for that all-important meeting next week, but once again, it looked like I would be going it alone...such is life, I suppose. I had also committed myself to preparing and delivering an elongated synopsis of the project from start to finish within the next week, and so once again it looked like I would have my work cut-out for me. Crossing the road in the main high street, I can see my car just across the street, and I am trying to see if a note has been attached to my windscreen –compliments of your friendly *traffic vulture*! I had been in Robert's office for over an hour, thus doubling my allowance of time to park, and so I wasn't too surprised when I saw the dreaded parking ticket flickering in the wind under my windscreen wiper..."Shit!"

However, upon closer inspection of the said ticket, not only did I discover that I had been booked less than five minutes after parking my car, but it also stated that a *standard car* was 'not eligible' to park in loading bays! I probably should have realised that these loading / off-loading bays were set-up primarily for the likes of box vans, transit vans and trucks, etc. but I guess I was hoping that I would get away with it. I cannot deny that I deserved the twenty pounds fine, but boy did I have a thousand better ways of spending that kind of money right now, than handing it over to the government. I also knew that I would have to hide the evidence from Caryl; otherwise I would be in even more trouble. The last time that I had been ordered to contribute funds to the 'Fixed Penalty Department' was back in the seventies, in the days when racing motorcycles along the highways and by-ways of Britain was the norm for me...along with getting caught by the police on several occasions, hence I continued donating my hard earned money on a regular basis.

My last parking fine was also many years ago, only on this occasion I thought that I could get away with parking my motorcycle in a disabled parking space in the centre of Cardiff, on a busy Saturday afternoon, whilst I popped off to the cinema for a couple of hours. Having recently survived a horrifying head-on collision with a car, where I sustained life-threatening injuries, along with being left with a permanent limp for the rest of my life, I figured that was entitled to park in a disabled zone –and so I did! However, when I returned to my bike and saw the proverbial parking ticket, nestled neatly under the seat strap of my bike, I was absolutely furious, to say the very least. In fact I was so annoyed that I decided to write a stormy letter to the clerk to the justices, vehemently refuting the summons, and refusing point-blank to pay a single penny of the fine. My letter of grievance had been hand-written, just to give it that personal touch, and I intended posting it first thing on the following Monday morning.

However, on the Sunday evening a friend of mine warned me about the possible consequences of sending such a letter, such as being ordered to take a secondary test, because of the permanent injuries I had sustained, adding that if I passed the test, then I might also have to display a disabled sticker on my bike henceforth, which would somewhat diminish one's 'macho' image of being a *rough-and-ready* biker. With these thoughts in mind I decided to substitute the letter for a cheque, and duly pay the courts for my crime.

Little did I know then that thirty-five years later I would write a book about my numerous exploits on motorcycles, which included no less than thirteen horrendous crashes, seven endorsements on my driving licence (long before 'points' came into force) and two driving bans. Mind you, the book would also tell the stories of three incredible motorcycle journeys around Europe and North Africa, which I undertook in the eighties, and so at least I had plenty of positive stories to tell to the world –as well as the negative ones.

CHAPTER 24

Talking To God

It had taken me seven very long months to get where I am today, and I could not afford to lose everything now. In eight days time I will be coming face-to-face with the head of broadcasting in Wales –the big 'He-bull' himself, Mr. Mick Chappell, and so I must make sure that I am one hundred percent ready for him. For the next five days I worked solidly on my presentation, literally wearing out the first ribbon on my typewriter with endless pages of facts, figures and dialogues…you name it –I wrote it. I then had the laborious task of minimizing all of my work, so that only the most poignant issues of my project were addressed, in minimal sentences, of course. On the Wednesday morning I hand delivered my synopsis-cum-itinerary for the meeting, along with a covering letter, to the BBC studios, expressing the urgency to the receptionist that Mr. Chappell is given the envelope "Today –please".

The *big day* had finally dawned, and today I hoped to find out what the *man at the top* really thought about my project. In no time at all, my car was loaded to the hilt with every item that could possibly relate to my African expedition, just in case Mr. Chappell should ask for something relevant to the journey that I hadn't managed to fit inside my briefcase. My left arm would be taken up carrying this humungous map of mine, of course, which was a trial in itself to lug around. After carefully manoeuvring my car in-between two parked cars, in a line of 'Visitor's' car-parking spaces, I suddenly realized that I would have to reverse my car out again, otherwise I would be unable to extract this seven feet long map from my car without bending, or possibly breaking it. My briefcase contained both of Ted's books (*Jupiter's Travel's* and *Riding Home*) along with my copy of the meeting's agenda, both books on Africa (compliments of *Stanford's of London*) and a cassette player, in readiness for my latest innovation.

Clambering up the concrete steps, whilst making sure not to trip over my *anaconda-sized* map, I entered the main reception area, before duly announcing my *safe* arrival at the reception desk. While my name is being channelled through the telephone wires to my host, I begin looking around the walls of the reception area for a socket that I can plug my cassette player into, in order to try out the latest addition to my pitching material, which is backing music for the opening scene to my documentary. A security guard is soon on hand to point me in the right direction, and so in no time at all I had set the cassette tape to the beginning of the iconic song 'Sitting on the dock of a bay', by Otis Redding. The idea of using this song had come to me quite by accident, whilst driving my car around Cardiff, and listening to my 'Love Songs' album on the cassette player. As soon as the song started, I knew that it would be the perfect theme music for the start of my journey through Africa.

The song would have to be faded-out before the line "I left my home in Georgia", of course, but that still gave me around thirty seconds to set the scene. My vision was for me to be sitting next to the motorcycle, on a hillside overlooking a ferry-port in the Mediterranean. Slowly, but surely the bustling harbour will become almost still, and the colourful ships will be nothing more than silhouette's in the background, as dusk settles into dark, and the setting sun diminishes over a distant horizon. Then, as the music fades, the bike and I will also become shadows in the moonlight, as the narration begins. After a few seconds, the initial scene will be replaced by a hand drawn map appearing on the screen, with moving indication arrows, outlining our proposed route from start to finish. Had I thought of this earlier on in the week, I would have also written the entire synopsis for the first episode, but it was too late to do anything about it now, for Anna-Marie, Mr. Chappell's secretary, has just turned up in the foyer to collect "Mr. Donovan".

After making myself known to the lovely lady, she immediately made a jovial comment about the amount of luggage I was carrying, and so I thought it best not to mention the additional items that were on standby in my car. Once inside the main lift, Anna-Marie depressed the button marked 'Three' and seconds later we were winging our way northwards, and so I quickly adjusted the knot in my tie (as one does), in readiness to meet the man himself. As we exited the lift, and entered the small corridor leading to his office, I was surprised to be personally greeted by Mr. Chappell, who said that he had been waiting for my arrival. I was even more surprised by the fact that this man looked so young to be in such a prestigious position. After initially shaking the man by the hand, I wondered if I should follow this up with a gentle bow, or perhaps a full curtsy would be more appropriate, after-all, in the eyes of the British Broadcasting Corporation this man was royalty...or so I had been led to believe.

Ushering me into his somewhat large office, which looked more like an elaborate penthouse suite that one normally associates with towering skyscrapers in New York or Miami, Mick left me on my own for a couple of minutes, in order to have a quick chat in the hallway with one of his producers. The decor of the place was extremely lavish, including two ornate coffee tables, their glass tops having been virtually obliterated by plate-loads of mouth-watering fairy cakes, assorted crackers and various cream cheeses. However, I could see that this mini-banquet of *naughty-but-niceties* had not been set out on my behalf, as there were already half a dozen empty wrappers adorning the cake stands, and a smattering of biscuit crumbs littered the tablecloths, as undeniable proof of my predecessor's indulgence. Mick returned with my itinerary clasped firmly in his right hand, before openly admitting to me that he had not had the time to read it through thoroughly, as if it was not an important issue –and maybe to him, it wasn't, but to me it was not a good start to my presentation.

Mick beckoned me to *take to the stage and perform*, as they say, but I had already realised that there was nowhere that I could hang my huge map, and so I simply rolled it out onto the carpet, before going into overdrive with my presentation, thus enlightening my host on the trials and tribulations of crossing this gigantic continent on a motorcycle.

Knowing that time would be of the essence, I mention Ted Simon and Derek Bryant briefly, before moving swiftly along to the all-important budget, referring back to the itinerary as often as possible, thus letting Mick know that he could always re-cap on my facts and figures at a later date, if and when we ran out of time. My rendition of Robert's 'Choices' plan sounded almost as good as his original performance –well, in my eyes it did anyway, and so I only hoped that it would encourage Mick into making a *positive* decision very soon. Throughout our twenty-five minutes together, Mr. Chappell stayed calm, cool and collected, saying very little until I had completely finished my spiel. However, this made me feel a little uneasy, as *audience participation*, whether the comments are good or bad, at least lets the performer know that the people are seriously listening to what the person has to say. In fact, Mick's only words throughout my entire presentation were to tell me when my time was up.

As I begin rolling-up my map, I suddenly remembered the music, and so I called to Mr. Chappell, who was by now walking towards the door to his office, in order to let his next client in. I told Mick that I could not leave his office until he had heard the opening soundtrack to my series, the audacity of my statement seemingly impressing him, rather than getting his back up, and so he calmly returned to his big, comfortable armchair, before beckoning me to continue. Whilst frantically searching his office for a plug socket, I began setting the scene for Mick, as I dare not lose his interest at this point, and I knew in my heart that this really was the *last chance saloon* for me. Plugging in my cassette recorder, I pressed the *play* button, and the song 'Sitting on the dock of a bay' started, Mr. Redding's entrancing voice soon lulling the pair of us into a dreamy world of exotic, far-off places. However, as wonderful as it sounded to both of us, Mr. Chappell expressed to me that he was 'on the clock', and so after the thirty seconds introduction, he politely requested that I stop playing the tape.

As I packed the final items into my briefcase, before clicking-down the locks, Mick thanked me for my time, before adding that he would contact James Gerwyn in the week, and that together they would look at my project "Sympathetically", his somewhat nonchalant statement immediately telling me that I had not won Mick over, as easily as Jim Howell's had thought that I would. My only consolation was that Mick said that he would let me know the outcome of their chat in a few days time, rather than me waiting a few weeks, as was the norm, for my answer, simply because we were running out of time to do the journey. To say that I was devastated would be an understatement. As soon as I arrived home I rang Jim, with a view to telling him about my increasing uncertainties

267

regarding the whole project, but I ended-up pouring my heart out to his secretary, who told me "Not to worry", as she knew both Jim and James Gerwyn very well, before adding: "James loves this sort of thing, and so I am sure that he will be right behind you every step of the way". The weekend came and went, and all I could think about was my forthcoming meeting with James, which I hoped and prayed would be nothing but positive vibes all around.

On the Monday morning I popped into the BBC studios, leaving a card at the reception desk for James and his wife, congratulating them both on the birth of their daughter. I also left a small gift for the proud couple, as a "Thank you" to James for all his help with my project. Today would be James' first day back at work, after his fortnight off, and so I hoped that he would be in great spirits, sharing the joy of fatherhood with everyone around him, just as I had done after the birth of my beloved daughter, Hayley. Jim had always insisted that James was "On our side from the word 'Go'" and why shouldn't he be; after-all, he had flown over 6,000 miles just to talk to Ted about his amazing journey, before spending fifty grand on making a documentary about how his book had influenced so many people's lives, and so I couldn't think of any reason whatsoever, why he wouldn't be every bit as enthralled about my project as I have been. However, I had this morbid fear that when push *came to shove*, James might call a halt to the whole proceedings, believing that I would not have what it takes to make a decent documentary series.

Having contemplated this over and over again in my mind, I now wanted to prove to James that I could write both interesting and exciting material about my forthcoming journey. I also wanted to show him how I could openly express my innermost thoughts, either in front of a camera, or through narration, and so I began preparing my role as a presenter, by writing a theoretical account of that opening scene, to accompany that wonderful song of Mr. Redding's. I had already spent many hours researching the various ferries and cruise liners that docked in Port Said, along with other Egyptian ports, but yesterday I found the perfect route for my Mediterranean crossing from Europe into Africa. Originally, I had intended arriving on the African content after ferrying from either Crete or Cyprus, but in the last twenty-four hours I had discovered a direct route from the European mainland across to Africa's northern shores, thus eliminating any island hopping whatsoever.

The ship that I now intended taking, sets sail from Venice, in Italy, heading southwards through the Adriatic Sea, before crossing the Ionian Sea near the southern coast of Italy. From here the ship will enter the Mediterranean Sea, before heading in a south-easterly direction to the port of Alexandria in Egypt –it sounded perfect. Although I was currently unable to find out any specific days, dates, or timetables for the crossing, I estimated that it would take approximately four days to complete the journey. Apart from this being a very relaxing way of 'crossing continents', it was also an additional *safety feature* of the project, because not only would it save me a small fortune in fuel costs, by not having to ride the full length of Italy, (which is over 1,000 miles long, if one includes Sicily in the equation) or 1,500 miles if I chose to ride all the way to Athens in Greece, but there would also be less wear and tear on the bikes components –especially the tyres, which are going to wear out soon enough on Africa's horrendously rough terrain.

As soon as I find out the name of the shipping company, along with a telephone number to contact them on, I will be asking them to sponsor a ferry ticket for both me and my motorcycle, of course, but in the mean time I could only pretend that I was at that ferry-port in Venice, waiting for my ship to set sail to southern shores, and so below is my rendition of what I would say to the camera at the start of my epic journey from Southern Europe, across the mighty Mediterranean Sea, through the heart of Central Africa, and on to my final destination at Cape-Town in South Africa:

"Venice, the 'City in The Sea', where gondola's rule the waves, and the local taxi comes in the form of a speedboat, ready to whisk you around the many attractions, such as the 'Lido De Jesolo', where you can top up on your tan, before taking a swim in the warm waters that surround this beautiful Italian coastline. From here sail on to St. Mark's Square, where pigeons eat bread out of your hands, and the local cafe-owners take 'plenty lira' out of your pockets, should you wish to indulge yourself in a cool glass of beer, or a fluffy cappuccino. After the sun has finally set over a glorious crimson horizon, why not spend a romantic evening with the one you love at one of the many sea-food restaurants overlooking the harbour, listening to the sound of classical music

serenading you in the background, whilst watching the decorative reflection of the bay lights, shimmering on the surface of the enchanting Adriatic Sea.

It has been over ten years since my last visit here, and the memories are flooding back to me, of piranha-type mosquito's, that chewed-away more of my flesh in one night, than I could lose by dieting in six months —and so yet another lesson was learned 'Never camp next to a slow moving stream —unless you are prepared to be eaten alive after dark, by Dracula's 'disciples of doom'. Needless to say, that my latest tent has a built-in mosquito net. In 1981 I came to these shores purely as a tourist, to see if this magical city was all that I had heard it was, and I certainly wasn't let down. However, this time I am here for an entirely different reason. In just over four hours, that huge vessel standing opposite me, which by then will be filled to the brim with cargo, passengers and crew, will be upping anchor, before making its way steadily out of the dock, while friends and family of all those on board, wave their final farewells from the quayside.

Heading south through the Adriatic Sea that separates Italy from Yugoslavia, our ship will continue sailing southwards through the Ionian Sea, and into the Mediterranean Sea, passing the shores of mainland Greece on our left, before dropping down in-between the islands of Crete and Cyprus, and finally coming to rest on Africa's north-eastern coast, in the 'Land of the Pharaoh's —namely Egypt. Me and my trusty steed, which has already carried me over a thousand miles from my starting point in Cardiff, South Wales, will be on that luxurious 'life-raft' for the four day crossing of continents by sea. At the end of this relatively small voyage is where my real journey will begin. Travelling alone, it is my intention to ride the length of the world's most exciting continent, from north to south, in approximately ten weeks. Considering that the overall distance is in the region of eight thousand miles, it means that I must cover around eight hundred miles a week on some of the roughest and most inhospitable terrain anywhere in the world".

At this point, a map of the African continent will take up the whole of the screen, and while I continue-on with my narration, an indication arrow will show clearly the route I intend taking from Alexandria, in Egypt, all the way down to Cape-Town in South Africa. My narration would go something like this:

"The route is straight-forward enough, although a number of alterations to my original course may occur, due to political, or topographical reasons, along

the way. Beginning my journey in Alexandria, I will cut right through the heart of Egypt, by following the River Nile for several hundred miles, before crossing Lake Nasser into Wadi Halfa, in Northern Sudan. Once inside the largest country in Africa, the bike, along with yours truly, will be loaded onto a train that will take us from Wadi Halfa, all the way to Atbara. This is because road travel is not normally permitted through this part of the Nubian Desert without a convoy of fully-armed soldiers. From Atbara I will take a south-easterly trajectory all the way to the Ethiopian border, in order to avoid the war-zones in southern Sudan. However, should this route also be closed, then I have two other options left to me. The first is to head in a north-easterly direction to Port Sudan, where there will hopefully be a ship I can board that is bound for northern Ethiopia.

From here I can then continue my journey south. Alternatively, if there is no ship going to Ethiopia, then I will have to board one of the more regular ships that go directly to Mombasa, in Kenya. This will mean missing out Ethiopia altogether, which will be a crying shame, of course, but unfortunately unavoidable. My other choice is to head further inland, crossing into Chad, before taking a south-easterly route through the Central African Republic, Zaire and Uganda, before finally ending up in Kenya. Although the latter route is a lot longer, either way will suffice. By the time that I reach Nairobi, Kenya's capital city, the roads should have improved considerably, thus making the second half of the journey much easier to negotiate -and subsequently a lot quicker to complete. From Nairobi, my trajectory will change once again, as I head in a south-westerly direction, through Tanzania, Zambia, Zimbabwe and Botswana, before crossing into South Africa, for the final ride into Cape-Town, and the long-awaited finish line."

As the map fades off the screen, so the scene will revert back to me sitting on the bike, and facing the camera, in readiness to make my final statement of the introductory scene:

"Well, it looks as if the cars and trucks are making their way to their respective terminals, and so I had better get myself down to the boarding gates, in readiness for our departure. See you all when I get to Egypt."

At this point I will close the visor down on my helmet, before starting-up the bike, and after clunking the gear pedal into first gear, I will ride towards the main ramp leading into the ship. The next scene will show the ship docking in Alexandria, where I think it will be a great idea to have Ted Simon standing on the quayside, waving and welcoming me to Africa's *far-flung shores*. The follow-up scene would show us sitting outside a local tea house, where we have a map of North Africa spread out over one of the tables, and Ted is giving me advice and instructions about following his route as closely as possible. After a brief chat, I would fold up the map, before shaking Ted by the hand and saying "I'll see you in Kenya then Ted". The pair of us would then ride off in different directions, as the scene fades-away into the distance, in readiness for the next stage of the production...the crossing of Africa.

By three o'clock in the afternoon I was all *done-and-dusted,* and so I duly drove over to the BBC studios, leaving a copy of the opening scene with the receptionist along with a copy of my itinerary from my meeting with Mick Chappell, for James' attention. Now all I could do was to *sit and wait.* I had not heard from Mick Chappell in over five days, and so I could only assume that he had already had his chat with James Geraint yesterday, and that all would be revealed to me at our meeting this morning. Once again I dressed myself up like a 'dog's dinner', before double-checking my face in the bathroom mirror, simply to ensure that there was not a single hair of stubble protruding from the pores on my perfectly shaven chin. It seemed like I had barely unloaded the boot of my car, before I was once again reloading it, and like all the previous times, there would be even more stuff being piled onto the back seat of the car, just in case I might need it?

Because I seemed to have come full circle with my efforts since I first started this project over seven months ago, I decided to take my *'Wheel of Fortune'* hub-cap with me –just for luck. Hiding it safely under the carpet in the spare wheel compartment, for fear of it being stolen, I drove off on what was by now a regular route for me, and one that I probably could have done with my eyes closed, it had become so familiar. As I drove through the streets of Cardiff, I wondered whether I would ever get the hub cap engraved as my official 'Wheel of fortune' –in other words as and when the documentary had been shown on television, or preferably when my beloved film had finally been released on the *Silver screen*. The movie idea certainly lay in the lap of the gods right now, and it seems like the making of the documentary would all be down to the *wit and wisdom* of James Geraint.

He was the first person I had spoken to over seven months ago when we first *started the ball rolling* with this outlandish idea of mine -and now, after months of chasing and pursuing the men who are sitting right at the top of the tree, it looked like James would be the one who will be making the final decision after all –how ironic is that? Standing outside the BBC studios for the umpteenth time in weeks, I tried to take stock of everything relating to the project, so that I would be able to come up with the right answers to whatever questions James would fire at me, but no matter how hard I tried, I could not overcome this morbid fear of mine that today was NOT going to be a 'Good' day? I tried to attribute the bad vibes I was feeling all around me to my nerves, but something inside of me kept telling me that it was more than that, and yet no matter how deep I delved into my innermost thoughts, I simply could not put my finger on it?

Walking up the *same* set of steps, into the *same* foyer, and speaking to the *same* receptionist, as always, I gave her the *same* name (Shaun Donovan) before confirming my appointment with the *same* person (James Gerwyn) just as I had done so many times before, and so how come this day seemed so different? Spotting a calendar by chance on the wall behind the receptionist, I suddenly noticed that the date on the calendar was the 16th June –the day prior to the fifteenth anniversary of my horrendous motorcycle crash back in 1977, when, but for the grace of God, my life could so easily have ended. Every year since then, on the eve of that fateful calendar date, I have either suffered from horrible nightmares about the crash, or been plagued with eerie premonitions about a similar thing happening again in my life, and so the 16th of June is the one day in the year that has always haunted me.

"So that is what this is all about", I say under my breath, having now convinced myself that my morbid apprehensions were purely an instinctive reaction, because of what day it was, and that they had absolutely nothing to do with today's meeting. The receptionist has just informed me that Mr. Geraint is running a little late with his schedule, before adding that I shouldn't be waiting for too much longer. Around ten minutes later I am greeted by a lady whom I have never met before, which is something different (at last) and she tells me that Mr. Geraint is waiting for us in the building that sits opposite the main studio centre. As we cross the road together, I tell her about my brother, Gary, who works in the scenery

273

department of the BBC, to whit she duly reciprocates by saying that her husband also works in the scenery department, and that he knows Gary very well. What a small world this really is.

Although our conversation could only be construed as *small talk*, it had certainly helped me a great deal in calming-down my nervous system, which was currently all over the place? After gaining access to the building via a key-code at the entrance, the lady also informed me that James had been temporarily promoted as the new 'Head of department', and so I sincerely hoped that this upgrade of James', albeit only temporary at the moment, had put him in a good frame of mind —especially towards struggling lowlifes like me? As we entered James' office, which was not as big as Mr. Chappell's, but it was every bit as elegant in its decor, I knew that this was it: the moment of truth was about to unfold, and I have to admit that I was now on tenterhooks.

CHAPTER 25

Video Diaries

James apologised for keeping me waiting, before adding that he could only spare me fifteen minutes of his *valuable* time, as he had an *important* meeting to attend "Over the other side". Assuming that James meant 'across the road', and not at a *clairvoyants convention*, I hastily unpacked my briefcase, whilst at the same looking all around the room for somewhere to hang my blessed map. It has to be said that I despise doing these 'quickie' interviews, where my concentration is marred by a mixture of tension and frustration, as I try desperately to cram all the aspects of my project into a very small space of time, inevitably leaving out the most important parts, until it is too late. Also, forgetting to ask the crucial questions at the right moment can mean the difference between getting positive or negative answers. As the floor space was limited, I decided that it would be simpler to drape my huge map over the leather sofa, and just do the presentation as best as I could.

Whipping through the proposed route in a matter of minutes, whilst simultaneously spreading the rest of my paraphernalia over James' coffee table, I then went on to tell James how much I had progressed since our last meeting, before asking him if he had read the itinerary that I hand delivered to the BBC studios yesterday afternoon? James said that he had gone through it thoroughly, which made me feel a lot more secure right now, as I had put my heart and soul into the document, and I truly believed that I had left no stone unturned. However, James' next statement shook me to the core. "So what exactly is it that you are asking me for Shaun; is it advice, assistance with the project, or funding; please be more specific about what you actually want from me?" I was absolutely gob-smacked, and words could not describe how I felt right now. Here was I thinking that everything had been moving forward nicely, but in true point of fact, apart from receiving a multitude of sound advice, a handful of inventive ideas, and numerous alternative suggestions, I was actually no further forward with my project than I was when I first started my quest almost eight months ago!

Had my mentors continually encouraged me from day one, purely out of the goodness of their hearts, simply because they did not want to shatter my illusions, I asked myself –or was it me who had let the side down badly, by failing miserably once the cards had been firmly planted on the table? However, I was not about to let James know my innermost feelings, and so instead of begging and pleading for his assistance, as if it was a matter of life and death, I decided to go straight for the jugular. "All three, I hope James", I said in a positive manner, knowing only too well that I had to get back to basics, and sell my wares to him as good as I had done in the beginning, otherwise this meeting would be over before it had started. I decided to go for the 'alternative close', which Robert had so kindly set up for me, in readiness for my meeting with Mr. Chappell, and so I duly presented my '3 choices' scenario, starting with choice number one –me going it alone. This was the cheapest option, of course, and so I knew that I would have much more chance of getting a "Yes" from James with this option, than from the other two, more-expensive options. After finishing my spiel, and having planted the ball firmly back into James' court, I sat back and waited for his *hopefully positive* reply.

"Let us look at this thing logistically Shaun. If we send you on this escapade on your own, there is no guarantee that you will make a decent documentary, is there? After all, you've got no filming experience whatsoever, very little idea of how to present a programme professionally, and not a single qualification in production, or sound recording techniques?"

I could not deny that James was right on all four accounts, and although a part of me really wanted to tell James that Ryan had promised to teach me all these things, before undertaking the journey, a voice inside my head was telling me that no matter what I came up with, I would be onto a loser, and so I simply took it on the chin, and accepted the fact that *the truth often hurts.*

However, unwilling to go down without a fight, which has always been my motto in life, I took the positive out of James' answer, and so instead of battling to save option number one, I moved swiftly on to option number two.

"I totally agree James, and that is why I truly believe that option number two would be a much better idea. Having a film crew following me throughout the journey means that the filming would not only be more professional, but the guys would also be able to get far more amazing landscape shots from the back of a Land Rover, than I could ever hope to attain from the seat of a motorcycle".

I daresay that James had expected me to go down this route, and so his next 'reality-check' once again brought me down to earth with a tumultuous 'bang'. "Have you any idea what the cost of sending a film crew to Africa for ten weeks would be Shaun…'astronomic' is the answer. I am sorry to say it, but option number two is completely out of the question!"

Even offering James a Land Rover and a full time driver for the duration of the trip fell on stony ground, I'm afraid, as James was having none of it. I was now out on a limb, and so if option number three didn't work, then, as my old papa used to say, I was "Up shit creek without a paddle".

"Okay James, so I daresay we are talking fifty –or possibly even sixty grand here, in order to complete the journey from start to finish, but if Robert Staines is able to bring the South African film company on board, then this cost could be halved. Surely twenty-five to thirty thousand pounds is not a lot to pay for such an epic production, is it, my friend?"

This was it; I was clutching onto the last few straws that I had left, before drowning in a sea of negativity. My brow was sweating heavily and my pulse was racing against my heartbeat, like there was no tomorrow, as I waited for James to give his final answer?

"I totally agree that it could be a great production Shaun, but to be brutally honest, I cannot see that a merger with a South African film company would work out with this kind of production; there is just too much bureaucracy involved. No, there is only one avenue left to you, that might possibly work, and that is 'Video Diaries'. This is a relatively new concept in film making, which is run by a separate department within the BBC, and so I will be more than happy to give them a call in the week on your behalf, if you like".

James went on to explain how amateur filming enthusiasts, who have a general idea of what they think would make an excellent documentary, are given basic instructions in the art of filming and sound recording, before being loaned a camera and a microphone for however long it takes for them to come up with the finished product, in readiness for cutting and editing, before being screened on television. However, this is only if their documentary *makes the grade,* of course. Because this is a relatively inexpensive way of doing things, in comparison to conventional filming costs, it has become a very popular way of making documentaries. Also, because the viewers are fully aware that these are effectively amateur productions, and therefore they are expecting to see a lower standard of video footage, both visually and audibly, there are rarely any complaints sent to the BBC regarding the standard of the documentary.

However, whether this would work for me I couldn't even hazard a guess at this point in time, and if truth be known, I couldn't really care either, for my mind was a total mishmash right now, and all sorts of things were running amok in my brain. Even if the idea was taken up by the Video Diaries department, it meant that I would have to travel up and down the country, in order to make it work, and lord knows I could barely afford the petrol to travel in and around Cardiff right now, let-alone the cost of covering hundreds of miles a week. James could see how disappointed I was, and so he tried to console me by saying that I was not on my own with my feelings of rejection, before adding that he had received over six hundred ideas for proposed documentaries this very week, of which he would only be able to accept ten of them, and the others would all have to be shelved. "And so you can see how many people are in exactly the same boat as you Shaun", he said sympathetically, as if advising me to simply 'dust myself down', and move on to the next stage in my life, but unfortunately 'rejection' is not one of my strong points.

As I began packing-up my belongings, James turned to me and said "If it any consolation to you Shaun, I have to admit that none of the documentaries were anywhere near as exciting as yours, but I have to be very careful on how I spend the licence-payers money." Unfortunately, this had the adverse effect on me, knowing that my proposal had been washed away so easily, or so it seemed, but then who was I to sit in judgment of my masters, after all, when it comes down to knowledge and expertise in the industry, I was nothing but a meagre lowlife –and I knew it. By the time I had packed everything away, James had already left to go to his next meeting, having left his secretary, Elisabeth, to see me out. Because I had rushed through my presentation at a rate of knots, I had forgotten to play the opening song to James –and I had also omitted to ask him what he thought about my write-up. It was just after eleven when I walked back into my home, and even the sunny smiles of my three beautiful children could not bring me out of my miserable state of mind.

Slumping into one of the armchairs in the living room, I asked myself a very pertinent question: "So what are you going to do now then 'smart-ass?'", I suppose I could always get a petition together, showing how all the 'normal' people out there are right behind me with the movie idea –or perhaps I should simply sell my story to one of the major tabloids, in order to get nationwide recognition of my efforts, I joked half-heartedly, my mind in a total daze about what to do next? A nice cup of coffee was the order of the day, as I always think more clearly once I have had my fix of caffeine. Having slurped the dregs from the bottom of my mug I decided to call Jim and tell him openly how I had failed miserably in my meeting with James. To my surprise, Jim was very optimistic about the idea, categorically stating that 'Video Diaries' was a very positive move forward, before letting me in on another one of his philosophical gems. "No-one has said 'No' to you since day one Shaun, which is a claim to fame in itself. Always remember the motto of the *Pontypool Front Row*, and you won't go far wrong, my friend *'We go up and we go down –but we NEVER go back'*. As far as I am concerned, we are still heading in the right direction."

Ryan shared in Jim's enthusiasm, saying that he has an acquaintance who has worked on two documentaries for the Video Diaries department in the BBC, before offering to set up a meeting for me with the person in question. "Although the man primarily works in London, he is always commuting back and forth to Cardiff, and so it shouldn't be too hard to pin him down" Ryan added quite confidently. However, having already experienced how difficult it is to get any kind of meeting confirmed in this industry, I have to say that I wasn't holding out too much hope. Thankfully, Robert managed to raise my spirits somewhat, by informing me that because Video Diaries were made by the BBC in London, it means that the documentaries are broadcast nationwide, as opposed to being initially broadcast in Wales alone. "Therefore, apart from covering a much broader range of the viewing public, it also means that publicity for the show would be far more widespread, thus making potential sponsors more inclined to donate their wares for the production Shaun", Robert added, his enthusiastic manner winning me over, as always.

Caryl had supported me right from the start, even though she was very sceptical about the project in the beginning, letting her man 'play with his new toy', as they say, until he became fed-up with it. Every time that something went wrong, or somebody had let me down badly, I expected her to utter those immortal words "Well, I hate to say 'I told you so Shaun', but"....but she never did say anything of the sort to me. Perhaps fate has ruled that this period in my life has occurred merely to teach me a lesson, for all the times that I spent moaning about my working-class lifestyle, rather than actually being a genuine opportunity for me to completely redirect my life? If this is the case then I must always remember the heights of euphoria that I have attained whilst putting this whole package together, and rather than seeing myself as a complete failure in life, I must simply learn by the mistakes that I made along the way, for one never knows when previous experiences, be they good or bad, could eventually prove to be truly beneficial in one's life.

If nothing else, then at least my trip out to California had helped me to lose a lot of my excess weight, because upon seeing the photographs that were taken at Ted's place, I was so taken aback with my current size and shape, that I decided to put myself on a strict diet at the beginning of the New Year. Eating only crisp-breads for breakfast, various salad dishes for my lunch, and a range of slimming soups with toasted brown croutons for my evening meals, was a complete contrast to the crap that I had been devouring for a very long time. Also, no longer would I finish off food that the children had left on their plates, nor nibble on biscuits and cakes with every cup of coffee I drank —and eating after eight o'clock in the evening was now a definite 'no-no'. In addition to completely changing my eating habits, I also decided to combine this with a 'get-fit' regime, by jogging around the perimeter of our housing estate at least twice a week, coupled with three or four 15-minuts sessions on Caryl's exercise bike every weekend.

My goal was to shed around two stone by the beginning of the summer, thus shedding about one sixth of my body weight, and enabling me to get back into clothes which I could ill-afford to replace. Anyone who has attempted this joint method of decreasing the bulk of one's carcass, will appreciate how hard it was for me to do, especially in the first few weeks, when all the temptations that one has been used to for several years, have to be completely ignored, for fear of 'giving up the ghost' before one has begun. On my first attempt at traversing all four corners of our vast estate, I had to stop for a breather at least six or seven times, my lungs desperately gasping for every breath of available air, whilst my brain was expecting me to pass-out at any given moment. By the time I reached the halfway stage, the inevitable 'stitch' had appeared in my stomach, relentlessly stabbing me in my tummy with each thumping step, until I finally completed this seemingly never-ending circumnavigation of Thornhill. C o m p l e t e l y exhausted, and dripping in sweat, I would inevitably collapse onto our front lawn, much to the amusement of my neighbours, who taunted me mercilessly about how unfit I was, adding that *a good days' work* would probably finish me off for good. I have lost count of how many times I considered giving up this ridiculous malarkey, as I continued telling myself how futile *running around in circles* was at my age, and so lord knows what kept me going? I guess it was the fact that deep down inside of me I knew that it would be so easy to turn around at any stage of the run, and bring everything to an immediate halt, whereas once I was in Africa there would be no giving-up –and certainly no turning back, and so this endurance test was as much for my mind, as it was for my body. The solitude of being all alone is also a major factor when one is running, because this is when one's mind goes into overdrive –well, at least with me it does. Keeping up the positive thoughts can be quite a daunting task, especially when one is puffing and panting like a fool, and the harder the run becomes, the easier it is to let into one's mind those unwanted negative thoughts, which is something that has always been taboo in my books.

Most of the people who go jogging today do not have to concern themselves with the above scenarios, of course, because ninety-nine percent of them have music blasting away in their eardrums, from all sorts of mini music-systems, whereas back in the early eighties the only option joggers had was to strap a *Sony Walkman* onto their shorts, or alternatively sing to themselves, as they pounded the beat. After three months of sacrifice, denial —and in some cases 'pain', I finally achieved what I had set out to do. Weighing-in at a fraction over ten stone, instead of a tad under twelve, I look and feel a lot better than I did this time last year. At this point I gave up using the exercise bike, but I kept up the jogging, running three miles every day for many more years to come. (*In fact, 15 years later, at the tender age of 47, I was actually jogging 'seven' miles a day, whilst living in sunny Cyprus —but that is another story... and indeed, another book!*)

Besides my daily jogging routine, I also continued playing five-a-side football for 90 minutes each Friday evening, at the Ely leisure centre in Cardiff. Everyone has a certain amount of vanity tucked away inside of them, and so let me be the first to admit that 'looking good' was my initial reason for trying to lose a few pounds of excess fat, along with getting fit, and feeling good about myself internally, of course. However, what I had not taken into account at the outset of my 'road to fitness' was how all this would affect me mentally. At the end of those first three months I had proved to myself that anything I set my mind in achieving, I was capable of accomplishing, no matter how long it would take me, or how hard it would be. This was just the encouragement I needed to know that I could conquer Africa single-handedly, and return home in triumph. However, the big question now was "Would I ever get the opportunity to do so?" Two days have passed since my meeting with James, and so he should have spoken to the Video Diaries department by now, or at least I hoped he would have done, and so I decided to give him a call, in order to put my mind at ease, one way or another.

To my great surprise, James, himself, answered the phone, and what surprised me even more, was the fact that he was literally reading through my budget papers when I called him. "What it has all cost me so far..." James said nonchalantly, simply to prove to me that he actually had the document in his hand at that very moment. James continued: "Since starting this venture back in 1991..." At that point I interjected, assuring James that I believed him, before asking him the $64,000 question; "What did they say at the Video Diaries department?" James hesitated with his answer, as if he was somewhat lost for words, before revealing all. "Listen Shaun; I have had a word with Mr. Chappell, and we have decided to keep your project 'in-house', rather than contacting any outside help from further afield –if that is okay with you?"

I was somewhat taken-aback by James' response, as his positive attitude was in complete contrast to many of the statements that he had made during our last meeting, but then I was also very pleased, of course, although I knew that there would be more to this than meets the eye (let's face it –there always is) and so I just sat back and listened to what else James had to say:

"We will be willing to supply you with a film camera for your trip, although I am afraid that we will be unable to assist you with any of the other costs that you have listed in your budget –I hope you can understand that?"

This was an almighty blow, especially after that first 'very positive' statement, and so in total innocence, I told James that he would not have to bother getting me a camera, as Ryan was already organising one for me, although I secretly hoped that James would now offer me some other form of contribution instead. However, James simply stood by his guns, saying that he was sure there would be no problem in the BBC supplying me with a top-of-the-range video-camera.

By the following Wednesday I have still heard nothing from James, and so decided to give him a call. As usual the phone was answered by James' secretary, who obviously did not recognise my voice, as she asked me who was calling? However, upon giving my name to her, I was shocked by her response:

"Well Shaun, if I have heard your name mentioned once in this office this week then I've heard it a thousand times".

Her somewhat outlandish statement was wonderful to hear, of course, albeit that my name had been used in the right context, of course. In a few seconds I would have my answer, as James' secretary continued-on with her spiel:

"Apart from Mr. Chappell, James has spoken to at least four other people about his 'gut feeling' regarding your proposed documentary. I also know that he is arranging to have a special type of VHS video camera tested for you in the very near future, and so I expect that you will be hearing from him shortly". James was unavailable to talk to right now, but I had already heard all that I wanted to hear, albeit through the grape vine, and so for once I was happy to say "No problem, I'll wait for him to give me a call when he is ready".

After replacing the receiver, I immediately lifted both Hayley and Carl up from the carpet, where they had been playing peacefully together with a bunch of toy cars, and holding one of them on each arm, the three of us then danced around the room together –much to my children's delight, of course, who probably thought that their daddy had finally *flipped his lid!* Having exhausted myself, and also my aching arms, to the point of collapse, I returned my beloved children to the floor, where they continued-on playing together, whilst I left a message on Caroline Watts' phone for her to ring me as soon as possible please.

As it is now nearing the end of June, Derek Bryant is due to arrive in London any day, so long as everything has gone to plan with his African crossing, of course, and so I do not want to miss the opportunity of having an in-depth chat with him.

CHAPTER 26

Taking A Break

It is now Friday evening, and I have just left my final message on Caroline Watts' answer-phone, in the hope that I will receive a call from her over the weekend otherwise things will be on hold for at least another week. My cousin, David, has offered to loan us his caravan for a week, and so all five of us will be leaving first thing on Monday morning, to drive to the caravan site in St. Clears, in Carmarthenshire, which is situated right in the heart of West Wales. The weekend soon passed, without hearing a word from anyone, but then late on Sunday evening, just as Caryl had finished ironing the last batch of clothes, in readiness to be crammed into our third suitcase for the journey ahead, the phone rang, and within seconds I was on the call. It was Caroline, simply calling me to let me know that I had not been forgotten, and also to put my mind at rest. She said that Derek will be staying in London for a couple of months "If and when he arrives in the UK", she added, thus letting me know that he was obviously still in Africa.

Intrigued by Caroline's last statement, and also by the tone of her voice, I asked Caroline if she would kindly enlighten me on Derek's current position, and also to let me in on the reason why he had obviously been held up on his journey. Caroline said that she had received a letter from Derek, saying that the truck had suffered several mechanical problems recently, so many in fact that he had actually considered abandoning the journey altogether. However, as luck would have it, he then received a call from Nissan, asking him if he would be willing to road test one of their latest saloon cars, to whit he said that he would be happy to do so, so long as he could drive the vehicle from Nairobi to Cairo. Nissan agreed. Caroline finished our conversation by saying that the letter from Derek had been dated 4th June and that it was post-marked 'Natal', and so she knew that he was at least two weeks behind his original schedule.

The following morning Liam got Caryl and me up early, as he was really excited about the prospect of staying in a caravan for the first time in his life. Personally, I would have much-preferred staying in my cousin's holiday apartment in sunny Torrevieja, on the Mediterranean coast of mainland Spain, which David had also kindly offered us for the week. Unfortunately, having no money to pay for five return flights, we had graciously accepted the loan of his caravan, as it was simply a case of *Beggars couldn't be choosers*. By ten o'clock the car boot had been packed to the gills, and Caryl was now cramming every last carrier bag of food into any remaining spaces that were left in the car. The children, who were now looking 'ever so posh', having been dressed in their best summer outfits by Caryl, had all been strapped into their respective seats, and just like us, they were really looking forward to a well deserved holiday. Sadly, the sun was not shining down on us this morning, the density of thick cloud that was currently hanging high above our heads having obliterated it from view, but nonetheless we were all in the right frame of mind for a day at the seaside.

Whilst Caryl had been doing all of the above, I was busy making one final phone call to James Gerwyn's office, simply to inform his secretary of my whereabouts for the coming week, just in case he needed to contact me urgently. I knew that this was highly unlikely, of course, but then at least it gave me the opportunity of possibly speaking to James one last time, before I went *off the radar* for a week. My luck was in, as James answered the phone to me, and he seemed to be in a very good mood. James said that one of his colleagues was currently in Paris, trying to obtain a video camera for me from Sony, and that as soon as he had completed his mission, James would have the letter that I had requested from him, confirming the BBC's inclusion in the project, drafted-up, in readiness for my return from my holiday. Knowing that this would make life so much easier for me with regards to pitching companies for sponsorship, I finally closed the book *on 'Bat out of Hell–The Movie'* (albeit temporarily, of course) before joining my beloved family in the car, as we set off on our summer holiday together.

Caryl had already made me *swear on a stack of bibles* (slight exaggeration there) that I would not mention 'Africa', the 'BBC', anything or anyone involved with my project whatsoever, during the coming week, otherwise she would leave me for good! For fear of her being deadly serious, I quickly put them all out of my mind, before sitting back in my drivers' seat and enjoying the beautiful countryside that lined our route to St. Clears. As we drove down the M4 motorway, Caryl began singing her version of Cliff Richards' famous song 'summer holiday', which although it is a very happy-go-lucky tune, Caryl's non-melodic version failed to gain any recognition from her *captive audience.* By the time we reached Swansea the sun was already fighting its way through the clouds, and as we headed into the centre of Carmarthenshire, the skies were now bluer than blue. The directions that my cousin had given me were spot on, and even though the car was getting on in years, it had performed well from start to finish, and so the journey went without a hitch.

By the time we arrived at the camp site all three children were sleeping soundly in the back of the car, and in no time at all we found David's caravan, which was located in one corner of the main field, next to a fast-moving stream. David had built a huge patio area adjacent to the caravan, complete with three low level surrounding walls, a set of steps leading up to the caravan door, and a built-in barbecue, which finished off the patio quite nicely. With all three of our off-spring being *away with the fairies,* as they say, unpacking the car was a piece of cake, and so after quickly sorting-out our respective gear for the week, Caryl and I were completely organised, and ready to enjoy our holiday. As beautiful as the setting was, my wife and I both knew that we would have to improvise on the surrounding area, in order to make it safe for the children, especially Carl, who we know will insist on going for a paddle in the stream the moment that he wakes up. Both of the picnic benches, which had been locked inside the caravan, I immediately placed in front of the two entrances to the patio area, thus eliminating any exit from these openings.

Caryl and I then turned all four of the sun-beds (that were also inside the caravan) on their sides, before placing them along the length of the underside of the caravan. We then propped them up by using a mixture of leftover slabs and house-bricks, which David had neatly stacked to one side, in order to prevent any possible exit, by going underneath the caravan itself. Liam was the first to wake, and so I carried him from the car, before handing him over one of the new *perimeter walls* to Caryl. I knew that Liam would not be amused by his enclosed surroundings, but for the safety of his younger siblings he would have to learn to live with it. Hayley was next in line to open her peepers, and so after releasing her from her *shackles*, I took her straight into the caravan, where Caryl was busy preparing food and drinks, in readiness for a quick lunch. This was the first chance that I really had to survey the interior of this huge eight-birth static caravan, and I have to say that I was more than impressed with what I saw.

There were three bedrooms in all, one with a double bed, and two with a single bed in each compartment, and so the sleeping arrangements were obvious ...or so we thought! Unfortunately, the travel cot, which we had borrowed for Hayley to use, turned out to be too big to fit in any of the bedrooms, and there was no way that we were going to leave her alone in the living room all night, and so once again I would have to give up the delights of sleeping in a large, comfortable bed, so that Hayley could sleep soundly with her mummy. Carl had only ever slept in a cot-bed beforehand, and so the pair of us could easily snuggle-up together in a full sized single bed, whereas Liam would have the third room to himself. Just inside the entrance to the caravan was a small toilet-cum-shower room, and even though there wasn't a full size bath in the caravan, the shower unit had a very deep base unit, and so it would be large enough to accommodate both of the boys at once.

However, if all three of the children needed to be done simultaneously, then whilst I was bathing the boys, Caryl would bathe Hayley in the large wash-hand basin –simple. To the right of the shower room was the kitchen area, complete with a sink unit with hot and cold running water, a four burner hob unit, a compact fridge / freezer unit, and a storage cupboard which was situated below the sink unit and the worktop area. The cupboard was full of cooking utensils and cutlery, along with a set of instructions on how to control the lighting and the heating. (*David had already told us where to find this*). Caryl had already stacked our vast supply of tinned food for the week on the three shelves that were located above the main worktop area, and the compact fridge was now bulging with cold meats, salad foods and various vegetables all waiting to be turned into healthy meals and pre-packed lunches.

As for the living room, and the dining room area, well this could not have been cosier, with plush patterned seating enveloping the walls and encircling the polished mahogany table, before coming to an abrupt stop at the beautiful feature-fireplace, which adorned one side of the caravan. Matching cabinets filled with an array of ornaments, and a selection of holiday souvenirs stood tall alongside an ornate looking electric log-effect fire. Joining the two cabinets was a highly polished mantle-shelf, sitting high above the fireplace, which had been adorned with a handful of cute little trinkets, along with a miniature grandfather clock, that chimed every hour, on the hour. The caretaker of the caravan site kindly took time out of his grass cutting duties to welcome us all to our temporary home for the week, before giving Caryl and me an insight into the local attractions, along with handing us a list of the forthcoming events that would be taking place during our short stay.

However, as much as we would have loved to partake of a number of these events, sadly our monetary situation would not allow us to do so, and so our daily routine would be somewhat basic. Breakfast would normally be served-up by me at around eight o'clock, and this would then be followed by a quick spruce-up in the bathroom for me and the children, whilst Caryl washed-up the breakfast dishes, before preparing our packed lunch. Then, while I dressed our off-spring, Caryl would duly do her ablutions in the bathroom, and by eleven o'clock we would normally be ready for the off. If the weather was good, then we would head for a nearby beach, where after a spell of catching crabs in the rocks, the children and I would then build *the sandcastle to end all sandcastles*. We would then begin digging-out various tunnels from four different directions underneath our fortress, whilst at the same time trying our uttermost not to have a *cave-in* before all of our hands finally met in the middle.

While all this was going on, Caryl, having commandeered the nearest deckchair available, would be working hard on topping-up her tan, before tentatively setting out lunch for us on our portable picnic table. On dark and dull days, Caryl and I would drive around the local areas, stopping-off in various little villages, to do a spot of ornament shopping, before consuming our lunch in the car –or if it was warm enough, we would head for the nearest park to enjoy a spot of picnicking on the grass. On the only rainy day that we had, we drove thirty miles to Tenby, one of Wales' most famous seaside resorts, where we also treated ourselves to a spot of lunch in a local cafe –just for the hell of it. Around four-thirty, give or take half an hour, the five of us would normally set off back to the caravan, where the children and I would have a paddle in the stream, or play football and mini-cricket on the lawn for an hour, or so. If the weather was cold, we would simply treat the kids to a Disney film on Sky's movie channel, whilst Caryl and I indulged ourselves in a good old game of cards.

Our evening meal would invariably consist of hot, freshly cooked food, which we would consume sitting inside or outside the caravan, depending on the weather. Hayley's ritual early-evening nap nearly always coincided with the boy's bath time, which was invariably between 6.30 and 7.30 pm, and so I guess you could say that we had a pretty good daily routine. The bathing would then be followed by a massive scramble to grab the driest, warmest, and largest bath towels to wrap each of our beloved children in, before drying their soaking-wet bodies as fast as we could. Getting all five of us in that shower room, which was less than five feet square, was certainly not an easy task, but with a lot of breathing in and shuffling around, we invariably managed it without a hitch. Once the children had gone to bed, Caryl and I would then finally put our feet up for the day, calmly relaxing and enjoying the sheer delight of Sky television –something that we could ill-afford to purchase in our own home.

In the middle of the week our camp site suffered a mains power cut, leaving all of the caravans in the dark for several hours in the evening. Luckily, Pam, David's wife, had invested in a set of decorative candles, and so at nine o'clock that evening Caryl and I enjoyed our first ever candlelit supper, proving to us both that true romance can never die. Caryl and I returned home, only to be greeted by a stack of letters waiting for us on the hallway mat, the majority of them being bills, of course. 'Final Demands' were in abundance, as usual, along with court letters, repossession orders, and fines for non-payment of standing orders. Bank charges for a multitude of sins, including overstepping my over-draught limit, having 'Insufficient funds' to pay my our monthly direct debits, and worst of all, a £35 charge for every letter that the bank had sent me, simply to inform me that they wouldn't be paying a £20, or a £ 15, or even a £10 standing order, were exorbitant, to say the very least.

All of my credit cards had either reached or exceeded their maximum limits, and so the monthly minimum payments would now have to include the overspills, which ran into several hundreds of pounds in total –and I barely had a penny to scratch my backside with. Worst of all were the mortgage payments, which we had been unable to pay for the last two months, and we both knew that a three month deficit was the maximum that the building society would tolerate, before repossession orders for the property would come into force. I knew that losing the house would be the final straw for Caryl, and the mere thought of my three lovely children being shunted from pillar to post in council estate properties on rundown housing estates around Cardiff, or maybe even outside Cardiff, in the valleys of South Wales, was eating me up inside. Fears of moving next to the dreaded *neighbour's from hell*, or even worse, living in gang-infested or drug-ridden areas, ran amok in my brain, and so I kept trying to put these dreaded thoughts to the back of my mind.

The mere thought of coping with the awful upheaval of having to pull Liam out of his wonderful nursery school, thus taking him away from all the friends that he had made over the last couple of years, was also breaking Caryl's heart. The Citizen's Advice Bureau (C.A.B.) had done all that they could for us over the last few months, but with each 'stay of execution' that they managed to obtain on our behalf, so another calamity would befall us, immediately throwing us straight back to square one. Talk about *seeing a light at the end of a tunnel,* well neither Caryl or I could even see a glimmer of one from where we were standing.

I was shouldering all of the guilt for our current 'abominable' situation, for it was me who had lost my job; it was me who had chased this impossible dream; it was me who had kept telling my wife that everything would be okay, when in reality, I had not even convinced myself that we would get out of this mess one day! If we lost the house, then I feel sure that it would also be the end of our marriage, and that Caryl would take the children back to live with her parents, which would surely break my heart in two. Whenever I was on a *downer* Caryl would always remind me that there are three million unemployed people in Britain right now, most of who are probably in the same financial position as us. However, her words of consolation (*which were no doubt being uttered around the country by supportive partners everywhere*) did little to appease this huge burden of guilt that was hanging heavily on my shoulders. Was I meant to have pity for them, rather than feeling sorry for myself, or should I just feel somewhat uplifted, by knowing that I was not on my own?

Whatever the reason was for Caryl's good intentions, I am sorry to say that her positively brave statements were being wasted on me, for I was on a downward spiral right now, and it looked like there was no turning back? Not being thankful for the many wonderful things that I currently had in my life, such as a beautiful wife, three adorable children, a very supportive family, a fabulous home, and a generally healthy body, is quite disgraceful, I suppose. However, when one is hurting so much inside and they truly believe that they have lost all of their dignity, their self esteem, and also their pride as an upstanding member of the community, then it is sometimes easy to forget how much joy and happiness they have all around them, and how nothing could ever replace one's own family –please trust me on this. It is only when I hear about wars being fought in other countries, or I read about the many tragedies going on in my own country, such as murders, rapes, violent crimes, child abuse, and so on, that I am reminded to count my blessings.

Also, whenever I see news reports or documentaries concerning the horrors of poverty, drought and famine in third world countries, I immediately forget altogether about my own problems, and all I want to do is whatever I can to help those poor people survive. I remember that fateful day when Caryl and I were watching an episode of 'This is Your Life'; where the person in the spotlight was Anneka Rice, star of the television show 'Challenge Anneka'. At the end of her tribute, Anneka was given a challenge to help scores of abandoned children, who were currently living in a very rundown Romanian orphanage. The BBC had chartered an aeroplane to fly around 300 aid workers, primarily professional people, such as carpenters, painters, plumbers, plasterers and builders, to Romania, as the orphanage they were in was in desperate need of a complete renovation. They also asked for volunteers to work as labourers, porters, cleaners and kitchen staff, asking anyone interested to call a hotline number as soon as possible.

I remember looking over at Caryl, the tears now streaming down my face, and she simply said to me "Go on –I know you want to help them Shaun". Immediately, I called the hotline number that was being displayed on the television screen, but by the time I got through to one of the many operators, I was duly informed that all 300 places had already been filled. I was devastated. This was the one chance I had been given to do some good in this world, and it had eluded me. I cursed myself for hesitating, and not getting on the phone sooner. "Maybe they will ask for more people to come and help in the next few days", Caryl said in a positive manner, her compassion for my disappointment standing as tall as ever. "Oh well, they say that 'God helps those, who help themselves, and so I suppose I had better concern myself with our numerous problems, rather than worrying about what other atrocities are being perpetrated elsewhere in this big, bad old world of ours", I said somewhat sarcastically.

I then rang the BBC, simply to inform them that their *entrepreneurial travel-writer* had returned home from his holiday, and that he was now raring to go again. I was half-expecting –or should I say 'hoping and praying' that all sorts of things might have materialised whilst I was away, but within seconds of my call being answered by James' secretary, my spirits had been dampened once again. "I am afraid that James is not in this morning" she said apologetically, having obviously realised how important time was to me right now, before continuing-on "You see he has been preparing for the interview that he has this afternoon, with regards to taking over the permanent position as 'Head of Department', and so he has put everything else on hold until further notice." Thanking the lady for her time, I then replaced the receiver, before shrugging my shoulders and mumbling a few choice words under my breath.

Today it would be Caroline Watts who would uplift my spirits, as she gave me great tidings of joy on the safe arrival of Mr. Bryant in the UK. "How are things going from your end Shaun?" she said enthusiastically, no-doubt expecting me to have nothing but good news for her, but all I could say was that things were still a little bit 'up in the air' right now "Although the outcome is looking more positive every day", I added.

Okay, so I was stretching the truth somewhat, but I was simply the piggy in the middle of this cat and mouse game, and it seemed to me that unless I began playing one individual against the other, then there would never be any commitment from anyone. Since day one I had been taking two steps forward and then one step back, and so what I wouldn't give right now, just to take one huge leap forward –with no steps back. Caroline seemed quite nonchalant about my answer, simply taking a deep breath, before asking me to keep her informed as soon as there was any movement forward from the BBC. Now I don't know whether one would call this *intuition*, or a *sixth sense* perhaps, but at this point a handful of tiny little bells began ringing inside my head. They were certainly not *alarm bells*, going off like screaming sirens from a police car that is in hot pursuit of a stolen vehicle, or an ambulance tearing-up the tarmac, whilst telling all the cars to get out of their way, as they need to get to the hospital as soon as possible, but more like the sound of jovial sleigh bells, jingling majestically in the distance, as if lulling me into a false sense of security, and yet subliminally warning me that all that glittered was not gold.

Caroline's hesitance, albeit for only for a few seconds, was telling me that she had something more to say, but that she had decided to retract the statement at the last moment for some reason, and I wanted to know why? I decided to ask Caroline a few questions about Derek's journey, just in case there was something that she had omitted to tell me about his African crossing, which might jeopardise him moving forward with mine? Unfortunately, Caroline was not very forthcoming, saying that she knew very little about Derek's expedition, before adding that as soon as she finds out exactly where he is staying, she will ask him to contact me, in order to set up a meeting "And then he can tell you in person all about his exciting journey from Cape Town to Cairo", she added, like a teacher telling her pupils at the end of a fairy tale "And they both lived happily ever after". Accepting the fact that it must have been my overactive mind playing tricks on me again, I ignored the jingling in my brain, simply thanking Caroline for her call, and also for the update on Derek's arrival in Britain, of course, before saying my goodbye's to her.

However, before I had the chance to finish my farewells, Caroline interrupted me, saying that she had something that she would like me to think about in between now and when I finally made contact with Mr. Bryant. Suddenly, those bells started ringing again, and I wasn't sure whether I really wanted to hear what she had to say to me?

"Derek will be returning to South Africa around the middle of October, and it is his intention to drive home, rather than to fly back to Cape Town, and so if you are considering beginning your expedition around this time, then perhaps we could arrange something with the BBC regarding transportation that would suit all parties concerned". It was no wonder that Caroline had hesitated in asking me this question, because all along I had been asking Derek for his help, his assistance and his guidance, in making my trip a success, and now here was his *right hand woman*, asking me for a 'favour'. For a moment it seemed like the tables had been turned, and yet the mere thought of a possible back-up from Derek for the entire trip, was a dream come true. So maybe those chimes I had heard in the background were actually 'bells of joy' and not a *warning sign* after all. Whatever they were, they were now silent, unlike my mind, which was now awash with a million thoughts of what I would do next –and what I would say to Mr. Bryant when I received that all-important call?

CHAPTER 27

Go Ten For Africa

The following morning I rang Mr. Gerwyn's office, in order to inform him of my forthcoming meeting with Derek Bryant, and also to see if James had been awarded the position of 'Head of Department'. James' secretary answered the phone, and she duly informed me that, much to everyone's astonishment, including James, he had not been given the appointment. However, she did have good news for me, saying that she had been given the go-ahead to 'open' the file on Shaun Donovan, which meant that everything was still well and truly 'On'. 'Overjoyed' would be an understatement in saying how I felt right now, for it seemed to me as if the tables had been completely turned around in my favour, although I did feel sorry for James, of course. He would obviously be feeling somewhat indignant about the job going to someone else, having held the temporary position for these past few months, in which time I am sure that he had proved himself to be worthy of the appointment.

I went through a similar scenario many years ago, when after ten years of holding the rank of an office supervisor, I suddenly found myself being demoted to a desk clerk. What made it worse is that the relatively new desk clerk, who had only been with the company for about a year, and who was currently working under my supervision at the time, was duly promoted to become my supervisor. This ridiculous role-reversal had everyone completely baffled, including my union representative, who failed miserably in trying to quash both of these ludicrous decisions, even to the point of threatening a mass workout by the labour-force if something wasn't done about it. Unfortunately, the decision was final, and the only walkout that came about, was when all of the participating N.A.L.G.O. union members, of which I was one, immediately opted out of the scheme, before transferring their memberships to M.A.T.S.A. I also resigned from the company on the day of the transition, before moving on to much better things in my life.

However, karma had been duly served it seemed, after bumping into an ex-colleague of mine several months later, who virtually begged me to *return to the fold,* before expressing his feelings to me in no uncertain terms: "The guy who replaced you is costing the company a fortune in lost revenue Shaun –he doesn't know his ass from his fucking elbow", he exclaimed in a somewhat furious manner. About a year later it transpired that the promotion and demotion decisions were nothing more an act of nepotism, as the 'New kid on the block' only turned out to be the nephew of the chairman of our company...now there's a surprise! Now that James was no longer the head of the department, I feared that he would lose his jurisdiction over sanctioning projects through to completion, as someone else now had the power to override his decisions. I was also concerned that this new *Head of Department* would probably know nothing about my project whatsoever, and therefore they would have little or no interest in an unknown traveller talking about his journeys across Africa.

Unfortunately, I had no alternative but to just sit back and see what happens over the coming weeks. Sifting through the deluge of mail one morning, whilst hoping and praying that there would be no *repossession notices*, or *eviction orders* in the pile, I happened to stumble upon a leaflet that had an outlining map of the African continent printed on the front of it -and standing out in big bold letters above this map were the words: 'GO TEN FOR AFRICA'. Upon reading the contents of the leaflet I discovered that a British overseas charity called 'Action Aid' were attempting to raise money, in order to send badly needed aid relief packages, along with a number of aid workers, to support ten countries in the African continent as much as possible. Supporters of their cause were being asked to do a total of 'ten' menial tasks, which could be anything from mowing ten lawns, to taking ten dogs for a walk; in fact anything that resulted in ten objectives being achieved. The fund-raising would culminate in a ten day event, which would be held at the beginning of October.

My mind immediately went into overdrive, but before putting anything concrete together, I called James, just to confirm that the BBC would be fine with me integrating Action Aid into the television production. James was very supportive of my idea, saying that the whole project was *my pigeon,* and so whatever I decided to do, and whomever I wanted to be involved with in my production, was entirely up to me. "The BBC will only be interested in the standard of footage that you can produce", he added. James also said that he was currently trying to obtain a video camera for me, although Sony seemed reluctant to part with one at the moment, but he told me not to worry, as he was still working on it, and he was very confident that he would get one in the end. James closed our conversation by saying that he would be going on holiday for most of August, before adding that he would have everything sorted for me by the beginning of September.

I so much wanted to share James' optimism, but time was already as *tight as a drum* and so I believed that losing all contact with James for a whole month would be nothing less than catastrophic right now. However, to appease my darkened thoughts, James then gave me the good news. "In the meantime I am sending you a copy of that letter that you requested, in order to assist you in obtaining sponsors, and if you bear with me a moment, I will read out what it says, to you, just to make sure that it meets with your approval, before putting the letter in the post to you this evening. I was so excited that I told James I would be more than happy to drive to the studios and collect the letter from either his office, or from the reception desk today, but James insisted that this would not be necessary, before asking me to just shut up (for once) and listen carefully to what he had to say:

"To whom it may concern, regarding Shaun Donovan –that is the heading at the top of the page", James said in an authoritative manner, before continuing on.

"This is to certify that Shaun Donovan has discussed his proposed solo motor-cycle journey across Africa with me as a production representative of BBC Wales Television.

The obvious televisual opportunities involved in his journey have been the subject of detailed discussions over a period of several months. BBC Wales is unable to commission Mr. Donovan directly, but it has undertaken to view video material shot by Mr. Donovan whilst in Africa, in order to assess its quality for possible broadcast on BBC television when he returns.

Before leaving for Africa, Mr. Donovan will also have the benefit of our professional advice and instruction with regard to documentary shooting techniques, since we are naturally anxious to ensure the material he comes back with is of the highest possible quality".

I am very pleased with the wording in James's letter, and also with the fact that James is happy for me to bring a third party into the production. On the back of the *African Aid* leaflet is an address to write to in Wrexham, North Wales, and below the address is a hotline number to ring in Bristol, should one require a sponsorship pack, or more details about the 'Go Ten for Africa' scheme. It is now the last week in July, and so if I intend getting this whole new idea up and running before the end of September, then I daren't waste a single minute, and so I duly rung the hotline number. Unfortunately, I am told that the man I need to speak to is currently out of the office, and according to his female colleague, he won't be back in work until after the weekend. However, after expressing the urgency of my call, the lady assures me that he will give me a call first thing on Monday morning. No sooner had I replaced the receiver, than the phone rang again, only this time it is my good friend (whom I have never met, of course) Caroline Watts.

Caroline asks me if I have a pen handy, before reading out a telephone number to me, saying that if I ring the number now, I will be able to speak directly with Derek Bryant. This was an opportunity I would not pass up for the world, and so, even knowing that this would add yet another few pounds to my ever-increasing phone bill, I duly dialled the ten digit number. Derek was obviously expecting my call, because he answered the phone almost immediately, and by the time that the introductions were over I already knew that I was dealing with a very professional –and somewhat reserved, gentleman. However, I was hoping that he would be willing to answer all of my questions, the first one being "How on Earth did you manage to get through Southern Sudan in one piece?"

"I didn't", Derek answered abruptly, as if to say "Next question please", but I wasn't going to give him that satisfaction, and so I just said nothing, in the hope that he would continue-on with his story. "They turned me back at the Sudanese border, and so I had to double-back through

Somalia", Derek added. I told Derek that I had seen recent reports on the television regarding the bloody war that was raging in Somalia, along with the horrendous drought which had left thousands of people suffering from starvation all over the country, and he confirmed the situation, before adding his own experiences to our conversation.

"It is absolutely awful out there Shaun; every two hundred yards, or so there is a dead body lying in the road, its stinking carcass being eaten-away by great swarms of flies. If one tries to imagine that not too long ago that pile of rotting flesh was walking around like any other normal human being, it immediately sends shivers down one's spine."

This was powerful stuff coming from Derek's lips, and it was certainly enough to deter me from going anywhere near Somalia. I then asked Derek what my chances were of being able to film in Africa, to whit he replied:

"You will not be allowed to do any kind of filming whatsoever in Ethiopia, but elsewhere you shouldn't encounter any problems...hopefully! Mind you, in Egypt I was charged twenty-five Egyptian Pounds on a number of occasions, just for the privilege of being able to carry a camcorder around with me. Should you wish to take your camera inside any of the tombs, or do recordings anywhere near the Pyramids of Giza, then you had better bring a lot of spare cash with you, because that can cost up to one hundred pounds a time."

Derek also warned me that I must keep a careful eye on my camera equipment at all times, as there are plenty of opportunist thieves, especially in and around the tourist areas, who would stop at nothing to own such a prized possession. I took Derek's advice and warnings very seriously, for I had already read the article in his book about keeping radio's concealed wherever possible, because people have been known to *kill* in order to obtain one. Road conditions were the next thing on my agenda, for this was probably one of my biggest fears regarding the whole trip, and I also wanted to have a rough idea of how long it should take me to complete my crossing of Africa.

"The roads in Egypt are pretty good all the way from Alexandria in the north, right the way down to the south of the country, but you will not be able to travel through Southern Sudan until late October, or early November, because there are no road routes, as such, due to the rains. However, after the rains have terminated, and the flooded roads have finally dried up, it will become passable again. With regards to the amount of time that it will take you to complete the crossing, well I have a simple answer to that question Shaun...whatever timescale you have allowed yourself to do the trip, simply 'double it' and you won't be far off!"

Derek was a mine of information, and I could have sat chatting to him all day, had each minute of our conversation not been costing me a small fortune, and so I decided that now was the time to *test the water*. "So what if I asked you to become involved with my project Derek, or possibly even to join me on my journey through Africa; is that something that would interest you?"

"The question is not so much as to whether I would be interested Shaun', it is more a case of 'would I be available at the time', for I am a very busy man".

I now wondered whether he knew that Caroline had already told me about his planned trip back to South Africa, but rather than *spilling the beans*, I decided to play the dumb-ass fool, and simply let him finish what he had to say.   "You see Shaun, this is what I do for a living, and let me be the first to admit that my services do not come cheap. For example, let's say that I receive a call from my agent tomorrow saying that a group of people are interested in crossing the Sahara Desert in October this year, and that they want me to act as a guide for them. Now if I commit myself to doing that trip, then there is no way that I would suddenly change my mind at the last minute and let those people down, even if it meant turning down a lot more money for a considerably bigger project, such as yours."

Derek's explanation had proved to me that I really was the *ignorant fool*, and I had to admire his sense of fair play, having now realised that the majority of his life is based around other people's wants and needs. The mere thought of not knowing whether you would be home for Christmas, or being unable to forward plan a birthday party for a loved one, just in case you suddenly have to go flitting off halfway across Africa, was quite daunting for me as a quote 'family man', and so this journey might be just what I need, simply to appreciate that good old *nine-to-five, Monday to Friday*, might not be such a bad thing after all? I now had to further my investigations, if I was to get any inclination of what his charges to accompany me would be, and so once again I *jumped in with both feet*. "So, hypothetically speaking Derek, if I was to call your agent tomorrow, and ask them for confirmation of your assistance in my journey, from start to finish, could you give me any indication of what the cost would be, say for ten weeks of your valuable time?"

"If I were to join you, then I would insist upon doing the journey by motorcycle, as well, and the bike would have to be provided by you, of course, as I never use my own vehicles unless it is absolutely necessary. With regards to my own personal costs, well they would vary, depending on which countries you intended crossing on your journey south. A safe bet would be to allow an expenditure of £100 per day, although some countries could be as low as £60. If I had my spreadsheet at hand, then I could tell you the average cost of each individual country throughout the continent".

I tried not to let Derek hear my gulping, as I quickly work out that £100 per day, multiplied by seventy days, is £7,000 —and that is *just for starters!*

"Of course, if you only want my advice, such as general information on individual African states, route planning along the best, and more importantly, the least dangerous roads, and what essentials to take with you on the trip, then by all means come to London and simply have an informal chat with me over lunch...I'll only charge you fifty pounds an hour for that."

"Well thank you for all that Derek; it was most informative, I must say...can I give you a call in a couple of days with hopefully some kind of proposition that will be suitable for all parties concerned?" I said politely, but basically meaning 'Don't call us, we'll call you!' It was only after putting the phone down that I realised his terms and conditions were quite plausible, taking into account the fact that he would basically be working sixteen hours a day, and also working away from home for several months. As for me supplying Derek with a motorcycle, well I was having enough problems trying to get one for myself right now, and so lord only knows how I was ever going to overcome that hurdle? My only hope now was that James' letter from the BBC would have the power to bring in sponsors from all directions?

Another weekend passed and another month disappeared into oblivion, as I said my farewell's to July, before welcoming-in August —the month of my birth. First thing this morning I received a call from Ronnie Riley, the man from Action Aid, who is responsible for the 'Go Ten for Africa' project. Ronnie said that he was currently calling me from his office in Somerset, but that he would be travelling up to Bristol this evening for a meeting.

"If you would like to meet up with me, then I will be at the Watershed Arts Centre, which is situated at the far end of the marina, not too far from the town centre and the Bristol Exhibition Centre. I will be standing in the main foyer at the bottom of the staircase, next to the cafe-bar, and holding one of our leaflets, so that you can recognise me. Would seven o'clock be alright for you?"

Ronnie had rattled-off the details so quickly that I almost burned a hole in my diary, as I tried scribbling-down the information as fast as he was saying it. If this man organised his charity events as precisely as he does with his rendezvous arrangements and directions, then he is the kind of man that I would like to be working with...providing that he wanted to work with me, of course? In the afternoon I typed out an introduction letter, before putting it in an envelope, along with a copy of the letter from the BBC. Even if Ronnie condoned my idea, he would still need the go-ahead from the hierarchy in his company, before being able to sanction the project, and so I hoped that the BBC letter would be strong enough to win them over. It was 6.45 pm by the time I joined the A48 heading eastwards, and I still had over forty miles to go before I would reach the city of Bristol, and so once again my schedule was running terribly late. By the time that I turned off the M4, and onto the M32, it had turned seven-thirty, and I was beginning to panic that Ronnie would be long gone by now.

In every major city I have ever visited, I have never failed in getting completely and utterly lost...and today would be no exception. Pulling into a bus lane, in order to ask for people for directions, and then being unable to pull out again, due to the density of traffic, was bad enough, but what do you do when you finally find the building that you have been searching so diligently for, only to discover that your pathway is obstructed by several million gallons of water? I had no idea that Bristol's civic centre had a small canal running through it, and so once again I had found out...*the hard way*. However, I soon spotted a bridge crossing over the canal, which was about one hundred yards from where I was now standing, having been lucky enough to find what I reckon was the very last remaining parking space in the whole of Bristol. Running as fast as my little legs would carry me, I soon traversed the distance, before crossing over the bridge and in through the doors of the Watershed Arts Centre. As I entered the main foyer, I could see two men chatting away together by the central staircase, the man facing me donning a receding hairline that had virtually reached the crown of his head.

Ronnie had already told me that he was *going thin on top*, and so I gathered that this could well be the man that I am looking for? Without further ado I walked swiftly towards the guy, with a view to introducing myself to him, and as I got closer to the staircase, I could see that he was holding one of the *Go Ten for Africa* leaflets in his right hand. After formally introducing myself to both gentlemen, I made my sincere apologies to Ronnie for being so late, to whit Ronnie's companion jokingly quipped that he was only keeping Ronnie entertained, whilst he was waiting for me to turn up. He then left Ronnie and me to continue our discussion in private, and so the pair of us headed off to the small cafe-bar, where Ronnie kindly treated me to a coffee, while I prepared to make him an offer that he couldn't refuse. Returning to our table with a tray of light beverages, Ronnie beckoned me to *spread my wares.* Pulling the envelope out of the inside pocked of my suit, I handed it over to Ronnie, saying that I hoped it would raise a sizeable amount of cash for the 'Go Ten for Africa' appeal.

As soon as Ronnie saw the BBC insignia at the head of James' letter, his eyes widened for a few seconds, and so I immediately knew that he would be suitably impressed with what he was now reading. Having read James' input, Ronnie then read the document that I had written, which was headed as follows:

"On the **TENTH** day of the **TENTH** month, Shaun Donovan will begin a **TEN THOUSAND** mile journey, spanning **TEN** African countries, over a period of **TEN** weeks, in an attempt to raise **TEN THOUSAND POUNDS** for **TEN** development projects in **TEN** areas of this third-world continent".

Underneath the main heading was a footnote, suggesting that we open the event on the first of October, with video footage of me setting off from Cardiff, which would then be followed-up by **TEN** days coverage of my journey throughout Europe, before showing a second video of me setting off from Alexandria in Egypt on the **TENTH** of October?

Ronnie was completely bowled over with the whole concept, from start to finish. In fact the term 'spellbound' came to mind, as Ronnie openly announced "It's perfect; just what we have been looking for to publicise the whole event". I had to agree, of course, as there is nothing better than television coverage to promote anything that is going on in this world. I also felt a tremendous amount of satisfaction in knowing that after nearly a year of hard bargaining at countless meetings, and an endless amount of phone calls, not only throughout the UK, but also to America, Canada and South Africa, my diligent efforts could end up saving 'hundreds', or maybe even 'thousands' of lives...who knows? The mere thought of being able to lift my head up in public and say that I had finally contributed to the well-being of others, rather than selfishly looking after no-one else on this planet, except me and my family, made me feel really good inside. With this thought in mind, I raised my coffee mug in the air, before making this rather pertinent toast "To a bright and prosperous future for us all". Lifting his cup of Rosy-Lee off the table, before gently chinking my mug, Ronnie echoed my thoughts by saying: "I'll drink to that Shaun".

Feeling great about everything right now, I decided to tell Ronnie the whole story, including my trip to America to see Ted Simon in person, of course, and to my great surprise, Ronnie said that he had already read Jupiter's Travels, and that he really loved the book. Déjà-vu! Every avenue I suggested going down regarding the project, Ronnie went one step further. Below are a few examples:

When I told Ronnie that I had a couple of contacts in the BBC in Wales, my learned friend said that he had several contacts in both the BBC in the West Country, and also in the HTV Television Network.

Also, when I informed Ronnie that I intended contacting British Gas Plc Wales for sponsorship, primarily because I had worked for the company for over a decade in the past, Ronnie once again took the helm "Leave that to me Shaun; British Gas has already sponsored several projects for Action Aid, and so I am sure that I can convince them to become involved with this one".

Ronnie then went on to say that he had a team of telesales operators back in the main office, who would be happy to contact every company in Wales for sponsorship if necessary.

With all this positive feedback, I did not want to appear negative in any way, shape or form, but I openly admitted to Ronnie that we could have serious difficulty in filming anything at all in several of the countries that I would be visiting. However, once again Ronnie came up trumps, by saying that the BBC had already filmed two documentaries for them in Africa beforehand and that because I would be travelling through four of the many countries that the charity is currently supporting in Africa, then he was quite sure that they would be more than willing to let us film in their countries, including doing interviews (with the assistance of an interpreter, of course) with the people in their individual villages. Ronnie was a real tower of strength for me, and the mere thought of being able to leave the *donkey work* up to someone else had simply lifted the world off my shoulders.

Ronnie was not only enthralled at the potential return that my project would create, both financially and also in name-awareness for Action Aid, but he was also encapsulated by the whole idea of the trans-African crossing, saying that given the opportunity he would join me "Like a shot". Unfortunately, Ronnie had various other commitments that made it impossible for him to even consider joining me on my trip, such as flying over to Ecuador in three weeks time, to work on a project that concentrates on saving the diminishing rainforests in South America. "So you will just have to tell me how brilliant it was" Ronnie added, with just a tinge of sadness in his voice. To compensate for his loss I told Ronnie that as soon as I had finished this crossing I fully intended shipping the bike over to South America, in order to begin the second part of the *Jupiter's Travels -20 years on'* documentary series. "Maybe we will link up with Action Aid again Ronnie, only next time, instead of raising money to save lives, we can help save the rainforests instead –and hopefully you and I can do a journey together?"

Ronnie loved the idea of this dream trip for us both, but for now we had to face up to reality and channel all of our energies into getting this project in place, before even considering looking at future projects. Our meeting had now reached its climax, and so I gave Ronnie a copy of the entire budget, so that he would be under no illusion of how much money it would take in order to make this project work. Ronnie studied each section in detail, flipping over the pages every few minutes or so, before suddenly coming to a grinding halt on page four. Glancing across the table, I could see him staring at the paragraph that was headed 'Capital required to cover my expenses whilst I am away in Africa'... and more importantly, the final figure of £ 5,000! For fear of this shedding a different light on the whole issue, I quickly intervened, before Ronnie's negative thoughts got the better of him. "Please don't concern yourself with my personal funding, my friend; for I already have that covered", I fibbed.

Ronnie then sat back in his chair, before gently folding his arms, and although he said nothing, I knew right then that he was not convinced with my statement at all. About thirty seconds passed, and by now Ronnie's *silent close* was certainly getting the better of me, as my inner-thoughts began begging him to say something...anything that was positive? Ronnie leaned forward, with both of his elbows now resting firmly on the table, while his open hands supported the weight of his chin. He then pushed the top half of his body even more forward, so that his face was within inches of mine, before uttering those immortal words "So what is it exactly that you are asking me for Shaun?" As honesty has always been the best policy (well, almost always) in my books, I am happy to tell Ronnie the truth, the whole truth –and nothing but the truth.

"All I am asking you for is your assistance in helping this journey reach fruition, and in return I will hopefully raise enough money to assist Action Aid in its quest for a better world. I have not come here to beg you for money, my friend, but to simply ask you for your professionalism and expertise in helping me raise enough funds, in order to get this project off the starting blocks. Any information that you can give me regarding African countries will also be more than welcome, of course".

At this point, Ronnie sat back in his chair once again, looking as if he was satisfied with my answer, thus releasing a mountain of pressure off my already weakened shoulders. I wanted to be completely straight with Ronnie, and so now that he had accepted the fact that I was not looking for a handout, I openly told him that if, at the end of the day, we were short of a few hundred pounds to complete the budget, only then would I ask his charity for a little financial assistance. In closing our meeting I told Ronnie that I fully appreciated the fact that Action Aid was a business, as well as a charity, and so if we both put our heads together on this, then I was sure that we could make it both profitable for the company and beneficial to the people. Ronnie can see that I am very genuine about what I say, and truly steadfast in my convictions, and so he asks if he can take my budget sheets with him, in order to show the people who always make the final decisions.

As we said our fond farewells, Ronnie gave me a firm handshake, whilst emphasising how impressed he was with all that I had done to date, adding that he is now one hundred percent behind my project, and that tomorrow morning he would be making a number of "Very important" phone calls. As Ronnie made his way to the main entrance, he turned to me and shouted "I'll give you a call tomorrow evening Shaun", which was just what I wanted to hear. In my books it had been a great meeting, and I couldn't wait to speak with Ronnie again on the morrow.

CHAPTER 28

Biker Off To Film The World

Unfortunately, that all-important phone call never materialised, and once again it seemed like I had been let down at the eleventh hour. Two more days passed, as I waited in vain for Ronnie's call, but there was nothing. In desperation I rang the Action Aid office in Somerset, which was Ronnie's home ground, and lo and behold the man himself answered the telephone. Feeling more than anxious by now, I openly asked Ronnie why he had not called me, knowing full well that I would be waiting on his call, to whit he claimed that he had tried to call me several times, but he had been unable to get a reply. The line then went quiet for a few seconds, and so I knew that it was only ever bad news that followed these deathly silences, for I had become somewhat accustomed to them by now. "You see Shaun, it is like this..." Ronnie said in a rather sympathetic tone, as if he was about to break the bad news to me as gently as possible.

"I have spoken to several of my colleagues and contacts these past few days, and they all say that there is just not enough time for us to put this whole concept together –and still make enough money out of it by October. A number of people are also quite concerned about the dangers involved in visiting Southern Sudan or Kenya, and so I am sorry to have to tell you that it is all off Shaun".

I was so disappointed, it was unreal, but rather than fighting my cause, which I knew would be a waste of both time and energy, I simply asked Ronnie to send my budget forms back to me, before calmly replacing the receiver. At this point my legs gave out and I slumped to the floor.

My whole body was in disarray, and my mind was in turmoil, as I tried desperately to regain my composure. Again I was at sixes and sevens with the whole project, wondering why I even bother continuing on with this uphill struggle for the return of my identity.

Thankfully, I have never quit at anything in my life, and so no matter how bad I am feeling right now, I know in my heart that I have no alternative but to get back up on my feet again, dust myself down, and go straight back to the drawing board. I have dozens of copies of the BBC's letter in my briefcase, so why not start the ball rolling by putting them to good use. I knew that my first and most important letter must be sent to Budget Rent-a-Car's head office, to see if they would be interested in sponsoring the motor-cycle for the project, and so I spent the next hour or so outlining everything on paper, before duly posting my letter off to the lady who is the head of the Public Relations Department in Hemel Hempstead. Jules has already informed me that the woman has recently returned from her honeymoon, and so I am hoping that she is in an *all-giving* mood right now.

A few days after sending the letter, I contacted the woman in question, only to be told that all of the company's sponsorship funds had been used up for the year, due to the amount that they had donated to the Olympics. However, she did say that I was welcome to contact the regional office in South Wales for assistance, as they have an entirely separate sponsorship fund from the English regions. This I did within minutes of replacing the handset, but the receptionist kindly informed me that Mr. Grayson, who was the gentleman that I would need to speak to regarding any kind of sponsorship deals, was currently on holiday for a week, and so I said that I would call back on the following Monday morning. However, when I told her about my relationship with Jules Gilmore, the young lady said that it would be best not to mention Jules's name on Monday, when speaking to Mr. Grayson, because him and Jules had not had a very friendly relationship for a very long time.

Apparently, Mr. Grayson had not only refused Jules sponsorship for various appeals on several occasions, but he had also turned down an advertising contract with the *Cardiff Devils* hockey team –a partnership deal which Jules had highly recommended. "Not taking Jules' advice on the latter option was something which Mr. Grayson now bitterly regrets, although he would never admit it to Jules, of course", the lady added. Unfortunately, Jules was no longer in a financial position to help me, primarily due to the recession, which had drained a lot of his resources, and so I knew that the only way that I was going to find a decent sponsor was to advertise for one. I decided to give Matthew Green a call, telling him that I was now ready to do the interview for our local newspaper. As luck would have it, my timing could not have been better, for Matthew admitted to me that at the end of next week he would be leaving the newspaper, in order to become an independent journalist, and so he would be happy to do the interview this week.

Matthew already knew a fair amount about the project, as I had sold the idea to him when we first met all those months ago, but I now had no choice but to put him fully in the picture with the situation, before asking him to express the need for further sponsorship in his article. I knew that the *South Wales Motorcycles* dealership in Malpas, Newport, would be delighted to have some free advertising in the newspaper, and so I arranged to meet the owner at three o'clock on Thursday afternoon, before confirming this with Matthew, who said that he would also bring a photographer along with him for some really great shots of me sitting astride one of the motorcycles. The following day Matthew gave me a copy of the article he had written, just in case there were any final adjustments that he needed to make to the storyline, before going to press. By now I had looked upon Matthew as a friend, rather than just another newspaper reporter whom I had happened to come into contact with, and after reading the excellent article that he had written about the birth of my wonderful daughter, Hayley, and the subsequent story about my dear mum's 60th birthday / *This is Your Life* surprise party, I had all the confidence in the world in him. Once again his write-up was 'spot-on', and I could not wait to see it in print.

Despite having two beautifully sunny days on Tuesday and Wednesday, Thursday turned out to be one of the wettest days in the whole of the summer. The heaven's opened around breakfast time, and the rain never stopped lashing down for the next six hours, the deluge of excessive rainwater literally gushing down the roads and pavements of Newport, before finally disappearing into the rainwater gulley's. Jules had kindly loaned me one of his official 'Budget Rent-a-car' jackets for the photo-shoot, primarily hoping that upon seeing their insignia advertised in the centre of a major *Welsh* newspaper, our friend, Mr Grayson might seriously consider the idea of becoming a part of our little venture, and thus offer me some kind of sponsorship deal. As I said before, I am not a liar in any way shape or form, but if there was a prize being awarded for the 'World's most prolific schemer'; someone who would stop at nothing, and let no-one stand in his way, in order to achieve his own personal goals, then I do believe that I would be a good contender for the title.

What I have learned during these last twelve months is that anyone who takes on the drudgery of trying to become a success, in any kind of business, deserves the right to be devious and cunning whenever necessary, and that so long as his (or her) motives are genuine, and their heart is in the right place, then they have the right to do whatever it takes in order to achieve their ultimate dreams and aspirations. Mind you, considering that I am currently *rolling around* in the category of 'The Most Unsuccessful People' in the country, I daresay that my moralising is quite worthless! This article could just be the starting point of a whole new career for me...who knows?

The photographer, whose name is Jamie, was waiting for me when I arrived at the dealership, and he said that before taking any pictures, he wanted to ask me a few basic questions about the journey, in order that he could write a caption underneath the main photograph that would coincide with Matthew's write-up. As soon as I had answered Jamie's questions, I made my way through the main showroom, before stopping at the bike that I intended using for the journey. After climbing aboard the Kawasaki 500 trails bike, I asked Jamie which angle he wanted to take the pictures from, to whit he just laughed aloud, saying that there was no way that he could take the photos inside the building, before adding that we would just have to "Brave the elements" outside on the forecourt. I turned to the salesman, who had been listening to our conversation with avid interest, and whispered "I was afraid he was going to say that!" At this point the owner appeared on the scene, and within minutes two of his sales guys were moving half a dozen bikes across the sales floor, in order that we could safely manoeuvre our *mean machine* out onto the forecourt.

Standing adjacent to the dealership was a large petrol station, and so I humbly suggested to Jamie that we take the photos underneath its huge canopy, in order to save us both from getting soaked through to the skin. Jamie was in total agreement, and so while he spent the next few minutes playing around with various cameras, filters, films and lenses, I duly pushed the bike from the doorway of the showroom, through the pouring rain, and countless puddles, until I finally reached the sanctity of the shelter. One of the salesmen had kindly loaned me a helmet for the photo shoot, and so I spent the next few minutes trying to fathom out how to secure the chin strap correctly? I was so out of practice with doing this menial task that I began to wonder how bad I would be when it came to doing a little bit of motorcycle mechanics at the roadside. For the next twenty minutes or so Jamie and I worked on all kinds of shots and angles, making sure that the name of the dealership, which adorned the main wall of the showroom, was in every photograph, as this was the condition that we had agreed with the owner.

With my one arm outstretched, holding onto the handlebars, Jamie also made sure that the *Budget-rent-a-car* insignia, which had been embroidered on the sleeve of my jacket, was ever present in each photograph, as like Jules, I sincerely hoped that the sight of the company name in the paper would hopefully encourage Mr. Grayson to become involved with the project. It was a very long shot, of course, but I was getting used to them by now. As soon as the photos were done, Jamie made a sharp exit, saying that he would pop a few of the pictures in the post to me within the next couple of days, before driving off in his car. This left me with the horrendous task of pushing the bike back through the puddles and potholes of this very uneven Tarmac surface, but then compared to what I would have to face in Africa, this would be like *a walk in the park*. Ironically, sitting opposite the toilets at the back of the petrol station was the rear entrance to the bike dealership, which was basically the workshop, where a handful of mechanics spent their days servicing and tuning motorcycle engines, along with repairing the damaged carcases, of course.

Unfortunately, the place was currently chock-a-block with vehicles, and so there was no way that I could slip in through the back entrance, thus avoiding my second soaking of the day. After finally handing the motorcycle over to its rightful owners, I told the guys to keep an eye out for tomorrow (Friday) night's issue, if they wanted to see our splash –and their dealership, of course, in the newspaper. My next port of call was Budget's offices in Cardiff, where I asked Gareth if he would kindly print me off another five 'Embassy letters', just in case I had no alternative but to cross into additional countries on my trek through Africa. Now it was time to see if my ex-colleague was right about British Gas? The gentleman in question (whom I shall call 'Bobby') had rather a high standing in the head office in Wales, and some time ago he had told me to write directly to the chairman, quoting Bobby's name in the letter, and so I was very hopeful about my chances of success.

When Bobby first told me that British Gas had a lot of spare cash that they were looking to donate to a worthy cause, I had jokingly remarked at the time that I would be happy to take it off their hands...and now here I am, hoping to do just that. On Friday, the fourteenth of August, 1992, the article, which included a photograph of me sitting astride the Kawasaki 500, was printed in the *early edition* of our local paper. However, in the *late edition* there was no photo. The heading read "BIKER OFF TO FILM THE WORLD". Immediately after reading the write-up, I tentatively cut out the article, along with the photograph, before placing it in an A4 envelope, along with a selection of documents that I would require for pitching prospective sponsors. The following day I compiled a comprehensive *sales catalogue*, in readiness to begin my *cold calling campaign*, in order to attract sponsorship for all of the other items that I would require for the trip -even though I had still not secured any kind of agreement on who would actually be supplying the motorcycle?

On Saturday morning I was in the middle of doing my *house-husband* chores, quietly singing away to myself, whilst chasing the children around the room with the floor sweeper...a game they simply loved to play, when I suddenly heard Caryl's voice shouting above the din of the children's laughter. Immediately I curtailed my sweeping, much to the children's dismay, as Caryl handed me the phone, saying that a gentleman from W.H. Smiths was on the line for me. "Shaun Donovan speaking -how may I help you", I say in my best *business* voice, even though my heart is pounding with excitement. "You ordered a map from us almost a year ago sir, and yet it is still here in the shop, waiting to be collected. Could you tell me if you still require the map, and if so, could you also let me know when you will be able to collect it please?"

My heart sank, although why I was expecting anything relating to my project in the first place, I cannot imagine, because up until now I had not pitched W.H. Smiths for any kind of sponsorship deal? I then explained to the gentleman that back in December of last year, one of his colleagues had ordered the incorrect map, and that I had already collected the correct map from his store the following month. Unfortunately, before I had the chance to go into one of my sales spiels, the gentleman simply apologised for the mix-up, before promptly cutting me off.

Shortly afterwards the second post arrived on the doormat, and I immediately noticed a 'British Gas' insignia on one of the letters. Tearing open the envelope like a man possessed, I unopened the neatly folded page, before slowly reading its contents, my heart now pounding with excitement, as a volley of positive thoughts continued swirling around in my head. 'Dear Mr. Donovan, thank you for your interest in British Gas. Unfortunately we will be unable to assist you, blah, blah...and finally we would like to wish you great success with your venture, blah, blah.' Obviously my friends' status in British Gas accounted for nothing, and so yet another *'Dear John'* letter could now be added to my ever-increasing portfolio of "Thanks, but 'no thanks'" letters. There was now only one other person who could be in a position to help me with my plight, but if he turned me down then everything would surely come to a grinding halt, for there was simply no-one else left that I could turn to. However, I know that certain members of my family will breathe one huge sigh of relief if the journey never materializes, for they had been dead-against me doing the journey from the start. I cannot dispute the fact that travelling alone through Africa is a somewhat irrational thing to do, but when a person has no alternative but to risk their life for their family, then chances have to be taken, otherwise all hope will be lost, and in my eyes that would be worse than death itself. The mere thought of having to leave Caryl and my beloved children for three months, or more, had been eating me up inside all along, and the dreaded thought of not making it home in one piece had tormented my soul from the moment of the journey's inception.

Before working as a Sales Executive for the Newport Ford car dealership, I had held the position of Sales Manager for an advertising company for about two years. The head office was situated in Gloucester, and I had run their offices in Cardiff, which primarily dealt with advertising vehicles for sale through various dealerships in South Wales, via a weekly newspaper. We also had a shop front for promoting private sales. During my time with the firm the owner of the company, Mr. Walter Finn, had been a very approachable person, who was always ready to listen to any ideas that his employees put forward, which is something quite unique for someone who owns a multi-million pound business corporation. With this thought in mind I rang him at his office in Gloucestershire, and duly booked an appointment with him for 11.30am on Tuesday, 18th August. On the Monday prior to our meeting I received the one and only phone call that the newspaper article had generated, only instead of it coming from a potential sponsor, it was from a fellow traveller, who immediately introduced himself to me as 'Nathan Smith'.

Nathan said that a workmate of his had showed Nathan the article, after Nathan had told him of his intention to do a similar journey across Africa, and so he wanted to know if we could get together sometime, in order to have a friendly chat, and also to exchange notes and advice about crossing the *Dark Continent*. "Two heads are always better than one, my friend," I said jokingly, before arranging to meet up with Nathan at his home around seven o'clock the following evening, knowing that I should be back from Gloucester (which is around 70 miles from Cardiff) long before that time. Tomorrow's presentation would have to be of the highest calibre, if I was to impress Mr. Finn, for we are talking about a very shrewd man indeed, who is extremely cautious when it comes to spending his hard earned cash on anything, let-alone donating thousands of pounds to some 'pie-in-the-sky' adventure. Knowing that I would have my work cut out for me, I duly enlarged the photograph from the newspaper clipping onto an A3 size sheet of paper, before obliterating all of the original decals that adorned the large front fairing.

I then created an artists' impression of what the bike could look like sprayed in Mr. Finn's company colours of a yellow background, with black pin-striping, accompanied with large white letters on an orange backdrop, for the company logo. The end product looked so professional that I felt sure it would strike a chord with Mr. Finn, and so all I had to do tomorrow, was to set an image in his mind of my bike flying across the plains of Africa, while at the same time his company name was flashing across the television screens of millions of families in the UK...it sounded just perfect. In retrospect, it was Mr. Finn who had taught me to sell advertising space in the first instance —and so why shouldn't I try and sell some to him?

CHAPTER 29

Going For Gold

The day has now dawned, and unbeknown to me at the time, but this day would go down in the history books as one of the worst days of my life, although the reason why I could never have imagined as I drove along the M4 heading towards the Severn Bridge. However, instead of crossing the River Severn into merry old England via this iconic landmark that separates both Wales and England, I would be taking the M50 motorway northwards, heading into the heart of the English countryside, just as millions of other vehicles had done prior to the building of this almighty bridge. *A second bridge, which was aptly titled the 'Second Severn crossing', opened to the general public on 5th June 1996.* As my appointment is scheduled for 11.30am, I have allowed around one and a half hours of travel time, in order to cover the seventy-odd miles from my home in Cardiff to the gates of 'Capitol House' –Mr. Finn's workplace.

Averaging around 60mph on the motorway, before dropping down to approximately half that speed on the country roads, I am making good time as I cross the border from Wales into England. The sun is shining brightly, and so I am in very good spirits, as the miles slowly but surely disappear under my wheels, and the minute's continue-on ticking away at a leisurely pace. Suddenly I run into a small traffic jam, thanks to a set of temporary traffic lights, and so I am not overly concerned, although I am still keeping a keen eye on the time. Within minutes the lights have turned to green, and one by one the cars in front of me begin moving forward, albeit at a snails' pace, until finally I am able to begin my wheels rolling, and become a part of the lengthy convoy. A very noisy generator is vibrating on the pavement, and several heavy duty plant machines are in operation on one side of the road, including a huge steamroller, which is now flattening the newly laid asphalt to a smooth finish.

As I reach the end of the road-works and return to my side of the road, there are no road markings, where the new asphalt has been laid over the past few days, but there *are* hundreds of small gravel stones splattered all over the Tarmac, which are being flicked up intermittently by dozens of sticky tyres, as the vehicles in front of me roll over them as tentatively as possible, so as not to damage the paintwork on their beloved cars. Because of this, everyone is still crawling along the road at around ten miles per hour, and there is currently no end in sight to this *new road*. After what seemed like an age, the *old road* finally reappeared in the distance and no-one was happier to get back on it than me. It was also a pleasure being able to use third and fourth gear again, and to finally take my foot off the clutch. Onward I drove, as the *'eleventh hour'* passed into oblivion, thus leaving me less than thirty minutes to reach my destination, park up the car, and find Mr. Finn's office...and to top it all, I now desperately needed to use the bathroom.

The closer I came to Gloucester, the more little villages I had to pass through, the traffic in each one seeming more dense than the last. I then had to stop at a petrol station in order to relieve myself before it was too late, which took another four to five minutes of my ever-diminishing timeframe, and so by the time that I finally turned onto the Bristol Road in Gloucester, I was already nearing the eleven-thirty deadline. Pulling over at the side of the road, I ask a pedestrian if he can direct me to Capitol House please, and to my pleasant surprise he tells me to drive straight along the road for about a mile, and I cannot miss the building, which is situated on the left-hand side of the road. As I continue driving along the road, I pass several familiar looking buildings, and I feel as if my life is flashing before my very eyes. On the left hand side of the road is a large gasworks, complete with two huge cylindrical towers the pair of them pumping out smoke like there is no tomorrow.

Sitting adjacent to these towers is what looks like a large warehouse-cum-office block, thus rekindling my memories of the ten years I spent working for British Gas at their spare parts stores in Cardiff. Just beyond the gasworks is the 'Target Ford' car dealership, which is not unlike the 'Newport Ford' car dealership that I worked for before being made redundant last September. On the opposite side of the road, and standing out like a sore thumb in my memory, is Gloucestershire's very own 'White Arrow' depot –a place I had had many dealings with when I worked for the White Arrow parcel distribution company back in Cardiff. Finally I reached, and the home of *PLC Publishing,* which apart from the aforementioned companies, and also working for one year as a painter and decorator, is the only other company I have worked for since leaving school over sixteen years ago. As I pull up outside the reception area of the building, the clock on the dash is reading 11.34 thus telling me once again that I am late for a very important appointment –shit!

Snatching my briefcase off the passenger seat, I quickly lock the car doors, before running hell-for-leather straight through the main entrance doors, whereupon a young man promptly leaves his desk in the main office to see if he can assist this *unannounced intruder,* who is obviously in a desperate hurry? After explaining about my appointment with Mr. Finn, before duly apologising for my lateness, the gentleman kindly escorted me up the main staircase, before asking me to wait in this small lobby while he announces my arrival to his great leader. As the man knocks gently on the large wooden door, before disappearing inside the room a few seconds later, I begin fumbling nervously through my paperwork, making sure that everything is in its right order, in readiness for my latest –and hopefully my greatest, sales pitch. Within a very short space of time I am greeted by Mr. Finn, who welcomes me with a beaming smile and a firm handshake, before leading me into the oak panelled boardroom, where he kindly offers me a seat at his own personal desk, before sitting opposite me in the comfort of his very plush leather swivel chair.

Directly opposite us is a very elegant and highly polished mahogany table, which is surrounded by eight matching hand-crafted chairs that have been strategically placed at equidistant intervals around the table. While Mr. Finn pops out to see his secretary, asking her to kindly make us a pot of coffee, I sit in total silence gazing at the empty chairs, whilst trying to visualise one of Mr. Finn's director's meetings in progress. I had never attended any of these regular *get-togethers* of the top brass in the company, but I have been told that they can become somewhat overheated at times, to the point where all hell can break loose on a bad day, but then that is what boardroom meetings are all about, I suppose. Upon returning to his desk, Mr. Finn and I discuss pleasantries for around ten to fifteen minutes, and so I am a lot more relaxed when it is finally time to get down to business.

Mr. Finn had an inkling of my project, after I discussed it briefly with his secretary when booking the appointment for today, but now Mr. Finn, who has never been a man to *mince his words,* wanted to know exactly why I have travelled all the way from Cardiff in order to have a personal audience with him. Knowing that I was now in a 'do or die' situation, I hit Mr. Finn with everything I had to give him, from the BBC's letter, to Action Aid's leaflet, from my meeting with Ted Simon to my chat with Derek Bryant, from sponsors to budgets, overland vehicles to motorbikes, merchandise to mileages –you name it, and I covered it. Mr. Finn was very intrigued by the whole set up, and so now was the time to *go for gold.* Reaching into my briefcase, I pulled out the advertising poster I had created, before unfolding it on Mr. Finn's desk, the photograph of me on the motorcycle, which was adorned in his company's colour's and logo, now standing proud for all the world to see. I watched Mr. Finn's eyes broaden at the sight of my *masterpiece,* as he gently lifted the poster off the table, before holding it above his head, as if in awe of its creation.

Phase one of this major operation, had obviously gone particularly well, but phase two was going to be the difficult part, for now it was time for me to ask the $64,000 question. Pulling out one of the budget sheets from a file in my briefcase, I tentatively went through the actual cost of purchasing the bike, along with the additional expenditure of adapting the frame to carry the excess fuel and water, and also the cost of the additional tools and various motorcycle equipment that I would need to carry, in order to cross Africa safely. Having now placed all of my credentials firmly on the table, I sat back in my chair, my heart pounding furiously as I waited patiently for Mr. Finn to say his piece. However, before he could utter a word, our negotiations are interrupted by the entrance of a very well dressed middle-aged lady, who, I have to say, is rather attractive to the eye. Donning a mane of long blonde hair that flowed majestically over her shoulders, before enshrouding the top half of her somewhat shapely figure, I can plainly see that this woman has obviously looked after herself over the years, although I am trying not to stare, of course.

As the lady gently places a tray of drinks on the table, Mr. Finn cordially introduces his wife to me -a lady whom I had only ever spoken to briefly over the telephone, and now here she was 'in the flesh', so to speak. Although it was very nice chatting with the lady for ten minutes or so, I felt that the crescendo of the meeting had been broken, and I knew that if I could not revitalise the impact of all my hard work, then I would fail miserably with my task of obtaining sponsorship from this man.

Pointing out the benefits of having his company's name splattered all over television screens across the country, I emphasized how much it would put his competitors in the shade, which I knew would strike a chord with Mr. Finn, for he is one of the most competitive men that I have ever had the pleasure of working with. Leaning forward in his chair, Mr. Finn began studying the picture, as if he was envisaging the bike flying his company colours across Africa –and across the UK, of course, before sneakily glancing over at the budget sheet for a few seconds, and then returning to the photo once again. Unfortunately, his prolonged silence, along with the fear of receiving a negative answer, was just too much for me to bear, and so I blurted out "I don't expect you to give me an answer right now Mr. Finn, but I would like to know as soon as possible please... if that is okay with you, of course?" I knew that I had broken the golden rule of the *silent close*, by speaking first, and that I had also given the client the option to *think about it* which is something that could quite easily lose me the deal.

Had I broken this rule, or given the above choice to a prospective buyer whilst working for Mr. Finn then he would have kicked my backside all over town, for I had now showed a flaw in my positivity –a weakness that could have catastrophic consequences if the prospective buyer already has any doubts whatsoever about the product in question. Before Mr. Finn could qualify my statement the phone on his desk started ringing, and so he duly answered the call, before swivelling around in his chair to face the window, as he began his conversation. At this point something inside of me said that I had overstayed my welcome, and so I immediately stood up and began packing up my briefcase in readiness to leave. Upon hearing the sound of the locks clicking on my briefcase, Mr. Finn immediately spun around in his chair, before raising his right hand and beckoning me to resume my place. At this point I heard a voice on the end of the line telling

Mr. Finn that he was already half an hour late for an urgent meeting, and so perhaps I was not doing so badly after all?

After replacing the handset Mr. Finn politely asks me if he can keep the poster, along with the budget figures, before assuring me that he will put my case to the board of directors at their meeting on Monday morning, and that he will have an answer for me within the next seven days. As we leave the room together, Mr. Finn heads off to his meeting, while I make my way downstairs to the main entrance on the ground floor. As I am about to leave the building, I suddenly spot my old friend –and ex-colleague, Calum, who was the regional manager for the company when I worked for Mr. Finn all those years ago. Calum has obviously gone up in the world, for he now has his own posh office in Capitol House, but as always he is happy to take time out of his busy schedule to sit and have a chat with me. Immediately I sell my wares to Calum, for he has always had faith in any of my undertakings, and I also know that if anyone can convince Mr. Finn to spend his money wisely, then it is Calum.

Calum is certainly intrigued with my journey, perhaps even a little jealous, for like me he has a very adventurous spirit, and I know that he would simply jump at the chance of crossing Africa on a motorcycle. However, Calum also warned me that Mr. Finn had just spent a small fortune on all of the latest computer technology equipment for his stores and offices, and so the annual 'spending' budget was looking somewhat empty right now. Nonetheless, I felt very confident as I said my farewells to Calum, before heading off in the direction of the company's main shop, which I had passed on my way to Capitol House, and so I knew that it would only take me a few minutes to get there. On the way to the shop I passed a *Beefeater Restaurant*, and so as it was nearly one o'clock, and therefore most of the guys would now be at lunch anyway, I decided to pop into the pub for a ploughman's lunch. By two o'clock I am inside the shop, tentatively looking at the dozens of photos of cars which adorned every window in the front and at the side of the building, pretending to be a customer who is looking to purchase a second-hand car.

As I gazed around at the hundreds of colour photos that now surrounded me, before reading a few of the brief descriptions printed underneath the photos of each individual car, I began to reminisce about my days in the Cardiff shop, which seemed like a lifetime away right now.

In the background I can hear one of the telesales girls offering to send out a complimentary issue of the company newspaper to someone who is already advertising their car through another media outlet, in the hope that they will read the company's special offer of 'Advertise your car until it is sold for a one-off payment of £40' and decide to take Mr. Finn's company up on its offer. On average, around seventy percent of the people contacted will accept the complimentary newspaper, simply because it is free, and for every 100 people contacted, around ten to fifteen people will end up taking out an advertisement with the company. And so the statistics are quite good, considering that it has only cost the price of a phone call and the postage cost for one newspaper (which will invariably be a leftover copy from last weeks' issue), along with five minutes of the telesales operators time, which in wages terms, is a pittance.

As I approach the counter, one of the telesales girls asks if she can help me, and I can see that she is poised to go through the whole sales spiel, which will include offering to send a photographer to my home, in order to take free photographs of my car, and to write-out the advertisement for me...and so on. It is a great feeling of power to have an unfair advantage over someone, and when I am in such a situation, I am always tempted to string the other person along, primarily to hear their personal sales pitch –and also to try and catch them out should they decide to throw in a few 'porky's' along the way. However, this time I have decided not to be so cruel, and so I simply ask the young lady if I can have a word with the manager please? Within minutes, Christian, the office manager, has appeared on the scene, and after a brief introduction, he says that he thought that he recognised my Welsh accent, but he just could not put a face to the voice? Christian was my counterpart when I worked in the Cardiff office, and although we have never met in person, we had often indulged ourselves in long chats over the telephone, and so once again it was a very pleasurable experience to finally put a face to a name.

I tell Christian all about my project, and like everyone else I have spoken to, he is absolutely enthralled with both the movie idea, and also the journey across Africa. The excitement that I create each time I tell my story is what encourages me to keep pushing ahead with everything, and although it may seem like I am selling my wares to someone else, deep down inside I know that I am selling myself on the merit of my efforts, more than anyone else, because without having belief in myself, I know that I am nothing. With so many people wanting to see my movie or my documentary reach fruition, if I fail in my efforts then I will feel like I have let them down just as much as I have let myself down. Having finished our friendly little chat, Christian then gave me a tour of the printing workshop, politely introducing me to each worker in turn, as I watched the guys preparing the latest edition of the newspaper in readiness to go to print. Christian rounded-off my tour by taking me to see 'Rodney' who is one of the company directors, and who is also the general overseer of the running of the whole company.

I had met Rodney on previous occasions, when he was my boss, and I have to say that he is a force to be reckoned with. Rodney is not a particularly large man in structure, nor does he have an excessively loud voice, but the man seems to possess this uncanny 'sixth sense' of reading people's minds, whether they are talking to him face to face, or even over the telephone. Should anyone ever try to pull the wool over his eyes, then they had better beware, because this man can sniff-out bullshit from fifty miles away, and so it won't be long before they are regretting every word they said. The mere thought of trying to get Rodney on my side was quite daunting, and I was actually more nervous about asking him for his opinion than I was when I was begging Mr. Finn for a great chunk of his money. However, after listening very attentively (*which was certainly a bonus, because if Rodney is not interested in what someone is saying, he will think nothing of cutting them short midway through their spiel*) to my idea, Rodney responded in a very positive manner, by saying how exciting the journey would be for both me and also my television audience.

However, whether I had convinced him to back me all the way was an entirely different matter, for this man is way too shrewd to give away even the slightest inkling of what he is thinking deep down inside, and so after thanking Rodney for his time, I duly said a polite farewell, before exiting his office. All along I knew that I would not get any kind of firm commitment from Mr. Finn or from Rodney today, but I had done what I came here to do, and that was to plant as many seeds as possible, not only with the heads of the business (which included Mrs. Finn) but also with Calum and Christian, both of whom were very respected by the powers above, and therefore their positive input, however minimal, could be just what it takes to tip the balance in my favour. By the time I finally left the office it was approaching the four o'clock mark, and so I knew that the rush hour traffic in the civic centre would be starting to build up. With this thought in mind I decided to take a different route home. Following the Bristol Road all the way down to the M5 intersection, I continued on following the signs for the M4, before heading due west in the direction of Wales.

However, it was only upon reaching the Severn Bridge that I remembered about the toll for crossing the bridge, which one only has to pay when entering into Wales –and not when exiting the country into England..."Shit." Trust me to do things the wrong way around, but as there was a minimal amount of traffic heading into 'Gods Country', and so the roads were pretty clear for best part of the way, I suppose it was worth paying the three pounds toll charge for a trouble-free journey back to Cardiff. Before heading for home, I also had an errand of mercy to complete, and that was to drop off a present for my nieces' eleventh birthday at my brother's house in Canton, which was a very joyful event to round-off my working day. It was a lovely sunny evening by the time that I finally turned into the cul-de-sac in Thornhill, where I live, and I was in an excellent frame of mind (which was certainly not a regular occurrence nowadays) and so I took full advantage of my 'happy-go-lucky' mood by singing in unison with the music that was now blaring out from the car radio.

As I cruised past a couple of my neighbours, who were standing and chatting together just a few doors away from my house, their conversation immediately stopped, and I could see the pair of them staring ominously at me, their glum expressions immediately telling me that something was not right? It was then that I spotted Caryl, who was standing at the top of our drive, and who was also looking as pale as a ghost, as she saw my car gently rolling towards her. Cuddled into Caryl's left shoulder, I could see my beloved daughter, Hayley, who was obviously sleeping, for she was completely motionless as I pulled up on the drive. "Don't tell me that you've locked yourself out again Caryl", I shouted through the open window, wondering if it was frustration in having to wait for me to return home, before she could enter the house again, that had caused her to give me such a frosty look. Caryl said nothing as I got out of the car, and as

I walked towards her she slowly turned her back on me, thus revealing Hayley's little face, which was looking all forlorn, and her beautiful eyes, that I love so much, had long since lost their wonderful sparkle.

I put my arms out to hold my darling daughter, and as she let go of Caryl's neck, I could see that her right arm was wrapped in a white bandage, which in turn was covered in a kind of string netting. Lifting Hayley gently out of Caryl's arms, I rested her limp body against my left shoulder, as she tucked her little head into my neck. Caryl then explained to me what had happened.

"I was in the kitchen around lunch-time, standing at the sink unit, washing a pile of dishes, while Hayley was playing happily on the carpet in the dining room next to me. I had shut and bolted the two saloon doors leading into the kitchen, as I was cooking a chicken in the oven for tonight's tea, but Hayley must have crawled under them somehow, and unbeknown to me she was now trying to get to her feet, using the kitchen units to assist her."

I began to shudder, and the mere thought of what Caryl's next statement would be, twisted a huge knot in my stomach, tighter than anything I had ever experienced before, as I tentatively squeezed my precious little baby's body, whilst gently massaging her beautiful locks of white hair, as Caryl continued-on with her story:

"Unfortunately, instead of putting her hand on what she thought was a cupboard, she put her hand on the glass front of the oven, which was still boiling hot, even though I had switched the oven off around twenty minutes earlier. Suddenly Hayley let out a horrific scream, which caused me to scream, as well, as I dropped the dishes in the sink, before turning around to see what on earth had caused Hayley to scream-out so loud –and that is when I saw what had happened. Immediately I tried to pull Hayley's hand away, but it was stuck to the boiling glass, and by now Hayley was screaming the place down".

My heart was in my mouth right now, and I felt sick to my stomach at the mere thought of the pain that my little girl must have been going through at that time.

"Finally I managed to prise her fingers off the glass, one by one, before using her wrist as leverage, to release the palm of her hand...it was awful Shaun."

By now Caryl was in tears, and so I tried to be strong for all of us, as I asked Caryl to continue on with her story.

"I lifted Hayley up to the sink, gently placing her hand under the cold water tap for about thirty seconds, before running with her to Connor's house, and thankfully he was at home, and so he immediately drove us to the emergency unit at the Hospital, which is where we have just returned from".

By now, Connor had joined us, probably to see how I was coping with the trauma of it all, as well as Hayley, for he knows only too well that what has happened to my baby girl will have devastated me to the core. I thanked Connor profusely for his assistance, to whit he jokingly replied: "That's okay buddy, hopefully you'll never have to do the same for me". I knew that Connor was only trying to alleviate the tension that currently surrounded my household, but I just wasn't in the mood for any kind of levity right now, and so I politely asked him to excuse me, before walking down the driveway, and disappearing into my house. Thankfully, my sister-in-law had kindly offered to collect my two lovely sons from school, and so they were now at her house, thus giving me the time I so badly needed to be alone with my beloved daughter.

As I cuddled into my little girl on the sofa, propping-up her dainty little head on one of the scatter cushions, my whole body was shaking like a leaf. Without saying a word, Hayley just looked deep into my eyes, her sorrowful face simply breaking my heart in two. It was as if she was asking her daddy what she had done to deserve such agonising pain, and I just didn't have any answer for her. It was just too much for me to bear, and at this point I broke down into floods of tears, as the pangs of guilt began tearing me apart. "Why weren't you there when your little girl needed you the most", I kept asking myself over and over again. "Because you were chasing some pathetic little dream, that's why –you fucking idiot", I answered under my breath. "While you were out having a nice little jaunt in the countryside, your baby girl's hand was melting away in the kitchen –so how does that make you feel Shaun!"

"If Connor hadn't have been home at the time, then how long would Hayley have writhed in agony until an ambulance finally arrived on the scene –ask yourself that one?" By now I had realised what a selfish bastard I really am, my tunnel vision having made me totally blind to what was most important in life, and that was my family's needs...and now I had paid the ultimate price for my utter selfishness. After around ten minutes or so (although it could have been an hour, for all I knew, as I was still in a semi-comatose state) Caryl came into the house, followed by my neighbour, Connor, who was brandishing a bottle of vodka in one hand, and three small tumbler glasses in the other. "What we all need is a good stiff drink right now", Connor exclaimed, before filling the glasses virtually to the brim with potent Smirnoff vodka. Unfortunately for me, I have never been partial to this Russian form of *diesel oil*, and so I only took a few sips, before giving the rest of mine to Caryl to finish off, as she probably needed it more than I did right now.

By the time that my sister-in-law had dropped off the boys, Hayley was fast asleep on the sofa, and so after the boys had gone to bed I decided to let Caryl in on my innermost thoughts.

"I just wanted to tell you how sorry I am for being a complete failure as a husband and as a father to both you and the children Caryl. Because of my failings I have decided that I am going to give up trying to make things happen on my own, for I obviously don't have the skill it takes to become a success in life. Next week I will head off down to the job centre, and I will be happy to take whatever shitty job they want to throw at me, regardless of how low-paid the job might be".

Because I meant every word that I said, I was hoping that Caryl would simply say "Good idea Shaun -I forgive you for letting us all down", and just leave it at that, but instead she decided to give me the biggest dressing-down of my life:

"How dare you give up on us like that; how could you even think of throwing away everything that you have worked so hard to achieve purely because of an unforeseeable accident that could have happened at any time whatsoever".

I was flabbergasted, to say the very least, with Caryl's attitude, although she was nowhere near finished with me as yet, and so I just sat back and said nothing, in readiness to take everything that she threw at me right on the chin.

"Stop being such a defeatist Shaun; what's done is done, and no power on earth is going to change that. Hayley's hand will heal in time, and our financial problems will also come to and end at some point in the future, that is for sure. If the doctor's told you that they were going to give up on helping Hayley to a full recovery, you would be the first to bitterly complain, and so what gives you the god-given right to give up on your family when we need you most?"

I had no answer to her question, of course, and I even managed to force a little smile, as I replied to her onslaught. "And they say that men are supposed to be the 'stronger sex!' Well, either the person that said that was talking a load of hogwash —or I had better start drinking Vodka!" The worst part of today's traumatic incident was now coming to an close, it seemed —well at least in our minds, it was, for I was no longer in a state of shock, and Caryl's outburst had certainly managed to cleanse my soul of the pain and guilt that I had been harbouring inside of me for so long —a deadly combination that had almost finished me off.

CHAPTER 30

End Of An Era

Even though I was now running very late, I rang Nathan Smith to say that I would still like to meet him this evening, once I had soaked my sweaty body in a steaming-hot bath, and tucked my three lovely children into their beds for the night. Nathan said that he would be staying in this evening anyway, and so we arranged to meet around nine o'clock. Thankfully, I know the Taffs Well area reasonably well, and so it did not take me too long to find Nathan's house, even though it was tucked away in what must be the smallest cul-de-sac in the area. Mind you, I have to admit that the Yamaha 600cc motorcycle, which was parked outside on the road, was a bit of a giveaway. Considering that Nathan stands around six feet, three inches in height, I am surprised that he lives in a house that has unusually low ceilings. The rooms were also quite small, and even though I am only five feet, four inches in my stocking feet, I was feeling a little claustrophobic as I sat on the Cotswold-type sofa in his living room.

In fact, all of the furniture in the room was somewhat countrified, and the place immediately reminded me of Ted's *dome* in California, apart from the fact that instead of having a wooden spiral staircase leading up to the second floor, Nathan's house could only boast a set of solid concrete steps, that I have to say were not too pleasing to the eye. The room was also very dark inside, primarily because there was only one small window to let in the light, and the bulb in the ceiling light was also of the *dim* variety. Wanting to break the ice as early as possible, I joked to Nathan that if the local zoo finds out that he has such an 'animal' parked outside his house,

330

then they are likely to come and kidnap the beast, before locking it safely away behind bars. Nathan agreed with me, saying that the bike is truly "A beast amongst beasts", adding that he is looking forward to crossing Africa on it at the end of October. For the next hour or so Nathan and I compared notes on what we had learned about crossing this vast continent, which included me showing him the two books which Stanford's had kindly donated to my cause.

Nathan reciprocated by showing me the two books which he had read from cover to cover, the first being 'Africa on a shoestring', and the second one was entitled 'The Sahara Handbook'. Nathan then added that if he took the first book with him on the journey, he would have to hide it, because in one of the earlier editions, the author had apparently made various derogative statements about a high ranking official in one of the major African states. (*Nathan did not say which one*). Because of this, the government had not only banned the publication of the book in their country, but they have apparently forbidden anyone from carrying a copy of the book with them whilst travelling through their country.

"Should I be found in possession of a copy at the border, then they can either refuse me entry into that country, or they will simply confiscate the book before letting me pass. However, according to one report that I read, they could also arrest me on the spot, which is apparently what happened to some poor bastard, who had to pay a ridiculous fine, before they would finally let him through" he added.

Every option seemed totally bizarre to me, but then that is what makes travelling through the *Dark Continent* so interesting and appealing to the independent traveller, I suppose. When I told Nathan about the vehicle I intended using to cross Africa, he was not impressed at all saying how small the fuel tank was compared to bikes like his, and also the purpose built 'Africa Twin', which had something like five gallons (twenty litres) fuel capacity. I immediately informed Nathan that I intended replacing the fuel tank with a much larger one, before assuring him that petrol was nowhere near as scarce as it used to be in Africa, well, according to various sources of information, it isn't? "Besides, look at the size of me Nathan", I hollered, before standing to my feet "I would have to wear a pair of high heeled shoes, just to be able to touch the floor on a bike like yours buddy, and so I really don't have much option, do I?" Nathan could see my point, but he was still not convinced that I was using the correct motorcycle in order to cross such a vast expanse of extremely rough terrain.

To add fuel to the fire, I then showed Nathan the route that I intended following, in the footsteps of Ted Simon, which he openly admitted was much too dangerous for his liking. "I'm starting my journey in Algeria, heading southwards through the Sahara Desert, before crossing into Niger and then on to Nigeria. From here I will then take a south-easterly trajectory, across into northern Kenya or Tanzania, whilst possibly dropping-off into a few other countries along the way", Nathan announced, his resounding confidence telling me that he already had his primary route worked out from start to finish, and that any diversions into additional countries would simply be a bonus. Nathan then suggested that we could possibly meet up somewhere along the line, but when I told him that I intended crossing Africa in only ten weeks, he said "Forget it", as he intended spending at least ten months in Africa, before selling his bike in Cape Town, in order to pay for his flight back to the UK.

At this point, I questioned Nathan about the *Carnet de Passage en Douane,* a legal document that supposedly prevents people from selling their vehicles in Africa otherwise they stand to lose their humungous deposit, which apparently has to be two and a half times the value of the vehicle in question? However, Nathan had obviously studied this document a lot closer that I had, because he began rattling-off a selection of clauses that are purportedly written into the document, which made me almost sorry that I had brought up the question in the first place? By ten thirty our interesting, and equally informative chat had come to an end, and as I said my farewells to Nathan I did wonder if we might *accidentally* meet up somewhere along the way? During the week that followed, I stayed as close to my phone as was humanely possible, as I waited with baited breath for my answer from Mr. Finn. However, having heard nothing from him for nine days, my impatience finally took over, and so I decided to ring his office. Without any apologies whatsoever for not calling me, Mr. Finn said that he had carefully considered my proposal, before openly admitting that he fully intended spending a considerable amount of money on advertising in the next few months.

"Unfortunately Shaun, your proposal has been *pipped at the post* by an offer from the main cinema advertising agency, and I have to tell you that I have already accepted their terms and conditions, and subsequently I have also signed a fixed promotional contract with them, confirming the agreement".

I was devastated –and I was also lost for words, as I knew that this would be *'the beginning of the end'* for my project.

"Don't be downhearted Shaun", Mr. Finn added sympathetically "I know that you will obtain a bike from somewhere and that you will do the journey exactly as you have planned it. Your enthusiasm and determination to succeed has already proved to me that you cannot fail in gaining recognition for your efforts, and I am very much looking forward to watching the first episode of your great adventure on TV".

"I wouldn't put any money on that Mr. Finn", I thought to myself, as I politely thanked Mr. Finn for his time, before replacing the receiver. What seemed ironic to me was that eleven months ago I had set out to make a movie for the big screen, and now it was the cinema that had curtailed my last chance of attaining any kind of goal for my efforts. In five days time it will be September, and James Gerwyn will be phoning me in order to make a final appointment, in readiness for my journey across Africa. Over the past few weeks I had plagued him to get me a camera from Sony, so that I could film all the wonders that Africa has to offer, and now I would be the one who is letting the side down, by not even having a bike to do the blessed journey! How I was going to break this news to James, well lord only knows, and so I thought that it might be a good idea to contact Jim Howell's first, to see what he would have to say about my project coming to a very disappointing end?

Jim admitted to me that he had been very sceptical all along about Budget paying for the bike, adding that whenever the BBC required sponsorship, they would always firstly contact the manufacturers of the merchandise required, as the chances of a donation are far greater from the original, and therefore the least expensive source. However, I knew that if I had personally contacted the concessionaires of Kawasaki motorcycles in the UK, they would have either ignored my letter altogether or I would have received the proverbial *"Dear John"* reply within weeks of writing to them. Then a thought suddenly struck me from *out of the blue* and so I asked Jim to put things on hold for a few more days, before saying anything to James. Connor, the neighbour who had taken Hayley to the hospital, had introduced me to his brother-in-law at Connor's daughter's christening a few months ago, and to my great surprise the guy only turned out to be an old friend of mine, (I will call him 'Martin') who I used to go bike riding with many years ago.

At the time Martin was working as a journalist for a popular motorcycle magazine, and we had exchanged phone numbers in the church, and so, after rummaging through my address book, I decided to give him a call. After enlightening Martin about my current situation, I asked him if he would be interested in doing a cover story for his newspaper, with a view to attracting possible sponsors.

"Actually Shaun, I am no longer working as a journalist, as I am now the editor of the TV Wales Magazine in Cardiff, and I can assure you that we would be very interested in doing a feature on your forthcoming motorcycle journey across Africa".

Feeling full of confidence, I asked Martin if he could possibly get me the telephone numbers of the concessionaires for the four major Japanese bike companies (Yamaha, Honda, Suzuki and Kawasaki) in the UK, in order that I could contact them individually, and duly ask each one of them in turn for a loan of one of their motorcycles for the journey. I had long since realized that the chances of any of them *giving* me a bike for keeps was non-existent, but if only one of them would offer to lend me a motorcycle for a period of three months, or possibly even a little longer, then at least I would still be in with a fighting chance of getting this show on the road. Within days Martin was returning my call, having not only obtained all four telephone numbers for me to call, but also the names on the individual people whom I should speak to in each company. Martin also advised me on how to word my sponsorship request, as he knew only too well how difficult it can be in obtaining any kind of *freebies* –especially from major companies such as these.

After making several calls, and leaving a handful of messages with various secretaries and work colleagues, I finally managed to speak to all of the people on my list, pleading my cause as 'professionally' as possible to each one of them in turn, just as Martin had instructed. The next couple of days came and went, without *hearing a dickybird*, but then on the third day I had my one and only positive response, as the Yamaha concessionaires confirmed that as soon as they had received confirmation from the BBC of

334

their involvement in the project, they would seriously consider loaning me a motorcycle for my journey across Africa. Feeling ecstatic about this massive step forward, I then set my sights on getting sponsorship for as many of the other items as possible that I would require for the trip. Unfortunately, this turned out to be the biggest nightmare of my life! If I heard the statement "You must be joking, my friend, don't you know that Britain is in the middle of a deep recession right now" once, I must have heard it a thousand times.

Jim was certainly right about going directly to the manufacturers, as trying to obtain any kind of *freebie* from local store owners was a definite no-no! Unfortunately, I just did not have the time to go searching through endless magazines and various phone books, looking for every individual manufacturer of biking and camping goods, before writing dozens of *begging* letters, and sending them all over the country, in the vain hope that a handful of them would adhere to my humble requests –not a hope in hell's chance. As much as I hated to admit it, I succumbed to the fact that getting everything sorted in time for an October take-off was now completely and utterly out of the question; in fact the way that things were looking right now, I would be very lucky to be anywhere near on my way by the end of November?

Leaving so late in the year means that I would have no choice but to do the unthinkable, and that is to miss out on Hayley's first birthday, along with spending Christmas away from my family, which would simply tear me apart. All of my plans were crumbling in front of my eyes, and there was nothing I could do about it. On 4th September I rang James, in readiness to admit to him how badly I had let him down regarding obtaining sponsorship items for the journey, but before I had chance to speak, James apologized to me, saying that he had still not managed to obtain a camera for me from Sony. After putting James fully in the picture with my current situation, I openly told him that unless the BBC would be able to assist me with substantial funding for my project, then I would have no other choice but to put off doing the journey until further notice. James simply replied that this would be "Impossible", adding that he could not even guarantee how much I would be paid, even if the footage I came back with was good enough to be transposed into a documentary?

In closing our conversation, I asked James to take my file out of the 'Current' tray, and place it in the 'Pending' cupboard, but certainly not to 'bin' my idea altogether, for I still believed that one day I would be 'banging on his door once again, only next time I would be far more prepared -and I would also have cash in hand, in readiness to make one of the greatest documentaries (and even a stupendous movie, perhaps) that the world has ever seen.

Telling all of the other people who had helped me along the way, that the dream was finally over, was not an easy task at all, and many a tear was shed during these conversations, as I admitted defeat to each one of them in turn. The majority of them were very sympathetic, some of them even assuring me that the journey will be done in time, only perhaps not this year. How could I have envisaged then that in only a few months time, my life would take a completely different turn, and that I would be off on my next great adventure?

CHAPTER 31

Going Under The Knife

Today is the 21st September, and I have been out of work for exactly one year. It is not the kind of anniversary that one would celebrate with relish, but one which I will personally never forget. Since I curtailed my project nineteen days ago, quite a lot has happened to both me and my family. Unfortunately, the majority of these events were unpleasant ones, and one or two of them were even quite painful. Below are a few examples:

One day, whilst Caryl was in the kitchen, doing her daily washing-up routine, Carl was duly enjoying one of his usual *climbing expeditions* in the lounge, when he suddenly lost his footing, falling head-first off the window sill, before nose-diving into the floor, thus knocking himself clean out for the third time this year. Bringing Carl out of unconsciousness consists of blowing as hard as possible into his face, in-between calling-out his name several times, in the hope that he will hear a voice in the background of his comatose state, and duly open his eyelids.

When they do finally open, only the whites of his eyes will be showing, and it will take a few more seconds before his pupils will roll down into their correct positions in his eye sockets. However, this is just the beginning, because as soon as Carl regains consciousness, he will then be unable to breathe, and as he begins gasping for every ounce of air, in a state of sheer panic, of course (which is probably the most frightening part of this horrendous ordeal) the blowing in his face must continue, until he is finally able to take a huge breath, before being able to breathe normally once again. Unfortunately, this is when the shock and pain of what has

337

happened to his little body kicks in, and so Carl will invariably burst into floods of tears, sometimes screaming out loud for quite a while, as his young mind tries to fathom-out what in the world has happened to him? This is one of the most traumatic situations for both child and parent to have to go through, and having personally experienced this scenario on two previous occasions with Carl, I know only too well that Caryl must have been *crapping in her pants* at the time!

Hayley was next in line, as she managed to lose her footing whilst learning to walk, before falling head first into the ceramic knob at the base of the living room radiator. Luckily, I managed to grab her just as her pretty little face made contact with the sharp cornered handle, whisking her backwards, and into my arms, before any serious damage could be done...or so I thought? As Hayley turned to face me, she gave me one of her beautiful smiles, as if saying "That was fun daddy", but as her lips moved, I suddenly saw a small part of her left cheek open-up, thus revealing a fleshy wound, as a few droplets of blood began rolling down her rosy red cheek. The razor sharp edge of excess plastic on the handle had sliced my baby's cheek open like a hot knife sliding through butter. Thankfully, Hayley did not seem to be in any pain, as the cut was so clean that it had obviously not registered with her brain, or at least not yet, because like paper cuts, they can take time before they begin hurting...a lot!

Immediately I showed Caryl the slice in Hayley's face, which was about seven or eight millimetres in length, saying that I had better take her down to the hospital, just to be on the safe side. "The nurse will probably put a few paper stitches across her cheek, and that will be the end of it" I said reassuringly to Caryl, as I dressed Hayley, before carrying her out to the car. When we arrived at the A & E (Accident and Emergency) department, I gave the usual mandatory details of how the accident had happened, to the receptionist on duty, before duly taking my seat in the waiting room. Hayley was sitting on my lap, happily playing with the features on my face, which is one of her favourite pastimes, and so her accident was certainly not classed as an emergency right now. As a matter of fact, the hospital was getting so used to my *almost regular* visits, that the receptionist even joked "we know your address Mr. Donovan; just tell me which one of your children it is *this time?*" After about twenty minutes, we were called into one of the cubicles, where the doctor studied Hayley's wound for a few seconds, before announcing to the nurse that it would definitely have to be stitched.

"Paper stitches, eh doctor; I've had a few of them in my time too", I joked, trying to alleviate the stress inside me that I had been hiding quite well, or so I had assumed.

"I'm sorry to say that the area is too volatile for paper stitches Mr. Donovan; every time Hayley eats any food, the stitches would simply bend, before eventually losing their adhesive and falling off. Hayley will require two or three cotton sutures sewn into her cheeks I'm afraid, as the wound is not superficial. However, as she is clearly in no pain right now, I have a few other patients that I need to sort out first, and so would you mind sitting in the waiting room please, until I am ready to do the stitches?"

This was the news I had secretly been dreading, hence I had played down the scenario to Caryl, otherwise she would be beside herself, knowing how much stitching-up wounds can hurt an adult, let-alone an eight month old baby –especially her little baby!' By now Hayley was feeling a little drowsy, as it was approaching her normal bed time, and so rather than playing with the other children, she chose to cwtch (Welsh cuddle) into my shoulder, while I rocked her gently in my arms, as we waited patiently for the doctor to call her name.

The rhythm must have been soothing for my darling daughter, because by the time her name was called, Hayley was now well in the land of nod, and even as I laid her down gently on the bed she continued sleeping to her hearts' content. The doctor openly admitted to me that the stitching was going to hurt Hayley a lot, albeit that it would only be for a very short period of time, as he reckoned that two stitches would be enough to do the job. He then gave me a small blanket, telling me to wrap Hayley up in it, with only her head showing, at the same time making sure that her arms were inside the blanket, before holding her tightly in my arms. This I did, while Hayley continued *'sleeping like a baby'*, as they say. As the doctor gently squeezed Halyey's cheek together, my daughter woke up, and looking deep into my eyes, she gave me that same loving smile that she has always given me, thus melting my heart in two.

At that point the doctor inserted the needle into her cheek, causing Hayley to scream out in agony, as she tried her uttermost to wriggle free, but I held on so tight to her, that she was powerless to move her little body more than a fraction, as the doctor continued-on with the task that he obviously hated perpetrating, as much as I hated watching –and my daughter hated receiving. More screams rang out through the ward, as a few more holes were pierced in my daughter's little face, before her suffering finally came to and end, and as the doctor snipped the chord with his scissors, I could see by the look on his *sweaty* face that he was equally as glad as we were that this horrible necessity was all over. By now Hayley's cheeks were streaming with tears, and as I unravelled her out of the towel, she looked directly into my eyes, as if asking me "Why did you put me through so much pain daddy?" I had no answer for her, of course, and after securing my beloved daughter in her car seat, where she promptly fell asleep again, I cried all the way home.

If the above wasn't enough to traumatise our whole family, then it was now poor old Liam's turn to get hurt, as he was about to *bear the brunt* of the school bully. It was Liam's first day at school, and he looked so smart as we walked together down the road, Liam donning his brand new school uniform, which Caryl and I had pawned our wedding and engagement rings the week before, in order to pay for it. As the morning continued-on as normal, Caryl unexpectedly received a call from the headmistress of the school, asking her if she could please collect Liam from her office, and take him to the hospital, as he now had a rather nasty gash on his forehead, which she believed would need stitching? Caryl rushed to the school, where Liam's teacher explained that Liam had been sitting on one of the benches in the playground, when one of the other boys had pushed him backwards off the seat, and he had inadvertently cracked his head on the floor, which had caused a gaping wound over his eye.

Unfortunately, this had not been done in a playful manner, and so the perpetrator's mother had also been called into the school by the headmistress, and the boy had been duly reprimanded for bullying, something which has never been tolerated by the schools' board of governors. Caryl and I had fought so hard to get Liam into this school, because it had such an excellent reputation, and after numerous letters and various meetings, over a period of several months, we had finally achieved our goal –and now my poor little boy had been hospitalized by the local bully, who apparently took great pleasure in picking on smaller boys than him.

With my beloved children having suffered so much in a month that Caryl and I will never forget for as long as we both shall live, I suppose it is only right that I should have my own story of woe to tell; after all, 'it is my book'. On the 7th September, I went into hospital to undergo the proverbial 'vasectomy', which would put a whole new meaning to the adage *'tying-the-knot'* –well, at least in my *watery* eyes, it did. There was only one other gentleman having the same operation as me on that fateful day (whom I shall call 'Adam'), and so we soon became *bosom-buddies* whilst sitting next to each other in the waiting room, the pair of us adorning our individual paper nightgowns and hats, and looking like a right couple of pussies! However, I take pride in the fact that I had come into the hospital for surgery on my own, whereas my new found friend was duly accompanied by his beloved wife, who spent half of her time affectionately cupping his right hand in her two hands, whilst looking deep into his eyes and trying her uttermost to convince him that the operation was totally straight-forward, and therefore it was also going to be completely painless.

"What an anniversary present", I quipped, before explaining to the couple that tomorrow, the 8th September, would be our 'eighth' wedding anniversary. "What a coincidence", Adam replied, "Because today is the 7th September, and it is also our seventh wedding anniversary". Well, if nothing else, at least this proved that my chance meetings with total strangers even in the most bizarre of circumstances, was still a major factor in my life. After about half an hour, one of the nurse's came into the waiting room and sounding as if she was *full of the joys of spring*, she openly announced: "Right then; which one of you two gentlemen would like to go first?" Before I had a chance to reply, Adam shouted out "Shaun", the index finger on his left hand now pointing directly at my face. As I have never had a fear of hospitals or operations, I was quite happy to adhere to Adam's 'unanimous' decision, before duly following the nurse into the operating room, which was about the size of your average doctors' surgery.

To cut (pardon the pun) a very long-winded story short, after prostrating my carcass on the bed, where a set of portable curtains were then rolled into place, just above my waistline, I laid back on my pillow, and let the doctor do his stuff. After a jab of anaesthetic, the area in question soon became numb, and so all I could feel was a lot of pulling and tugging, as the doctor and I chatted away freely for about twenty-five minutes…and that was it! After covering my private parts with a very light cotton sheet, the nurse then wheeled me through the small corridor, until we reached the double doors leading back into the initial waiting area. As one of the other nurses opened the doors for my bed to be pushed through, I could see my counterpart heading for the exit door, while his wife was holding onto his waist *for grim death*, whilst doing her utmost to try and drag him back to his seat. This was an opportunity that I could not pass up, and even though I couldn't feel an iota of pain, I just had to make Adam believe that I was in absolute agony.

As the nurse pushed me through the opening, I smiled and winked at her, before screaming out loud, whilst at the same time tossing and turning on my portable bed, as if I was writhing in agony…it was hilarious. By now Adam was halfway out the door, as his wife and a couple of the nurses tried their best to pacify him in the small porch-way, but it was only when I called out his name, before bursting into tumultuous laughter, whilst admitting that I was only fooling around, did he finally return to the waiting room. My name was mud, of course, but even the nurses appreciated my frivolity, saying what a coward he had been all along, and that he deserved a *swift kick up the backside* –metaphorically speaking, of course, for being such a baby. To compensate for my trials and tribulations, I was treated to a cup of tea and two slices of toast, whilst waiting for the initial effects of the anaesthetic to wear off, and after about an hour or so I was duly given the all clear to get dressed and go home.

As the nurse handed me a package, which contained a dozen antibiotic tablets, before explaining the hourly dosage to me, I asked her politely if I would be able to drink alcohol whilst taking these tablets, to whit I received a resounding "No –not a chance".

"In that case, I hope you don't mind if I start the course of tablets tomorrow nurse, for sitting in my fridge at home right now, is a four pack of painkillers, that I am sure will suffice me once the anaesthetic has worn off completely this evening". The nurse gave me a simple smile, before promptly giving me my marching orders. As luck would have it, I never did suffer an ounce of pain when the anaesthetic wore off, or for that matter, from that moment forward. However, ten years later, when I went into hospital to have a 'vasectomy reversal', well, that was an entirely different matter –and a truly enlightening story that is told in another book. Last year, Caryl and I went out for a meal to celebrate our anniversary, but this year we stayed in and watched television, for our romantic evening together. It took two more weeks for the boredom and frustration of doing nothing positive with my life, to finally get to me.

Now if one couples those two ingredients together with all of the other emotions I have already described in my book, before throwing a terrifying shock into the mixture, then perhaps those of you, who have also experienced such a traumatic year in your lives, will fully appreciate the admittance of my final story. Carl was sitting quietly in one of the armchairs, happily drinking his bottle of milk, while I was sorting out a video for him to watch on the television. Caryl was washing dishes in the kitchen, and Liam was in the bathroom upstairs. Hayley was peacefully playing with a toy on the living room carpet, and so everything was quite tranquil in the Donovan household...for a change. The child gate at the bottom of the stairs was safely closed. As our staircase is of the *open-plan* variety, I could both see and hear Liam, as he descended the stairs, but thought nothing of it, as I continued rewinding 'Postman Pat' back to the beginning of the tape. As soon as it had finished, I pressed 'Play' on the video recorder, but as I did, I suddenly spotted Hayley's little body standing at the top of the stairs.

Immediately, I ran to the bottom of the staircase, but just as I reached the gate, I saw my daughter trip over the small bar on the top gate, before tumbling head-first down the stairs. For a second I was in shock, as Hayley somersaulted down the steps towards me, her little body being bounced around like a rag doll in a washing machine...I was petrified. However, and lord knows how I did it, but I managed to catch Hayley, just as she was about to go crashing into the steel gate at the bottom of the stairs. A second later, and Hayley was now screaming the place down, as Caryl came rushing into the living room, to see what the hell had happened? Running over to the sofa, I gently placed Hayley on the cushions, before checking over every inch of her body for broken bones, as I quickly explained to Caryl what had happened. At this point Caryl became hysterical, fearing the worst, as her baby girl continued sobbing uncontrollably.

After what seemed like a lifetime, but was probably only a few minutes, Hayley stopped crying, her little body suffering no more than a handful of bumps and bruises, thank goodness. At this point, Liam came over to the sofa, innocently asking his daddy what had happened, and that is when I exploded: "It's entirely your fault Liam; you left the bloody gate open at the bottom of the stairs –and now look what has happened; Hayley could have broken her neck in that fall –don't you understand that son," I screamed in his face, at the same time clenching my fists in frustration and anger, as the sheer thought of what might of happened overwhelmed me completely. Liam just burst into floods of tears, not only because his daddy had never shouted at him like that before, but also because he realized what might have happened to his baby sister? This was *the straw that had finally broken the camel's back* and I knew it. Quickly, I grabbed Liam, hugging him tightly and apologizing profusely for shouting at him so nastily, before assuring him that it was 'NOT' his fault; it was MY fault for not double-checking the gate after he had come downstairs.

Unbeknown to me at the time, but Caryl had been wiping one of the carving knives in a tea-towel when all of this hullabaloo had kicked-off, and now that same knife was only inches away from my throat, as she threatened to use it on me, if ever I shook my fists at any one of our children again. It seems like I wasn't the only one who had finally snapped, albeit that Caryl was totally in the right, of course -and I was most definitely in the wrong. Deep inside, Caryl knew that I would never harm a hair on any of my children's heads, and that I had simply clenched my fists in sheer frustration, not directing them at anyone, except myself. However, as protective as all mother's are to their children, my wife was not going to put up with anyone screaming at her kids…especially their own father. Within ten minutes the Donovan household was peaceful and tranquil once more. Hayley was fine, because like all babies, she had 'bounced', and so it was obviously shock that had caused her to cry out so loudly, and Liam and I were giving each other one of our massive cuddles on the sofa that the pair of us loves so much.

Caryl was happily humming away to the theme tune of Postman Pat, and Carl was busy munching his way through a packet of crisps. Later on that evening, when the children had all gone to bed, I said to Caryl that she might have done us all a big favour, had she used that knife on me, because our only monetary asset was my life insurance, and so I was effectively worth a lot more to her 'dead', than alive! Caryl simply told me to "Shut up", before going to the cupboard in the kitchen and opening-up a sealed packet of cigarettes that she had kept hidden away for over five years, having given up smoking the day that we had decided to plan for parenthood. Thankfully, Caryl only smoked one cigarette, no-doubt to calm her shattered nerves, but the following day, when the vet informed my wife that her beloved cat, whom she had taken in as a stray many years ago, would have to be euthanized, Caryl swiftly lit up another cigarette, thus ending her many years of abstinence from those deadly cancer-sticks. *(As you might have already guessed, I am an 'anti-smoker').*

Having read the above, I guess you can understand what a crazy three weeks it had been in our lives, and to put the final icing on the cake, we only recently discovered that Carl was having learning difficulties in his pre-school classes, and the doctor's had also told us that he will probably require intensive speech therapy, unless he starts speaking clearly within the next six months. Carl was also tested for Autism at one stage, but thankfully this had been ruled out after several sessions with the experts in that particular field. Tomorrow I will begin my second year in the ranks of the unemployed, and on Monday I will be visiting the job-centre, in an attempt to find permanent work for the future, albeit that the salary will have to be extremely high, of course, in order to cover my mortgage, which is currently standing at over £1,000 per calendar month! Even though I have officially been out of work for 365 days, I cannot say that this has been an uneventful year, because like everyone else, my family has *had their ups and downs*, but this will certainly not rank among one of the best years of my life to date.

Caryl and I have long-since come to terms with both our own personal situation, and also the current economic climate, which can only be described as 'Dire' right now, and so we have come to accept the fact that all we can do now is to *ride-out the storm* and get through the struggling years as best as we can. I never answer the telephone unless I have to, because I know that it will be one of the many debt-collection agencies, insisting that I give them a date of when I will be making my next payment, even though they already know, by simply looking at the monthly expenditure sheet that I sent them months ago, that their requests -and even their threatening demands, will *fall on stony ground*, which is always the case when one is absolutely penny-less. I can handle opening all of the *official* letters, even-though they disturb me, and really upset Caryl, but when one of your credit card statements tells you to make a minimum payment of over £1,000 by the end of the month, even-though I sent the card back to them over a year ago, well frankly, even I begin to shudder at the thought of how much debt we are actually in?

In a couple of weeks time I will have my usual 'careers interview', at which point I will be openly asked if I truly believe that I am doing everything in my power to gainfully seek employment, even though they know as well as I do that the chances of a man who is in his thirties, and who has been unemployed for over a year, attaining full-time employment is virtually 'Nil'... or at least that is what the statisticians are saying, and they are supposed to be the people in the know? With regards to my journey, well maybe I will try again for sponsorship next year, after-all 1993 will be exactly twenty years since Ted started his journey around the world, and perhaps things will be looking a lot better in twelve month's time...who knows? In all fairness to Caryl, who has stood by me *through thick and thin*, she simply advised me to go back to my movie idea, but Robert Staines had already advised me against doing this, saying that it could take years before anything materialized, and he knows only too well that time is definitely 'not' on my side.

Robert insists that I must stay strong with my current idea, because having already covered most of the groundwork it would be far easier to reinstate the project at a later date, and much cheaper for the broadcaster's, of course. "I could always put an advertisement in the Financial Times for an amateur camera-man, along with a budding sound-recordist to come with me on my journey –providing that one of them is stinking-rich, and looking to spend lots of money, of course", I joked to Caryl, thus proving that *after all is said and done*, I still had my sense of humour. All joking aside, I knew that I could not continue-on in this world without being a part of the *working fraternity*, but I also knew deep down inside that my hands were well and truly tied. Doing another 'This is your Life' for my mother is out of the question, of course, and even if I had any more *entrepreneurial* ideas –which I don't, then Jules has already told me that he doesn't have any spare cash to fund anything anyway, and so what the heck am I going to do, in order to keep my brain active for the next few months?

Suddenly, it came to me. For the last year I had been keeping a rigid diary of my project, along with the events that had occurred in my life over the past twelve months, or so, should my movie or the documentary series ever reach fruition, at which point I would then be able to put my full story in writing, as a follow-up book? So why don't I transpose those notes in my diary into a book right now, in readiness for that day to come? If nothing else, then at least it will keep my mind occupied until something else appears on the horizon.

CHAPTER 32

A New Beginning

On the second of October I began writing in earnest, spending between five and seven hours every day, plugging away at my typewriter, until I had reached my goal of doing forty hours a week, primarily to convince myself that I was grafting as hard as any other working person. With each full page taking approximately two hours to complete, my goal of twenty pages per week was set. My second year in the ranks of the unemployed was now fully underway, but at least I felt like I was doing something positive with my life, even if it was *unpaid labour*. Whether my completed manuscript would ever be considered as a possible publication was an entirely different matter, of course, and one which I had decided not to build any hopes upon, for fear of further disappointment. Sifting carefully through my current diaries 'The Dream Begins', 'Follow the Dream' and 'Dream to Reality', I estimated that the completed manuscript would be around 200 pages, and so I set myself a target of ten weeks to complete the book.

If nothing else, then my writings would represent a nostalgic reminder of one of the hardest and yet undoubtedly one of the most exciting, years of my life. What the following twelve months would have is store me and my family, I could not even hazard a guess, but with the recession in the UK deepening with every day that passed, to the point where in certain cases it was being compared to the Great Depression of my grandparents day, I was very concerned, to say the least. With my project on hold until next year, and yet another dismal winter now looming on the horizon, I can do nothing more than just sit-it-out, in the hope that that some sign of economic recovery will appear soon, thus encouraging potential sponsors to be more forthcoming with donations towards my cause. Throughout August and September the weather had been simply atrocious, the rain lashing-down mercilessly across the country, thus drowning-out all hopes of having an 'Indian Summer'.

The children had become extremely bored with their incarceration, to the point of being overbearing on times, as their days in captivity continued one after the other. By mid-October all five of us had had more than enough of our *imprisonment*, and as Caryl's frustrations reached an all-time peak, she suddenly blurted-out a statement that would change the course of our lives forever.

"Do you know Shaun, I think that I would take a chance on living abroad, in somewhere like California, where the sun always seems to be shining, and where we could enjoy a lot more freedom in our lives, rather than being cooped-up inside the house all day long!"

Caryl's words had really shocked me, for she had always opposed my ideas of living in a foreign country beforehand, even though she loved feeling the warm sunshine on her body almost as much as I did. However, for her to make such an outlandish statement, I knew that things must have gotten really desperate for her. I could only hope that she meant every word that she had said, for nothing would give me greater pleasure right now than to be living with my wonderful family in a country that boasted seventy-plus degree temperatures all year round. Sitting on the sofa next to Caryl, I decided to challenge her words...not aggressively, but firmly enough for her to know that I meant business.

"If you are deadly serious about moving to a healthier climate Caryl, then I will do everything within my power to make it happen. However, certain sacrifices will have to be made, as I am sure you can appreciate, and you must realise that once I have put the wheels in motion, there will be no going back.

"Well, I wouldn't mind trying it...just for a couple of years, I suppose" Caryl hesitantly replied. Even though Caryl's statement lacked the commitment that I really wanted to hear, it was enough to send my mind spinning into overdrive, as I began conjuring-up ways and means of Caryl and me achieving our latest goal. The last time that we had seriously considered moving *lock, stock and barrel* to another country, was back in the early eighties, and we actually came very close to achieving our objective. Like many of the millions of Brits who visit Spain's sunny shores for the first time, on my initial holiday to Ibiza I had fallen head over heels in love with the whole atmosphere of sun, sea, sand and sangria –not to mention the relatively cheap cost of living compared to the UK, of course. The following year, I went to Magaluf in Majorca for the first time, and after visiting a multitude of bars and restaurants, the majority of which were packed to capacity, from dawn to dusk, I could see how much money was being made by the proprietors of these places, and the entrepreneur in me immediately took over my brain.

Our closest friends, Amber and Jake, had also set their sights on buying a restaurant abroad, and so the four of us decided to pool our resources, before hunting in earnest for an affordable bar / restaurant that we could purchase as a joint venture. Jake was really the backbone behind our project, as he was a qualified chef, and so he was duly appointed as the 'restaurant manager', who would be solely in charge of all the cooking facilities, food preparation and the ordering of our food-stocks. As the most outspoken person in our group, I was duly appointed as the bar manager, whose job it would be to welcome and serve the customers with their initial drinks, along with introducing any stage performers that we might hire at weekends, or during the busy summer months. I would also be in charge of ordering all of the beverages, along with any bar snacks, etc. The ladies would be the respective manageress's of the bar and restaurant, and while Jake and I were *doing our bit*, they would be spending the majority of their time serving the guests with food and drinks at their tables, along with assisting Jake in the kitchen whenever possible.

And so I guess you could say that we had it all worked out, although we fully appreciated the fact that we would probably have to hire additional staff, such as a *pot washer*, a cleaner, and probably an assistant chef to help Jake in the kitchen. Having scrutinised various magazines and newspapers for several weeks, I finally came across an advertisement for a small restaurant that was currently up for sale in Puerto Pollensa, in northern Majorca, and so I duly wrote to the owner, asking him for more details regarding the property. The selling price was £22,000, which was a lot of money back in the early eighties, but this not only included the bar / restaurant area, but also separate living quarters, that included two bedrooms, which were situated upstairs on the second floor. It sounded perfect. Now all we had to figure out was where to get the money from, as the four of us had very little ready cash available between us. As luck would have it, one of the major loan companies were offering 50% mortgages on properties overseas, and after making various enquiries with one of their financial advisers, we were told that as there would be two couples involved in the purchase, we could (unofficially) use one-another's loan as the 50% capital required for each mortgage –how brilliant was that?

At the time, Amber was the area manager for Thomson Holidays, and so she would arrange flight tickets (at a considerably reduced rate) for us to pop over and see what could be the start of a whole new life for the four of us. Everything was going according to plan, and Caryl and I were really excited about our new great adventure together, but then came the crunch! Amber rang me in work to say that Jake had been offered the position of 'Head Chef' in a hotel in South Africa, and that he had already accepted the position, even though Amber had refused to go with him. According to Amber, the job was very well paid, and apart from having his own private accommodation, he would also have the use of a swimming pool and a gymnasium, along with many of the hotels' other excellent facilities. Jake would also avoid all the pressures of trying to make a new business work, such as slogging his guts out from sun-up to sundown, seven days a week, for the first few years, with no guarantees at the end of it all, and so who could really blame him for taking the easy way out.

Reluctantly, we all agreed that Jake had made the right choice, but without him in the equation, we could not hope for this venture to work, and so sadly we kissed goodbye to 'The dream that never was'. In the years that followed, Caryl and I revisited Majorca on several occasions, including spending our honeymoon in the same complex that I had visited with the lads on my very first trip to the island a few years earlier. We also heard that Puerto Pollensa was now a thriving holiday destination for the rich and famous, and that the restaurant we had considered buying was making a small fortune... which was just our luck! However, the flame had now been lit for another dream to begin, only this time it would be even harder to make it work. Apart from having to sell the house, in order to pay off a huge mortgage, I would also have to find a full time occupation in a foreign country, and be legally allowed to work there for twelve months of the year, in order to support my wife and children.

The task was enormous, but the lure of the challenge was even greater –and everyone knows what a sucker I am for challenges. However, the burning question, which was spinning around in my head right now was which country should we head for –or more to the point, which continent? Caryl had point-blank refused to move to a non-English speaking country, primarily because of the children's education, and so her adamant decision had ruled out the majority of the countries in Europe, unless we manage to find something on the Rock of Gibraltar, of course? America seemed the obvious choice, and California the ideal destination, having visited this enormous state beforehand, and fallen in love with their wonderful lifestyle of sand, sea and surf. However, I had always had a hankering to visit the east coast of America, and there were certain factors that would have to be taken into consideration before making the biggest decision of our lives, which are as follows:

California is situated on the west coast of America, which is approximately 6,000 miles from the UK, whereas the east coast of America is only half that distance (around 3,000 miles) from the UK's western shores. Being closer to the UK means that flights across the Atlantic Ocean will be somewhat cheaper than flying all the way over to the Pacific coast, and so we will hopefully be able to afford to visit our families in the UK, and likewise they visit us, more often if we chose the Atlantic side of America. A climate comparative to that of Southern California, as Northern California can be somewhat cooler in the winter months was a must, and so we knew only too well that the best option would be Florida –'The Sunshine State'. Just like California, the weather in the southern half of Florida is a few degrees warmer than the top half, and whereas the eastern coast is surrounded by the relatively cool waters of the Atlantic Ocean, the western shores are lapped by the warm, turquoise seas of the Gulf of Mexico...'Paradise found'.

Both California and Florida have Disneyworld Theme Parks, for the children's ultimate pleasure, and according to my research, the cost, quality and also the standard of living in Florida far outstrips the UK...and by all accounts it is also leagues ahead of its American counterpart. A major advantage of living in Florida would be the abundance of direct flights to and from the UK, including Cardiff Airport, during the summer months, whereas direct flights to California are very few and far between. (*Last year I flew to San Francisco via Houston, in Texas, if you recall*). And so our decision, albeit monumental in the great scheme of things, was not really a difficult one to make. The stage was now set, even though we had not even reached the first rung of the proverbial ladder, as yet, and so I began planning and preparing my first move, which was to sort out all the legalities of living in the good old US of A. A friend had already warned me that the American Embassy in London is continually bombarded with hundreds of people, who are looking for working permits and temporary visas, and so turning up at the embassy without an official appointment would be fruitless.

Remembering the horrendous problems that I encountered whilst trying to sort out a working visa at the Australian Embassy in London a few years ago was enough to put me off for life. And so I decided to ring the US Embassy, only to be greeted by an answer-phone message, giving me a hotline number to ring. This in turn gave me a list of instructions on how to go about applying for immigration forms for entry into the United States of America. I have always despised these type of calls, as confusion soon abounds in my mind, as to which digit to press on my phone, in order to move on to the next *correct* stage of my enquiry. Call me *old-fashioned* if you like, but I still say "Give me human communication to technological wizardry any day of the week." Battling with bureaucracy has also never been one of my strong points, and this whole blessed issue seemed to be so long-winded, that the fear of failure began rearing its ugly head, which was something that I must avoid at all cost.

Even though last years' dream was by no-means over, this immense burden of guilt for not only letting my family down, but also for wasting all of Jules Gilmore's investment in my project, along with probably diminishing his trust and belief in me, was still hanging firmly over my head, whether I liked it or not. Thoughts of the above continually plagued my mind, but it also made me more determined to succeed with this latest venture, for deep inside I knew that a lot more would be at stake than just friendships, should I fail with my efforts this time around. I began racking my brain, trying my uttermost to think of someone who could assist me with my plight? If I was considering emigrating to California, then Daniel or Ted would probably be the ideal candidates to assist me with my efforts, but "Who the hell on earth knows anything about that wonderful state that hangs off the south-eastern edge of the North American continent", I kept asking myself?

Suddenly a name came to mind, and I could only wonder in amazement at why I had not thought of him beforehand? "Of course –I know the perfect man for the job... 'Leo Cummings'", I exclaimed, my heart racing with excitement. Lee had been born and raised in Florida, I believe, and I initially met him at the Hilton Hotel around eighteen months ago, when I first started working in the car sales industry. The company that he worked for had sent him over to the UK to lecture and promote Ford Motor Company's latest way of financing brand new cars, and I was one of around thirty sales executives from all over the UK, who had been selected to attend his opening seminar in Portsmouth. At the end of our first day in the classroom, Leo and I had chatted for a while in the lounge area of the hotel, and like all true American's, Leo couldn't wait to tell me how wonderful his homeland was –especially Palm Beach, where he had been living for the past few years. Naturally I was green with envy at Leo's lifestyle, and he knew it, and so it wasn't long before he was offering me a room for the night if ever I was in the area, no-doubt assuming that I would never take him up on his generous offer, of course.

However, I was just about to do just that...that is if I can ever remember where the heck I had put his business card? I knew that it had to be in the house somewhere, and so I began searching in earnest for that small piece of card that could possibly change my life forever. Diligently, I searched every pocket in every item of clothing that I owned, from three piece suits, outdoor jackets and full-length overcoats, to working shirts, Polo shirts, tracksuits, jeans and shorts –in fact anything that had a pocket was ransacked... but to no avail. Having now eliminated every item of clothing from the fitted wardrobes in our bedroom, I began scrutinizing every piece of furniture in the rest of the house. Chests of drawers were turned-out, kitchen cupboards were emptied, and my filing cabinet was literally stripped bare of every document it contained, in my quest for Leo's all-important contact number –but again I drew a blank. Hope was fading fast, as I collapsed into my armchair, my mind and body feeling totally exasperated and exhausted with my completely wasted efforts.

My briefcase was next in line, as it not only encompassed most of the paperwork relating to my project, but it also contained the majority of my memorabilia from my trip to California last year. Every envelope was opened in turn, each document was flicked-through at a rate of knots, and a stack of advertising leaflets were shaken individually, just in case that blessed card was lurking inside any one of their many folds, but there was no sign of it. And then a thought came to me. Caryl and I have always kept our most important documents, such as our marriage and birth certificates, our passports, driving licences, etc. on the top shelf of our lockable wall unit in the living room. In the past I had often put any business cards that I had received inside the plastic wallet that holds my driving licence, simply for fear of losing them if I put them in one of my pockets. Taking the key off the main key rack, I opened-up the cabinet door, before extracting my licence from the unit.

After a few seconds of fumbling-around inside the wallet, I pulled out a selection of business cards, and sitting right in the middle of them was Leo Cumming's personal card. I was ecstatic. The card had been well preserved inside the wallet, and so Leo's name and telephone number stood out boldly on its face, as if they had both been waiting to be resurrected from the depths of a darkened wallet, to the light of a new born adventure. Within seconds I was dialling the international code of '001', before carefully pressing each number in turn, making absolutely sure that I never missed out or misread a single digit, for fear of getting a totally wrong number. The few seconds delay in connecting my call seemed like an hour, as I sat there praying that Leo had not moved from his last address, or worse still, that his phone number had been changed, or his telephone line had been cut-off for one reason or another. Any one of these scary scenarios could terminate our dream once and for all.

After a few seconds I could hear the single dial tone, which is synonymous with American calls, unlike the dual ring tones that we have here in the UK. Within seconds an answer-phone message was apologising for the intended recipient of the call being unavailable at present, before asking me to leave my name and number –and assuring me that the person in question would call me upon his return. Even though no name was mentioned, a voice inside my head was saying that I had successfully made contact with Mr. Leo Cummings himself. Apart from Leo's business card,

I also found another telephone number, along with the name 'Peter', both of which had been hand written on the back of a piece of paper. I thought deeply for a few seconds, before realising that 'Peter' was the chap from Liverpool, whom I had met on the plane going to Houston last year. Peter had told me then that he only intended staying in the United States for six months, and so he should be back in the UK by now.

However, when I rung Peter's home number, simply to ask him how much he had enjoyed spending the winter months in Nevada and Colorado, his father, who had answered my call, gave me a reply that I certainly wasn't expecting:

"Well if he ever comes home, then he might be able to tell you, my friend, but my wife and I haven't set eyes on him since he left the UK about a year ago. The last we heard was that he was now living in San Diego, in Southern California —well, according to the postcard we received from him a few months ago, anyway".

He then added that in all fairness, Peter had rung home numerous times over the last year, and that he had recently given them a telephone number where he could be contacted, which he was happy to pass on to me.

"If you do manage to get hold of him, please give me a call back and let me know how he's getting on over there?" was his final statement, as we said our farewell's to each other, before replacing my receiver.

A part of me wondered why this man didn't simply call his son in America, rather than ask a virtual stranger to pass on a message to him, but then perhaps money was just as tight for them as it was for me right now?

Having two sons of my own, who would wholly rely on their fathers' protection for many years to come, I also accepted the fact that one day they would be men, and therefore they would no longer require me to watch over their every move. At that point, would I be happy to leave them to their own devices entirely, or would I still want to interact with them in many facets of their lives...or more importantly, would they want me to intervene in their personal activities, or would they just tell me to "Mind my own business". However, having a lot more important things on my mind at this present moment, I decided to concern myself with these issues at a much later time in my life. Peter was not at home when I first called him, but a Yugoslavian (*this was long before the final break-up of the country in 2001, although the 'Yugoslav Wars' had already started in 1991*) student, who answered the telephone, told me that Peter had actually moved out of the flat only this week, but that he would be calling back in the next day or so, in order to collect the remainder of his clothes that had been left in the wash, along with his motorcycle, which was in another part of the building.

She also said that Peter had moved to an apartment not far from this one, and so if I would care to give her my telephone number, then she would be happy to pass it on to Peter. How Peter had managed to stay in America for so long was a mystery to me, and so I really hoped that he would call me back soon, and put me out of my misery. Now it was time once again to collect my beloved son, Liam, from his school –a task that I absolutely relished, as seeing his lovely smiling face light up as soon as he saw me, before chatting with him about what he did in school today, as we walked happily along the road together, meant the absolute world to me. As I reached the top of our drive, I could see Caryl driving towards the house, and so I simply gave her a wave, as I was already running a little late, and so there was no time to stop and tell her all about my phone calls, or the subsequent messages I had left for both Peter and Leo, otherwise my life would not be worth living if my *wee man* was waiting all alone for me outside the school entrance when I arrived.

As luck would have it, the teachers were running late today, and so the tables had been turned, and now it was me who was doing the pacing up and down outside the school gates. Within minutes Liam came running through the double doors, before bounding into my arms, and so after spinning him around, and duly plonking him on my shoulders, the pair of us then headed off for the local newsagents, where chocolate bars and ice creams were the order of the day. It seemed uncanny that even though I felt as if all the pressures of the world were upon my shoulders right now, in the presence of my son I could so easily slip them to the back of my mind. For years I had lived with this vain / imaginary self belief that for some inexplicable reason I might be one of those 'special' people, who has been sent to this earth to do something that no-one else has ever done before, or perhaps to be someone like no-one else has ever been before? It would not be anything earth-shattering, just something that I had managed to achieve, which had never been accomplished before. Even when everything in my life went horribly wrong, I still held on to this firm belief that one day I would come out on top, albeit that the current statistics, both financially and psychologically, were certainly telling me otherwise.

Having survived over a dozen crashes on motorcycles, I even joked to myself about being immortal –especially having lost more lives than a cat could ever dream of having in one lifetime, and yet here I was, still here to tell the tale. I also toyed with the fact that I had been put on this earth purely to father my children, and that one day, one of them, or maybe even all of them, might end up becoming famous, or better still, 'legends in their own lifetime?' I could only compare these mindless imaginings to those of the proud mother of her newborn child, who is dreaming of her off-spring becoming a famous doctor, or a wealthy lawyer perhaps, even though the umbilical cord has yet to be cut by the midwife. In reality I know that I am just your average run-of-the-mill guy, who wants the best that life has to offer for both me and my family, but like the story of the hero and the coward, both of whom are afraid of what lies ahead of them, I know that it is what the hero does that makes him a 'Hero'. I humbly believe that I have the audaciousness to be a winner, and the tenacity to refrain myself from ever becoming a loser...I only hope I am right.

What drives me so passionately about becoming a success, is thinking about how my parents had scrimped and scraped all their lives, in order to make ends meet, without any just reward for their valiant efforts, and I guess I have always harboured a grudge against the trials and tribulations that the working class of this world have to suffer all of their lives, in the hope that one day they will make it to a decent retirement age. As for the people who live in poverty, and those who are homeless, living on the streets without a hope in hell's chance in this world, while others are languishing in the lap of luxury, well don't even get me started on that one, or I will never get this book finished. I have always been determined that I would never fall into the same trap as my parents. Surviving with little food on the table or leaving gas fires turned off in the freezing cold weather, in order to save on the fuel bills, is something that I never wanted my kids to suffer, and the mere thought of Caryl having to walk miles from one supermarket to the next, in order to save a few pennies on the weekly shopping bill, like my mother had to do all her life...well heaven forbid!

I have always said that I came into this life with nothing, and that I would probably exit it the same way, but then life does have a strange way of showing you how wrong you can be, and more importantly perhaps, it also makes you realise how lucky you are to have what you've got. Being born to a set of wonderful parents that loved me unconditionally, and having two big brothers who both cared for me dearly, is more than millions of other babies have to begin their lives with, and I sincerely hope that by the time that I finally have to leave this planet, I will have made hundreds of friends, all of whom will like me in their own special way, as much as I will like them. I believe that one of the most questionable things in our lives is 'maturity'-where does it begin, or more importantly 'Where does it end?' We all know that a five year old will do things that he or she wouldn't dream of doing when they are say, fifteen, and that a fifteen year old will do things that would simply make him or her shudder, if he or she was asked to do them again when they are twenty-five.

By the time we reach the *ripe old age* of twenty, the majority of us consider ourselves to be somewhat mature in life, having finally said our fond farewell's to those wild teenage years. Many of us will also believe that we are relatively wise to the ways of the world, and therefore more than capable of making the right decisions with all of the important issues that are about to be thrown upon us. Some of us even have *delusions of grandeur* about what we want and expect out of life, having already set out a number of goals to achieve, along with a detailed timescale of where and when they will reach fruition. Unfortunately, life will come along in the meantime and reshuffle everything completely around, in some cases turning our whole world upside down. As for my own personal *maturity level*, well at the good old age of thirty, I suddenly found myself completely disagreeing with the things that I had said and done when I was twenty –and now that I am approaching thirty-five, (*I wrote this back in 1994*) I am starting to challenge the beliefs that I had, and the decisions that I made when I was twenty-five?

To support my theory that I was just playing 'The game of life', and not *going completely off my trolley,* I started questioning several of my friends and relatives, who were now in their forties, fifties and sixties, and to my pleasant surprise, all of them confirmed that what I had experienced over the years was certainly not unusual, and that I was simply experiencing the *'Circle of Life'.*

"If you think you are going crazy now Shaun, then wait until you get to my age", said one of my uncles, who was in his seventies at the time, and had probably seen and experienced more changes and upheavals in his life than one could shake a stick at! If I am lucky enough to reach his tender age, then I only hope that my life will improve considerably over the next forty years, and that my mind will be able to function on a normal wavelength throughout that time, otherwise I could end up spending the rest of my days being tied up in a straight jacket, before being locked away in some isolated ward, with thick padding on the walls and steel bars on the window.

I had my palm read several years ago, and I remember the woman telling me that I had a very long lifeline, but that my life would be marred with several *unsavoury interventions*, and so apart from the numerous traumatic events that have already occurred in my life, perhaps I had better beware of what the future holds in store for me? If this quest for a new life, in a new land, fails miserably, then I cannot hazard a guess as to what will be my next challenge in life, for without challenges, I am nothing.

CHAPTER 33

Building Bridges

As Liam and I entered the house, Caryl told me that she had just received the weirdest of phone calls? She said that a man with an American accent had rung the house asking for someone called 'Darren', but when she informed him that no-one of that name lived here, and therefore he must have dialled the wrong number, he duly read out the number which had been left on his answer-machine, adding that the caller had asked him to contact him urgently? "I didn't know what to say, and so I just insisted that the caller must have given him the wrong number, before putting the phone down on him", Caryl added, shrugging her shoulders, before turning away and walking into the kitchen. Immediately, I realised that Leo had mixed me up with my colleague, Darren, whom I had mentioned on the tape recording, and so after explaining the situation to Caryl, albeit after the fact, I duly rang Leo again, only to be confronted by that very same answer-phone message, and so this time I quickly explained about the misunderstanding with my wife, before apologising for the mix-up, and asking him to please call me again

The following day Leo did call the house again, and for the second time he spoke to Caryl, only this time there was no confusion. Leo informed Caryl that he would be flying into the UK at the end of the following week, and that after landing at Manchester Airport on the Friday evening, he would be making his way to a hotel in Stoke-on-Trent, where he would be conducting his latest sales seminar, which would begin on the Monday morning. After giving her the name and the telephone number of the hotel he would be staying at, Leo told Caryl that if I wanted to have a chat with him, then I could contact him any time on the Saturday, or the Sunday. Leo's timing could not have been better, and apart from being able to ring him on a UK number, which would be far cheaper that calling America, it meant that I would now have the opportunity of speaking to him face to face. That evening my efforts were even further rewarded when I received yet another long distance call, only this time from Southern California. It was my good friend Peter, returning my call –apparently within minutes of receiving my message from his Yugoslavian friend.

Peter spent the next twenty minutes telling me all about how he had worked his way southwards, spanning the length of America's Pacific Coast, by doing all sorts of carpentry and joinery projects, including re-varnishing a number of boats along the way. Peter then went on to tell me all about his latest flat, before adding that he would only be staying there until just after Christmas, because him and a group of his American friends would then be heading off to South America for around six months or so, on their latest 'great adventure' together. Peter's enthusiasm inspired me greatly, as the memories of my wonderful days in San Diego, where Peter was now living, came flooding back to me. It seems that my impatience knows no bounds, and so instead of ringing Leo over the weekend, as I was instructed, I decided to call him at his hotel on the Friday evening.

Unfortunately, Leo's flight had been diverted to Gatwick Airport, in London, and so he had no choice but to drive all the way to Stoke-on-Trent, having literally arrived at the hotel only minutes before I called him. "I was just about to catch up on a lot of lost sleep when you rang Shaun" were Leo's first words to me, letting me know that my call was not a welcome one, and so, after extending my profuse apologies to him, I said that I would call him again tomorrow afternoon. However, Leo insisted that I continue, as he was obviously intrigued as to why I had contacted him after all this time.

"I must talk with you as soon as possible Leo", I said in my attempt at creating urgency for him to see me.

"Well you *are* talking to me", Leo answered, sounding a little perplexed with the forcefulness of my statement.

"No, I don't mean over the telephone Leo; I need to come and see you at your hotel, sometime next week, my friend...if that is okay with you, of course", I blurted out, as if it was a matter of life and death.

However, Leo was too tired to play guessing games, and so he insisted that I tell him right there and then why I wanted to make a 300 mile round journey just to talk to him? I told Leo how I desperately needed his help at seeking employment in the US, although I thought it best not to enlighten him right now on the fact that I would also need him to put me up at his home in Florida for a few days in the forthcoming weeks. Leo accepted my plea, inviting me to pop over to his hotel one evening during the week, before adding that he would be flying to Germany the following weekend, in order to spend a few days at the 'Octoberfest' beer festival. A close friend of Caryl's kindly offered to give me a lift to Birmingham on the following Wednesday, adding that she could bring me back to Cardiff the next day, on the proviso that I could make my way to New Street Station by seven o'clock in the morning, as she had to be in Cheltenham for a business meeting before lunch.

As much as I really appreciated her generous offer, the thought of having to make my way to Stoke-on-Trent from New Street Station, before returning to the station again in the early hours of the morning was enough to put me off the idea. Couple this with the fact of having to pay for the cost of a nights' accommodation in a bed and breakfast establishment, which is something that I could ill-afford, and I knew that I had no choice but to decline her offer –even though it was a kind gesture on her part. My only other alternative was to drive to Stoke-on-Trent in our family car, and then, after chatting with Leo for an hour or two in the evening, sleep the night in the car, before driving home early the following morning. The round trip would still cost me around twenty-five pounds for fuel, which again, I could not really afford right now, but there was simply no other way around it. If Leo did offer to put me up at his home in Florida, then everything would be worthwhile, as I would then only need to cover the cost of a return flight to Florida, along with enough spending money to last me a week. There was no going back now –I had to do it.

The big day soon arrived, and apart from having to have a tyre changed, which ended-up delaying my departure time by an hour and a half, both the car and I were in ship-shape condition, and raring to go. 'Would this be the start of another glamorous adventure in my life, consequently leading to a prosperous and happy new lifestyle for me and my family', I asked myself –or was this just going to be another great pipe-dream of mine, waiting for the bubble to burst at any given moment, thus putting one more nail into my coffin of defeat? "Well you are never going to find out the answer to that question, unless you try", I whispered to myself, as a kind of kick up the ass, before taking off down the road in the direction of the M4. Thankfully, the road ahead was clear, and the ubiquitous road-works seemed conspicuous by their absence, as I headed up the A449, my mind awash with what I would say to Leo as and when we finally get together again? By seven-thirty I had passed the turn-off for Birmingham, and now I was heading for the junction where the M5 meets the M6.

The light had long-since faded, which is always a bind to me when I am driving, as my eyesight in the dark is certainly not the best, especially when cars that are travelling in the opposite direction have their lamps on full beam, thus dazzling me into oblivion, as they continue-on whizzing down the motorway, often at an ungodly pace. Apart from the normal fatigue, which almost every driver encounters on long distance journeys, I also had to combat a stream of adrenaline rushes that were continually flowing through my body at a rapid pace. This was due to my nervous system, which was having a field day right now, along with the sheer excitement of knowing that I would soon be at my destination. Having turned off the motorway, I am now following all of the signs for Stoke-on-Trent, my heart beating faster and faster as I begin counting down the miles to the city centre. As I approach the main road leading into the centre of town, I quickly check in my mirror, to make sure that there are no cars behind me, before swiftly pulling over to the side of the road, in order to ask a pedestrian that I have just spotted, for directions.

Unfortunately, like my good self, the gentleman is a visitor to the city, and so he is unable to help me at all...unlike the next person I accost, who is only too-pleased to inform me that the hotel is "Within spitting distance" (his words, not mine). By eight-thirty I am standing in the main reception area, chatting freely with the very helpful receptionist, who has kindly called Leo in his room, in order to announce my arrival. Within minutes my host is descending the huge open-plan staircase, and I recognise him immediately. Dressed in a plush dinner suit, with a beautifully pleated white shirt, and a black dickey-bow to match, Leo looks more like a celebrity out of the BAFTA awards, who is just about to receive his Oscar for 'Best Actor' in a major movie production. In comparison, I felt like the *wreck of the Hesperus*, my attire primarily being made up of a pair of black corduroy trousers and a denim shirt. With his arms outstretched, Leo gives me a huge welcoming hug, as if we had been best buddies for years, before gently ushering me off to one corner of the bar, where we began chatting in earnest.

I questioned Leo vigorously about Florida, literally bombarding him with questions about anything and everything, including the people who lived there, the difference in climates from the north to the south, the general working conditions, housing, schooling, current crime-rates –and even the aftermath of hurricanes. In fact, you name it, and I will bet that we covered it. Having exhausted Leo of his entire knowledge of the state of Florida, I then told him all about my great adventure in California, and my subsequent quest for success, which he genuinely found interesting, saying that I should write a book about it all. Leo then told me about the company that he worked for, adding that they were currently looking to recruit British lecturers, who would travel around the UK promoting finance packages to car dealerships, just like Leo was doing right now. Because I had worked in the car industry, Leo humbly suggested that I apply for one of the positions on offer, as the money was good, and it also involved a lot of travelling, which he could see I enjoyed immensely.

As good as it all sounded, I had now set my heart on living in the sunshine, and once I have set my mind on doing something, there is no going back. Far from wanting to shatter my illusions on living in 'The Promised Land', Leo warned me that the worldwide recession had also hit America –with devastating effect in certain parts of the country. "Mind you, with all due respect to the UK, even living in a recession in Florida, I would say that our standard of living is still invariably higher than living in Britain in a normal economic climate", he added. By this time I was feeling as if I had known Leo as a close friend for years, rather than just a business acquaintance that I had spent five days in a classroom with eighteen months ago, and so now seemed the ideal opportunity to ask him my two most important questions.

"If I came over to Florida Leo, would it be possible to stay with you for just a few days, while I searched for full time employment? Also, if you knew of anybody who may consider employing me, would you be kind enough to introduce me to them?"

I knew that this was probably a tall order for Leo, but right now he was my best hope of finding a job in Florida, and he knew that as well as I did. With a gentle nod of his head, Leo said that he would assist me in any way that he could, although he also warned me that he invariably worked away during the week, and so the only time that I would be able to catch him at home was on the weekends. An hour had now passed and I gathered that Leo must be looking forward to eating his evening meal in peace, as he humbly suggested that I begin my long drive back to Cardiff. However, when I informed him that I intended spending the night in my car, he almost choked on his drink, jokingly saying that anyone who considered sleeping in a car during autumn time in the UK must be nuts! Having already questioned my sanity a thousand times over during this past year or so, I was inclined to agree with Leo. Nonetheless, within minutes I was reclining the driver's seat of my car, as far as it would go, before snuggling-up in my continental quilt. Feeling very satisfied with my evening's work, I duly dropped-off into the land of nod.

Surprisingly, I slept rather well, apart from the occasional turnover during the early hours of the morning, which was unavoidable, I suppose. At seven o'clock, the sun rose over the horizon, and so after licking my right index finger, and gently removing the last remaining crystals from the corners of my sleepy eyes, I duly piled my blankets and pillows onto the back seat of the car, before returning the drivers' seat to its original position, in readiness for the off. My first port of call would be the nearest petrol station, where, after filling up the fuel tank, I would ring Caryl, in order to let her know that all had gone well last night, and that I was now on my way home. Once again my timing was impeccable, and my foolishness supreme, as I began following the signs for Birmingham and the M5. It is now eight o'clock, and I am approaching one of the largest and most congested cities in Britain...at the height of this mornings' rush hour. Every inch, in every lane has now been taken up by cars, trucks, busses, vans and motorcycles, as they gently edge past one-another, in a vain attempt to finally get some breathing space on the road.

I had to do likewise, of course, but the whole situation seemed hopeless. I could see the directional signs above my head, telling me which lane I should be in, in order to take the ring road around the city, but the chances of me getting into that lane before taking the relevant exit, seemed futile right now. I simply had no control over my destination, and so like a lamb to the slaughter, I just continued-on, following the rear end of the car in front of me. Looking in my rear-view mirror, I can see the frustration on the face of the driver who is sitting directly behind me, who is undoubtedly doing likewise with the ass-end of my car. For the next half an hour our never-ending convoy of vehicles continued chugging along at a snails' pace, the tailback of traffic becoming greater with every minute that passed. Suddenly there is a break in the traffic, and so I swiftly change lanes, whilst trying my uttermost not to *cut anyone up* in the process. Up ahead is a multi-exit roundabout, and the sign at the side of the road says that one of those exits will put me on the road to 'South Wales'.

"Free at last", I shouted, as the broken white lines on the Tarmac in front of me became visible again, and the number-plate in front of me that my eyes had been focused on for the last thirty minutes or so, finally disappeared from my life forever. Being able to use third and fourth gear again seemed like an enormous privilege, and as soon as I hit that fifth gear, I gave a huge sigh of relief, safe in the knowledge that I could finally give my left leg the respite it deserved from all that declutching. My right foot was also as *happy as a sand-boy* now that it was pressing firmly down on the gas pedal, as opposed to the monotonous task of chopping and changing with the brake pedal every few seconds. By ten-thirty I am once again sitting in the comfort of my armchair in the living room, my children now swarming all over me, all three of them making up for missing their usual cuddles and bedtime stories from *daddy* the night before. The first stage of this major operation was now complete, and so without further ado I must move swiftly onto the next stage, which is booking my flight to Florida.

Thankfully, our great friend, Amber, was still the area manager for Thompson Holidays, and so I gave her a quick call, asking her to make all the arrangements on my behalf, as I knew that she wouldn't let me down. I was also fully aware that I would have to fly from either Manchester or Gatwick Airport, as there were no direct flights going to Florida from Cardiff Airport during the winter months. However, thanks to Amber's due diligence, which included *pulling a few strings*, of course, I managed to get a return flight for only £99... happy days. According to the paperwork, I would now be travelling as a 'company representative', and as such I could only officially confirm my flight seat four days prior to the departure date. Amber had already warned me that if all of the seats on the aircraft have been booked in advance, then I would have no choice but to reschedule my flight to a later date. However, because the cost of the flight was around one third of the official price, I was more than willing to take that chance.

Amber also said that if I did miss this 'business connection', then a later flight could cost me around twice the price. Paying two hundred pounds for a flight was simply out of the question, unless I was willing to sleep on park benches and beaches for the duration of my stay in Florida! A holding reservation was made for Tuesday 27th October, but as Amber would be out of the country on that date, she duly left everything in the hands of her colleague, Maria, who would keep me up to speed with the situation. As there were currently forty-odd empty seats on the plane, with less than seven days to go to 'blast-off', both Amber and Maria were quite confident that there would be no problem in me catching the flight in question. I wanted to share their optimism, of course, but with so much resting on this one chance, I was sceptical to say the least. Actually, apart from a possible long weekend with Leo, accommodation is something that I had not taken seriously into account, even though it would be a major aspect of the trip.

Getting to America was all that seemed important right now, and so I would worry about such *trivial* matters once I knew that my flight had been booked and paid for. However, fate sometimes has a funny way of creeping up on you when you least expect it to, and that is exactly what happened next. The following morning, whilst I was reading my daily newspaper, I happened to stumble upon an advertisement that could possibly solve a large chunk of my accommodation concerns in one fell swoop. A reputable estate agency in the UK was now expanding their portfolio by promoting prestigious holiday homes in various parts of Florida, and their current package was offering two nights free accommodation to anyone who was willing to undertake a two hour sales presentation at one of their many housing developments that were dotted around the sunshine state. Flights were not included in the package, but as I already had that part of the equation in hand, I thought that I would give them a call, and find out what the exact criteria was?

Having already done battle with various timeshare salesmen in Europe over the years, without succumbing to their clever tactics, or their amazing financial strategies, that would put Wall Street to shame, I figured that I could easily stand toe-to-toe with their American counterparts. The gentleman I spoke to was the marketing manager for the company, who introduced himself to me as "Mr. Norman", and so I gathered that he had no intentions of us being on first name terms for our opening conversation. Mr. Norman was very enthusiastic about the various homes that the agency had for sale, which included a selection of 'Bovis Homes', a name synonymous with top quality properties in the UK. He also took great pride in telling me that their company had several housing developments, spanning from one end of the sunshine state to the other. He then went on to say that their three major sites were in Orlando, right in the centre of Florida, Newport Richey, which is situated on the north-western coast, overlooking the Gulf of Mexico, and Fort Myers, which is located deep down in the southern half of Florida.

This was just what I wanted to hear, and so now that he had exhausted his spiel, I was about to go into overdrive with mine. "According to your advertisement in the newspaper you are currently offering two nights free accommodation on the proviso that your guest undertakes a two hour presentation at any one of your developments, is that correct?" I asked him quite forcibly, knowing full well that he was going to give me the affirmative answer. However, before he had a chance to begin explaining the terms and conditions of the offer, which might possibly hinder my sales pitch, I followed-through with the rest of my spiel. "Excellent, well in that case I would like to visit all three of the developments that you mentioned, staying for two nights in Orlando, two nights in Newport Richie, and two nights in Fort Myers, so that I can decide exactly where I would like to purchase a property from you...I daresay that would be fine with you, my friend?"

Mr. Norman hesitated before answering my question "Um –yes, I suppose that would be okay?" he said in a somewhat reluctant manner, knowing full-well that I had purposely put his back up against the wall in a way that he couldn't refuse. Mr. Norman then said that he would send me a selection of brochures relating to all three of the aforementioned sites, before adding that I would also receive a package from the US within the next couple of days, giving me a general insight into the company, along with introducing me to some of the people who worked at the various sites in Florida. Rounding off our conversation, Mr. Norman asked me to call him as soon as my flight details were confirmed, in order that he could arrange for one of his colleagues in the US to meet me in the *Arrivals* area in Orlando Airport. Having crossed yet another bridge, I decided to ring Maria, to see what the position was regarding the flight seats…it is now 5 pm on Thursday, 22nd October.

Maria informed me that there were now only 'six' seats available on the flight that Amber had booked me a seat on only a few days ago, which was not good news at all. However, Maria then assured me that as it was now approaching closing time, the same amount of seats would be showing on her screen in the morning. It is now 8.45 am the following morning and I am standing outside the travel agents, waiting for it to open at nine. Within minutes of the door being unlocked, I am sitting alongside Maria, the pair of us now staring at her computer, and waiting patiently for the flight details to appear on the screen. "There they are Shaun; six seats still available, just as I had told you they would be…I gather that you have the money with you, in order to pay for the flight Mr. Donovan. I couldn't get the cash out of my pocket fast enough, and within seconds there were five crispy twenty pound notes resting neatly on her table, set out like a hand in a card game, so that there would be no mistake.

By now Maria was already on the phone to the main booking office, giving all of my details to her colleague on the other end of the line. All seemed to be going well, but then Maria suddenly stopped in her tracks… "Oh, I see, in that case I will have to come back to you" she said in a negative manner, before replacing the handset. "What's up?" I asked, my voice trembling in anticipation. "Well, according to my colleague at the central booking office, because there is a weekend involved in your four day booking window, coupled with the fact that this is a bank holiday weekend, it means that we will have quite a good chance of selling all six of those seats on Saturday, and so she is unable to confirm your seat until the close of play tomorrow evening.. I'm really sorry". I was totally gutted, but I also accepted the fact that this was one of those odd occasions in life where there is absolutely nothing that anyone can do about it, and so I just left the money with Maria, on the understanding that she would book my seat on the flight, the very second that she was able to.

In all fairness to Maria, she did offer to book me a seat right now, at the reduced price of £199, which was still a £100 saving from the original price of £299, so long as I was willing to pay an additional £40 for flight insurance, something that was included in my 'company' flight, and so as much as it hurt me deeply, I had no choice but to politely decline her offer. On Saturday morning I took my children to the park, and while they played happily on the swings, slides and see-saws, I sat silently on one of the many park benches, my eyes transfixed on the black clouds that now filled the skies above my head, while my body shivered in the cold wind that was now whistling through my bones. All I could think about was how warm it was around this time last year, when I was swimming and sunbathing in the eighty degree Californian sunshine –and how warm I would be next week, should everything go ahead as planned with my flight to Florida. Knowing that Maria would be keeping an eye on the flight situation on my behalf, I decided to pop over to Caryl's parent's house, in order to write a few more pages in my *'Diary on the Dole'*, as I so eloquently called my manuscript at the time.

However, as soon as I arrived at their house, Donald, Caryl's dad, duly informed me that Caryl had rung him to say that Maria had rung the house regarding my flight, saying that I must contact them urgently. I rang the office immediately, and Maria informed me that there were now only four seats left on the flight, and so it was now 'Make your mind up time'. I told Maria that I would ring her back within the hour, before sitting down to work out my finances…for the umpteenth time this week. Caryl and I had already sold our beautiful garden furniture for £300, even though we had paid over double that amount for the elegant table and eight reclining chairs (complete with very thick cushions) just over a year ago, and my brother, Gary had kindly loaned me £100 for the cost of the flight. My dear mum had also cashed-in two small insurance policies, in order to raise anther £ 50 towards our cause, and Caryl's parents had kindly loaned us a further £ 150, thus giving us the sum total of £600, which would have to pay for everything from start to finish.

If I purchased a 'normal' flight for £240, including insurance, and then added another £35 to cover the cost of my transportation to and from Heathrow Airport, it means that I would be left with around £325 to last me a fortnight. With such a relatively small amount to survive on, it means that my initial idea of renting a car for the second week would have to be scrapped, and I would have to rely on the Greyhound Bus services to transport me around the state. What an awful predicament to be in, but this was our future at stake, and so I duly rang the travel agents, in order to find out the latest on the flight situation? Maria tried the availability for four seats, but after a few seconds it was rejected, and so she then tried for three seats, but again her request was turned down. "Try for two seats; there must be two seats left, surely to God; there have been four available for days?" I said commandingly, my heart pounding like an Olympic runners' at the final stages of a 10,000 metre race.

Maria was already on the case, of course, knowing only too well how desperate I was to be on this flight.

"I'm sorry Shaun, but 'Two' has also been returned to me as unavailable, and so I think that you will just have to accept the fact that you have lost your seat, because they are almost always booked in pairs. Nonetheless, I will still try and see if there is one seat available, of course?"

I can hear the digits clicking rapidly in the background, but not being able to see the screen was nothing short of a nightmare for me. Suddenly the clicking stopped, and there was nothing but total silence, as the pair of us waited in deep anticipation for those *all-important* words to appear on Maria's screen?

"Well, I don't believe it Shaun, of all the luck; there is one seat still available" Maria said nonchalantly, as if we had all the time in the world to spare. "Well don't just tell me about it...book the blessed thing", I hollered down the phone, my politeness having long-since gone out of the window.

"I am, I am", Maria retorted "But I can't do anything to make the machine go any faster Shaun; please try and be patient for a moment".

I really wanted to apologise to Maria for my rudeness, but I was so frustrated right now, that all I could think about was getting on that blessed flight —and the consequences I faced if everything fell through at the last hurdle. A minute had probably passed by now, perhaps even less, but to me it seemed like a lifetime.

"That's it Shaun...your flight is confirmed" Maria said with a massive sigh of relief...for both our sakes. I was so pleased that if Maria had been sitting in front of me right now, then I definitely would have given her a huge kiss —without question. I would now concentrate all of my efforts on the journey that was in front of me, and put last years' trip to California to the very back of my mind...well, for now, anyway. However, I would still take my half-finished manuscript to Florida with me, in order to show the sales representatives that I meant business, as I had already told Mr. Norman that the purchase of one of their properties would rely solely on me receiving a contract from a major book publisher. I had decided not to mention the fact that I would also be looking for full time employment in Florida, as I already knew how difficult this was going to be, although in hindsight that might have been a more plausible reason for me looking for somewhere to live, rather than simply purchasing a holiday home.

With regards to my employment prospectus, I had also photocopied a handful of reference letters and testimonials that I had received from previous employers, along with photocopying a dozen copies of my C.V. in readiness to show to prospective employers in the US. This is where I really hoped that Leo would *pull out all the stops* for me, by arranging meetings and interviews with anyone he knew in the car industry, because let's face it, with a job like his he should know dozens of car dealerships in Florida. My flight was due to take off around midday, and so once again I arranged to stay at Anthony's flat in Caterham the night before. Ironically, it was less than a year ago that I had said my fond farewell's to Anthony and his family, fully expecting not to see them again until the next family wedding, christening or funeral popped-up, and now here I am, less than twelve months later, preparing to turn up on their doorstep once again.

Unfortunately, my ever-faithful sponsor, Jules Gilmore, wasn't giving away any freebies, such as complimentary hire cars to and from the airport, this time, although in all fairness, he did offer me a car at a reduced rate for the return trip, but sadly the price was still way too high for my ever-decreasing budget. Taking the bus was also out of the question, for not only was there no direct route to Caterham, but the timetables were so tightly knit, that failing to make the right connection on time, could result in me arriving in the middle of the night, which was something I was not about to risk, as it would not be fair to Anthony if I had to wake him up in the early hours of the morning. Another option, of course, was to have Caryl drive me all the way to Caterham, before doing the return trek back to Cardiff, which was a round trip of approximately 350 miles. However, the mere thought of having our three children belted in the back seat of a car for approximately six to seven hours, was enough to instantly rule-out that idea.

Besides, I doubted that our poor little Vauxhall Astra, as reliable as it was around town, would be able to survive such a lengthy non-stop motorway journey. And so it seemed that my only choice was to use good *old faithful and reliable* 'British Rail' for my commuting. Thankfully, the return fare was not that much dearer than going by bus, and the connecting trains go directly to Caterham Station. There is no doubt that it would be a much more comfortable ride on the train, than it would have been by taking the bus, as there is nothing more claustrophobic than being squashed-up in a window seat, with only the headrest of the seat in front of you to stare at once the sun has dropped over the horizon, whereas on a long-distance train ride, there is a normally a good chance that there will be tables separating each row of seats, where one can sit in comfort, either reading a book, or simply tucking into a bag of goodies along the way. Upon my arrival at the station, I was under orders to call Anthony, and he would be there within minutes to collect me and take me to the flat.

On Sunday afternoon I packed my case for the journey. Because of the additional charges for suitcases to be put inside the hold of the aeroplane, an amount which would severely diminish what very little spending money I had left to take with me, I had decided to travel the length and breadth of Florida with nothing more than a hand luggage bag. As I would also be travelling by bus, instead of driving a car, having hand luggage only was probably a better idea anyway, as dragging a huge suitcase around with me for two weeks would probably have broken my back! Three-quarters of the clothes that Caryl had washed and ironed for me the day before, were now dutifully returned to the closet, and my beloved briefcase, that normally housed all of my important documents, would now have to be substituted for a large *zipper* wallet. The camera that I had borrowed for the trip, would have to stay hanging around my neck during the daylight hours, and my good old bum-bag, that I had used for our journeys around Europe all those years ago, would once again be put to good use.

Holding all of my essentials, such as my ready cash, which was a mixture of both American dollars and Sterling, a handful of travellers cheques, my railway and flight tickets, my driving licence and passport –and my debit and credit cards (even-though there was bugger-all left on either of them) the bum-bag would remain strapped around my waist for the duration of the trip. With my case now packed to the gills, and my tick-list of essential items having been double-checked by me...before being *triple-checked* by Caryl, I was ready for the off. "But where was I really off to? And what was I going to do once I got there?" I kept asking myself...but the answer was always the same. Only after I had landed in Florida, would the reality of what I was actually doing hit me square in the face, and I knew it. "How will it all end, and will Caryl and I finally be happy again, like we had been for many years in the past –and will our lovely children eventually enjoy a world of sunshine and happiness in the 'Promised land?'"

These were the questions that I had asked myself a million times over since that fateful day when Caryl first mentioned about moving abroad a few weeks ago? Even though both our family and our friends had supported us from day one, a part of me was still telling me that I was living in a 'Dream-world', and I just couldn't bear the thought of letting everyone down again, like I did with the movie idea –even though I still like to believe that it will be a winner one day. Snapping-out of my depressive state for the umpteenth time, I reassured myself by looking on the bright side of things, just as I have always done in my life. "Stop thinking of a dozen reasons why it won't work Shaun –and think of one reason why it will".

CHAPTER 34

Florida Here I Come

On Sunday evening Caryl left a message through Amber's letter-box, telling her all about the problems we had encountered obtaining a flight seat, just in case there was anything she could do about the price? Amber rang back the next morning to say that she was unable to do anything about the cost, but that we need not pay the balance immediately, as she would call in and collect the rest of the money whilst I was away. Amber also said that I only needed to ring the travel agents to obtain a booking reference and my flight ticket would be waiting for me at the airport. At 3.20pm we pulled-up at Cardiff Central Railway Station, and while Caryl and the children stayed-put inside the car, I began fumbling-around with the hatchback lock, which has always been somewhat temperamental, in order for me to be able to retrieve my suitcase from the boot area. My train was due out at 3.25pm, and so I had to get a move-on, because if I missed this one then I would end up getting to Caterham about midnight! Two taxi drivers, who have just pulled up behind me are grunting and muttering (probably unspeakable) words under their breath at me, because I am currently blocking their taxi rank area, but I was going as fast as I could and there was simply nowhere else to park.

Having only the time for a quick kiss to all parties concerned, I then raced to the nearest window to buy my ticket, before ascending the steps like an athletic runner to the required platform. Thankfully, my train was waiting patiently for me on the tracks, as if knowing that I was on my way, and so I immediately jumped aboard. After a few loud whistles and

376

a tumultuous slamming of doors we were off. I remember Ted telling me that his journey never really ended, and here was I, wondering when mine was actually about to begin? Caryl's statement, about wanting to live in a healthier climate, was the true beginning of this latest chapter in our lives, although the start of my newest adventure could be leaving the house today I suppose, or maybe flying out from the airport tomorrow…who knows –and does it really matter anyway, I ask myself? As the train picked up speed, and the noise of its engines burned-out the conductor's voice, as he bellowed-out "Any tickets please?" I sank back into my chair, reminiscing about an old Chinese tale I used to listen to when I was just a young teenager.

It is taken from a musical album I bought on Chinese philosophy, which was primarily the soundtrack from the original 'Kung-Fu television series that was iconic back in the eighties. It is certainly not the type of record I would buy nowadays, I must confess, but twenty years ago it had inspired me into achieving great things, and somehow I had linked this one particular story with my immediate quest for fulfilment of life. If I remember correctly, the story goes something like this:

A Shaolin priest is walking around the courtyard of his monastery, when he stumbles upon one of his students' (Grasshopper), who is sitting next to a small pond. The young lad has a handful of stones and he is aimlessly throwing them, one after the other, into the water. The priest leans over and taps the young disciple on his shoulder:

"Where does your pebble walk to grasshopper?"

"It walk's, its journey is to nowhere!

"Each journey begins –and it also ends".

"Then the ending is the bottom of the pool".

"Yet does not the pebble entering the water, begin fresh journeys?"

"It seems unceasing"

"Such is the way of life; it begins; it ends; yet fresh journey's go forth. Young man when I was a boy I fell into a hole in the road and was broken and could not climb out. I might have died there, but a stranger came along and saved me. He said that it was his obligation, for help that he had once received he must in turn help ten others, each of whom would then help ten others, so that good deeds would spread throughout the land –like the 'ripples from a pebble in a pond'. I became one of his ten –and you became one of mine. I pass this obligation onto you".

Although I hadn't played that particular album for over a decade, I had listened so intently to the words as a young lad, that they had been etched in my memory forever. (*I am writing the words from memory right now, and this is another twenty years down the line*). My thoughts were abruptly dispersed by the guard, as he orders me to vacate the *first class* compartment, as my ticket is not valid for such comfort "But I suppose it was worth a try", he added, grinning all over his face as he pointed to the carriage door with his grubby, nicotine-stained finger. "Don't worry my friend, I'm used to being shown the door", I joked, keeping everything in a light-hearted mood, before trotting-off to the 'economy-carriage', to join three drunken (male) rugby supporters, who were singing to their heart's content. At Newport Station the 'karaoke kings' got off the train, and so peace was resumed. I then spent the next two hours reading the 119 pages that I had written so far about my rather extraordinary first year on the dole.

However, as much as my life out of work was certainly different from the norm, I was desperately hoping that this chapter in my life would soon be over, and that it wouldn't be too long before my family and I were enjoying a fabulous lifestyle in the 'Sunshine State'. At 5.30pm the train pulled into Paddington Station, and soon afterwards I am being whisked by one of the many tube trains on the London Underground to London Road Station, where my final connecting train will take me directly into Caterham, and so what could be simpler? Suddenly the train comes to a screeching halt in the middle of a tunnel and apologies are being echoed-out through the speaker system, along with an announcement that "Due to circumstances beyond our control" (which is what they always say at times like these) "Route changes will now be coming into force for the next couple of hours" "Here we bloody go again", I hear one of my fellow passengers blurt out to anyone who wants to listen, and so I just roll my eyes in agreement.

To cut a very long-winded (and also a rather confusing) story short, I ended up travelling on the *Northern Line –going south*...if that makes any sense? I have already missed my connecting train, but I have been assured by a fellow commuter that "There will be another-one along soon" –famous last words, or what? As it happened, my friend was quite correct, and the good news is that there are only two of us in the whole carriage, which is a complete contrast to last *crushing* journey, where the carriage was *packed to bursting*. At the next stop a guard is waiting for the train to arrive, and as soon as the doors open my latest travelling companion and I are politely asked to move to another carriage by the guard in question, who claims that the engine carriage they are using is not powerful enough to carry the last three carriages and so they intend disconnecting them. 'Why can't anything run smoothly for me' I keep asking myself...without ever getting the answer that I am looking for?

At 7.25pm I finally arrive at the Caterham terminal, exactly four hours after leaving Cardiff, and so I immediately rang Anthony, Caryl's cousin, who turned-up within minutes. Before I know it we are pulling up outside his private flat, and it feels quite surreal that it was almost a year ago that I was climbing the back staircase with Anthony's son and daughter, and yet now it seems like it was only yesterday? The living-room was just as I remembered it, with its polished hardwood floor, the centre of it being hidden by a large Persian rug, and the brown, leather sofa, along with matching armchair's, are still set out in an arc formation around a huge log-burning fireplace.

"You'd better ring Caryl, Shaun; she might be worried, because you should have been here over an hour ago", Anthony said, before going into the bathroom for a quick *spruce-up*. "Then you and I are going out on the town, my friend", he blubbered, his speech now being muffled by a hand towel, which he was swishing around his face at a rate of knots. I knew that this would be *par-for-the-course*, as Anthony is a man after my own heart when it comes to enjoying a social pint out with the lads, especially if they are family members. Carl was first to pick up the extension line in the bedroom, and so I listened to a flurry of gobbeldy-gook for a minute or so, before Caryl had a chance to grab the phone downstairs. After *making my peace* with the missus, and having a quick word with my beloved son, Liam, it was my turn to have *a wash and brush-up* in the bathroom, as I was feeling a little sticky and sweaty after all the travelling and running-around that I had been doing today. As soon as I had freshened-up Anthony and I headed out onto the town. At first we supped a few ales in the pub where Anthony's daughter, Chloe was now working, and then *my host* kindly treated me to a slap-up meal, along with a few glasses of red wine in a nearby restaurant...all on the company's expense account, of course.

Although the wall clock shows 7.10am I am no hurry to get out of my bed, because the time is now 6.10am, due to the clocks going back an hour for the winter period. "I just hope the airport have remembered to turn the hour back", I chuckle to myself, before turning-over for another forty winks...which inevitably turned into two hours! Making myself a cup of coffee in the kitchen, I can hear people running around downstairs in the office, as Anthony's motor-factor business is in full swing by now –and so it is *all hands on deck* for the workers, as they begin another gruelling day. "Only *dole bums* like me can afford this jet-set lifestyle", I say sarcastically, my mind still in turmoil about whether or not I was really doing the right thing –or was I simply chasing yet another unrealistic dream? Phoning home didn't do me any favours either, as Liam tells me that he had had a bad dream about me going away without him, before asking me to "Come home" because he was missing me. Caryl also said that she was missing me, which made it even harder for me to say my final farewells, before putting down the phone.

I was missing my family too...big style, but I had to do this now, otherwise Caryl and I would lose all hope of having a better lifestyle someday, and what would happen to us then, well I daren't even think about it. I had to put the thoughts of failure to the back of my mind, or I could easily crack under the strain of it all, and then our chances of a happy life would be gone forever. The mere thought of a friend putting his arm around my shoulder, before uttering those immortal words: "Well, you gave it your best shot –and that's all you can do", made me shudder –and it also made me even more determined than ever to succeed. After breakfast Anthony handed me an envelope, which contained a letter of recommendation from him, which I immediately put with my other testimonials, character references and résumés. Wrapping both of his huge hands around my *little paw* (by comparison) and looking deep into my eyes, Anthony wished me "All the very best of luck", which I knew had come straight from the heart.

Of all Caryl's family, Anthony is the only person I had ever poured my heart out to (invariably after a few of the *amber-nectar's*), but I had always felt a great bond between us, and I also knew that I could trust him implicitly with regards to keeping our conversations strictly between the two of us. Anthony was not only a good listener, but he was always on hand to give advice, which is something we all need at some stage in our troublesome lives, and for that I would be eternally grateful to him. Anthony was fully aware of our dire financial situation, saying that Caryl and I were purely *victims of circumstance* "Circumstances which are, unfortunately, way beyond your control Shaun", he added sympathetically. Both of Caryl's parents, along with my mum (*my dad had passed away five weeks after Caryl and I got married*), had always helped us as much as they could, but none of them wanted to interfere with our lives (or our marriage) and for that I will always have the greatest respect for them.

Jack, Anthony's service manager, kindly offered to give me a lift to the airport, and so once again I bade a fond farewell to the *'Triple C'* service station and its occupants. In total contrast to last year, the airport was nowhere near as busy, and so within no-time at all I have checked-in for my flight to America's *Sunshine state*. At 11.15am I made my way to departure gate number 22. On the way over I popped into the newsagents to purchase a newspaper, along with a couple of batteries and a new film for the camera. I then used the last of my change to phone my brother Paul. You see we have a long standing (rather cruel) trick which we play on each other, whenever one of us is flying off to sunnier climates. As soon as Paul answered the phone I immediately held it up in the air, in order that he could hear the announcement "Will the passengers of flight BA2527 to Orlando, Florida, please make their way to gate number 22" booming-out around the terminal.

"You bastard" I hear him jokingly shout down the phone, and so I have a quick chuckle with him, as I know that he will be very envious of me flying-out to Florida, even-though he knows that I will be *surviving on a wing-and-a-prayer!* No-sooner had I parked my butt than I was being ordered to make my way to the aircraft...here we go again! Just like the airport, the plane was half empty, and so how I was supposed to have had the last available seat, well lord-only knows? However, the important thing is that I am here, and I am more than ready to begin this new chapter in both mine, and hopefully, my family's lives. Our vessel that will be carrying us across the Atlantic Ocean is a Boeing 767, and it is nice to know that this time I won't be changing aeroplanes before arriving at my final destination, as that long-haul flight, followed by a second relatively long-distance flight, really *knackered-me-out* last year. Within half an hour we are airborne, and soon afterwards I am tucking into my first in-flight meal of stewing steak, potatoes, peas and gravy, followed by strawberry trifle and cheese and crackers –yummy.

By now we are way above the clouds, of course, having entered that haven of clear blue skies and endless sunshine. Although we are travelling at around five hundred miles an hour, it seems like we are barely moving, and that we are simply *floating on air.* Rather than flying in a direct line to Florida, we are firstly heading northwards, crossing over Scotland, before flying due west for two-and-a-half thousand miles, at which point we will be dropping down over Canada into the United States, where our aeroplane will then head directly southwards for another thousand miles, before entering Floridian air-space. From here it will be just a hop, skip and a jump (well, about two-hundred miles actually) to Orlando Airport. At 6.45pm our captain announces that we are now travelling at around 21,000 feet above sea level, and shortly afterwards he says that if we look out the right-hand side of the plane we will be able to see the city of Boston below us "And on the left is Cape Cod", he adds. Long Island and New York were next in line, followed by Jesapeke Bay, Atlantic City, Norfolk and Warmington, Kiddihawk, Charleston, Georgia and Jacksonville, before finally making our descent into Orlando Airport.

Our pilot had been a mine of geographical information, and I now knew how the Americans must feel when they are being driven around London in the back of a taxi, with the local *Alan Whicker* cabbie giving them all the spiel on England's capital city. Nonetheless, I enjoyed every minute of our nine hour *crossing of continents*, and as our plane taxied along the runway, waiting to come to a halt, I, for one, couldn't wait to put my feet back on American soil once more. Actually, this should have been my second visit to Florida, but unfortunately fate had played yet another crucial part in my life, by dealing one more severe blow to an already weakened and disgruntled human-being.

About three years ago Caryl and I had around £3,000 in our personal bank account, and so we decided to use £2,000 of that money, to buy a decent family car. With the *grand* we had left we wanted to take Liam to Disneyworld for a holiday. A two-week package holiday in Florida cost around £399 for adults and £199 for children at the time, and as Liam was under two years of age, he would travel for free, of course. Caryl and I also knew that Amber would give us a decent discount off the original price, and so we would put that money towards the cost of the Disney passes -simple. However, I then saw an advertisement in the local newspaper stating that 'Free holiday's for two people' were being given away with every car bought from the main Nissan dealership in Cardiff –and one of the destinations on offer just happened to be 'Florida'. To qualify for the 'top holiday' a minimum of £3,000 had to be spent towards buying one of their new or used cars for sale. And so I figured that putting an extra *grand* into a better car, along with getting our *dream holiday* as well, was a sound idea.

After negotiating a deal with the salesman, that included him confirming with the travel company that Liam would definitely be included in the package, Caryl and I vacated the showroom as the proud owners of a 1.3litre Ford Escort Ghia. The three of us also had the 'holiday of a lifetime' to look forward to, which, if truth be known, I was more excited about than the car. However, within a few weeks all of our dreams would be shattered, as I sat reading my daily Sun newspaper over breakfast one morning. In the centre pages was an article about the Nissan dealership in Cardiff, stating how it had been ripped-off for £22,000 by a bogus travel company. The article went on to say how 'embarrassed' Nissan were at having to tell everyone that the dream holiday's which they had duly booked in all good faith, had never existed in the first place! I was mortified, and I was dreading having to tell Caryl that the holiday was off, along with telling my little boy that he wasn't going to see Mickey Mouse or Donald Duck after-all!

"A spokesman for the company said that Nissan would 'lend a listening ear' to all of their dissatisfied customers", was the final statement in the article, but I wanted a bloody-side more than a 'Samaritan's lug-hole' as compensation for this horrendous travesty of justice. After lengthy negotiations with Nissan, Caryl and I were refunded £600 off the initial cost of the car, which I believe is the amount that Nissan had paid the company for the *non-existent* holiday?

Ironically, the three of us ended-up with a last-minute 'cheapie' package tour to Majorca, instead of fourteen days in a sub-tropical paradise, but I swore right there and then that I would not break my promise to my son, and that one day I would take him to Florida, to see the magical world of Disney. Still, that was then and this is now -and so just forget about the past Shaun concentrate on the present, and look-forward to whatever the future may bring.

CHAPTER 35

The Sunshine State

As the plane drew to a halt, and the air hostess opened the door, I was surprised to see that we would not be descending a set of *moveable* steps, but walking through a tunnel that would lead us directly into the main terminal building. This was something I had never encountered before, but now there were even more surprises in store for me, as a special shuttle train whisked away all of the passengers to the passport control area, before being deposited by two huge lifts in the luggage reclamation area –amazing! Clearing customs was a very simple process (probably because I had no contraband in my baggage) and so within minutes I am out in the massive car park, breathing in lung-full's of that fresh, Floridian air. The local time is now 4.40pm, and the temperature is a stifling 82 degrees Fahrenheit –whoopee-doo!

Florida, which in Spanish means 'Land of the flowers' is more affectionately known as 'The Sunshine State', simply because it averages between 2,500 and 3,000 hours of annual sunshine. Apart from being the flattest state in America, Florida also boasts the longest coastline, which stretches 2,170 Km (1,350 miles). Besides its huge landmass, the state of Florida also includes over four and a half thousand islands, which is more than any other state, except Alaska. It is also the only state that borders both the Gulf of Mexico and the Atlantic Ocean. The northern half of Florida enjoys a sub-tropical climate, whereas the southern half of the state basks in a tropical climate, similar to that of Hawaii. The Florida Reef is the only living coral barrier reef in North America, and it is also the third largest coral barrier reef system on the planet. Approximately fifty of the world's billionaires live in Florida, and many famous writers, such as Ernest Hemmingway and Tennessee Williams have made their homes here. There is also a large Cuban expatriate community.

Florida's economy primarily relies on tourism, hence it has more theme parks than anywhere else in the world, along with agriculture (the state is famous for its orange groves) and transportation. Tallahassee is the state capital, whereas Central Florida is also known as the 'Lightning Capital' of the US, as it experiences more lightning strikes than anywhere else in America. Florida also has more tornadoes than any other American state, and the state has also been struck by over one hundred hurricanes in the last one hundred and fifty years, Hurricane Andrew causing more than $ 27 billion worth of damage, when it struck in August 1992, which is only two months prior to my arrival here. Classed as a 'Category 5' hurricane, it struck the Bahamas first, before ripping through Florida and Louisiana with 280 kph (175 mph) winds that completely destroyed 63,500 homes and left 65 people dead in its wake.

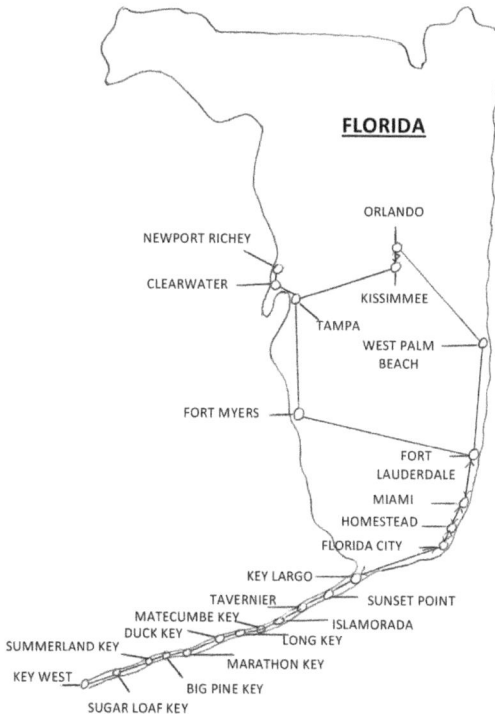

FLORIDA

ORLANDO
NEWPORT RICHEY
CLEARWATER
KISSIMMEE
TAMPA
WEST PALM BEACH
FORT MYERS
FORT LAUDERDALE
MIAMI
HOMESTEAD
FLORIDA CITY
KEY LARGO
TAVERNIER
SUNSET POINT
MATECUMBE KEY
DUCK KEY
ISLAMORADA
SUMMERLAND KEY
LONG KEY
MARATHON KEY
KEY WEST
BIG PINE KEY
SUGAR LOAF KEY

Suddenly I spot a gentleman who is carrying a piece of card with the name 'DONOVAN' written on it in bold letters, and so I walk over and introduce myself to the man in question, in the hope that he is my lift to the hotel?

The guy tells me that he had only been given the message to pick me up about an hour ago, which concerned me somewhat, but when we reached the hotel and he asked me who was supposed to be paying for the room, I was horrified, and so I figured that now would be a good time to put this guy completely in the picture with the deal that I had arranged back in the UK –like it or lump it! The guy (who I am going to call 'Roman') said that he felt really embarrassed by his company's lack of communication, before handing over his credit card to the receptionist, who swiftly swiped eighty-three dollars out of his checking account, before returning the said card to its *rightful owner* with a big, beaming smile, of course. Roman and I then arranged to meet at the swimming pool at ten-thirty the following morning, before he went his way and I went mine. Slinging my travel bag over my shoulder, I proceeded to the main lifts, as I had no intentions of climbing the four staircases to my room.

Using a kind of credit card to open the door, instead of the usual metal key, was yet another new experience for me, and a very acceptable one too, as it is so much easier, and considerably lighter than carrying a lump of crinkle-cut metal that has invariably been attached to a key fob which is the size of your average doorknob! As soon as I entered the room I could see why it was so expensive, because apart from having two –yes 'two' double beds, the place also had a huge sofa-bed, a massive television set –and even a massaging machine should the mood take you? There was also a built-in safe in the cupboard and an air-conditioning system, which I immediately switched on, as it was quite humid in the room at present. I then decided to have a quick shower and a change of clothes, as by now I was feeling rather sticky and sweaty. Unfortunately, the air conditioning unit made so much noise that I turned it off within minutes of getting out of the shower

It was still only seven o'clock in the evening in the state of Florida, but my body-clock was saying midnight (GMT being five hours ahead, of course) and so I was now feeling rather tired, and probably a little-bit jet-lagged too. However, there was no way that I could go to bed this early, and so out onto the town I ventured. By now it was dark, of course, and so I kept walking in a straight line, counting the blocks, as I passed each one in turn, as I dare not lose my bearings back to the hotel, otherwise lord-knows where I might end up? Thankfully it wasn't long before I found a secluded little bar where I was able to order a nice cool pitcher of lager... all to myself. With the majority of pitchers containing around three pints of lager, I figured that one should be enough to see me through the night. Just as I had expected, it wasn't long before one of the locals came over to investigate the stranger who drank all alone, whilst writing notes in his little black book all night.

Surprisingly, it was a young woman, in her late teens I would say, and not the usual peak-capped, pot-bellied pool player, who showed an interest in the *shady-dude from out-of town.* The young lady was very polite with her inquisitive questioning, and so I was quite happy to tell her what I was doing in Florida, along with where I was staying, how long I would be staying here for, and all about my family back home...the normal stuff.

It was only after staying in Florida for a week that I realised that this woman was not a nosey individual, but a typical friendly native of the land, who wanted nothing more than to welcome strangers to her country –and also to help them in any way that she was able to. In the weeks that followed I would meet several more of these wonderful, 'special' people.

Tonight, this young lady's assistance would be in the form of a little piece of advice, as she warned me not to return to my hotel by the same route that I had taken to get here, as apparently there had been a shooting in the area only a few days ago, in which a local guy had been killed! Florida was certainly living up to its *other* reputation, as being a dangerous place to go out at night. However, for fear of getting permanently lost in the back streets I did not adhere to her advice, and like the true coward that I am I ran all the way back to the hotel, not stopping to look back for a second. After shutting the door to my hotel room, I immediately clicked on the latch and hooked up the safety chain, my lungs still gasping for air, as I whispered "Safe at last", before falling onto the bed, and laughing hysterically at how ridiculously exaggerated I had made this whole situation, my overactive imagination having been the biggest culprit, as always.

When I awoke in the morning the room was still dark, and so I naturally assumed that it was still very early, but boy was I in for a surprise? As I pulled-back one of the thick brown curtains the sun's rays came blazing into the room, like a giant laser beam from nowhere, instantly brightening-up my entire bedroom, and also my new day ahead. Opening up one of the patio doors, which lead onto a small terrace, I am met with a barrage of hot air that instantly warms up my entire body. It was just as if a giant hairdryer had been turned against me, and it was simply wonderful. Under a ceiling of clear blue skies I took a deep breath, expanding my chest to capacity, as I inhaled as much as I could of that clean, fresh, Floridian air. Apart from feeling extremely healthy, and completely relaxed, I also felt totally unhindered by the obstacles that life had thrown at me. My problems, troubles and worries were all at bay as I sat on this petite veranda, gazing down at the rushed world below me.

Aches and pains were non-existent, except for a small twinge under my rib-cage, reminding me that my heart still yearned for Caryl and the children to be at my side. "Will this ever be our home?" I asked myself –and "Can I really make it work?" –or am I living on dreams that simply don't exist in my world? Oh how many times I would ponder over those questions during the course of the next two weeks, but for now it was time to concentrate on the present, and concern myself with the future as and when it arrives. After breakfast I sauntered down to the poolside, before stretching out on one of the dozens of sun-beds that surrounded our Olympic-sized pool. It was only nine-thirty, and so I still had an hour to kill before meeting Roman.

A continuous stream of soft music and romantic tunes emanated from a dozen speakers that were attached to various flagpoles and also dotted along the outside walls of the hotel, gently lulling the sun-worshippers into a world of eternal tranquillity. As Gilbert O'Sullivan sang about being 'Alone again –naturally', I remembered how I had cried when my father died, and at that precise moment the very first cloud that I had seen since stepping off the aeroplane, hung solemnly over my head, instantly blotting-out the sun's rays, and sending my mind back to the darkest day of my life. It seemed to have appeared from nowhere, and now it was just sitting there, defiant and immovable, and conspicuous by its loneliness in an otherwise cloudless sky. Could I have conjured-up the beast with my dark thoughts, as if reminding me that life was not all sunshine and roses, or had I simply closed my eyes long enough for it to have crossed the horizon? I have to believe the latter, for fear of assuming that the world revolves around *Shaun Donovan*, and not the other way around?

I suppose we are all guilty at some stage of assuming that events or occurrences which took place in our lives, did so purely for our own, personal gratification, rather than believing that they just 'happened'. Our whole lives revolve around coincidences; some good and some bad, and yet if we had the time to survey and analyze every movement we made, along with every place that we visited, we would probably find that the chances of say, bumping into an old friend were more likely than unlikely. Indeed, we have probably passed by people that we know, a thousand times, without actually realising it, because they were either walking on opposite sides of a street, or travelling in a car perhaps –or any number of scenarios, where the people in question might have simply been looking the other way...who knows? (*In today's world sending texts, or checking-out the internet on mobile phones whilst one is walking along the street, or even crossing a road, has to me a major contributory factor for people passing by one-another without realising it?*)

As a child I was told that I had not been born with a brain, but a cash register, because my mind was continually working out facts and figures, most of which related to my own personal wealth, of course, which rarely amounted to more than a few pounds at a time, but to me it was a fortune. However, as the years have progressed, I would say that my brain is now more like a computer (albeit on a much more primitive scale, of course) because it only accepts the perfect key combinations for it to function properly! In other words, if my mind refuses to believe that what it has just heard is positively correct, or my brain cannot immediately see the logic in a statement that has been made, either verbally or in writing, then it buzzes-around like a *micro-chip on heat*, as it tries to fathom-out the correct answer to a question –or how to solve the problem that it has been presented with...it is all very confusing?

As I lay there, wallowing in self-indulgence, my stream of ultra-violet rays were once again blocked, only this time it was not by a cloud, as I was about to find out. "Good morning", said a man's voice, in an unmistakable American accent, and so I opened my eyes, only to see Roman standing there, his arms folded proudly across his chest, and his face beaming with a perfect smile. He was early, and I wasn't ready for him, but he simply told me not to rush, as the pair of us walked back into the hotel lobby together. I told Roman that I wanted to ring my wife, to let her know about my safe arrival in the US, but he duly warned me about how expensive it would be to phone her from the hotel, adding that I would be much better off sending her a few postcards, before assuring me that they only take three days to arrive at their destination in the UK, and that they were very cheap to post. Looking seriously at my money situation, and not really knowing what lay ahead, I decided to take his advice, even though the stamp machine in the hotel ripped me off for forty-five cents every time I used it.

After ten minutes or so I am dressed, and soon afterwards the pair of us is winging our way to a place called Kissimmee, which, according to Roman, is just on the outskirts of Orlando, to look at a selection of two, three and four bedroom houses. The area of Kissimmee was initially named 'Allendale', after confederate Major J.H. Allen, a gentleman who operated the 'Mary Belle', a cargo steamboat, which was the very first to sail along the Kissimmee River. Kissimmee was given its new name in 1883, when it was finally incorporated as a city. Ranching was the main source of income prior to the opening of Walt Disney World back in 1971, when tourism boosted the city's economy to a much higher level than it had ever experienced before.

However, cattle's ranching still plays a major role in the area, especially in nearby Osceola County. On August 14th 2004 (a dozen years after my visit here) Hurricane Charley blasted its way across the city, with winds in excess of two hundred miles per hour, severely damaging homes and uprooting trees along its path of destruction. Only three weeks later, Hurricane Frances came billowing through the city, causing further disruption and chaos to the area –and this was followed by the devastating Hurricane Jeanne in the last week of September, which swept across Florida, along with hitting Puerto Rico, The Bahamas, Haiti, The Dominican Republic and the US Virgin Islands.

Orlando, which has been appropriately nicknamed 'The City Beautiful', is the county seat of Orange County. Orlando sits right in the centre of Northern Florida, and it is the state's largest inland city. During the Florida Land Boom in the 1920's Orlando flourished, and property prices soared, but within a decade all of this came to an abrupt ending, after several hurricanes devastated Florida in the late twenties. This was then followed by The Great Depression. Today, Orlando is known as 'The theme park capital of the world', which initially began with the opening of Walt Disney World in 1971, and since then the city has become one of the most visited destinations on the planet. Orlando also has one of the busiest airports in the United States, its theme parks, hotels, bars and restaurants playing host to around seventy million visitors a year from around the globe. The majority of Orlando's attractions are located on the famous 'International Drive', and the city also hosts an abundance of conferences and conventions throughout the year. Orlando is officially fourth in line as the city in America that most people would like to live in.

As we drove down the freeway I explained to Roman that at this point in time I was only interested in looking at the properties, on the understanding that I would realistically consider purchasing one as and when my book had been bought by a publisher. What I omitted to tell him was that I was only halfway through writing it! Mind you, I also said that if I saw a house which really *took my fancy,* and my wife liked the photographs that I took back with me, then the pair of us would pop over in the next few months, with a view to placing a holding deposit on the property in question. *How the heck I was ever going to afford to do that I knew not at this point in time, but it really wasn't that important right now.* After around twenty minutes or so Roman pulled over into this brand new housing estate, and the pair of headed directly for the sales office, where a lovely young lady not only gave me a complete overview of the planning and development envisaged for both the housing estate and the surrounding area, but she also bombarded me with a list of social amenities, such as sports clubs and leisure centres, which would be built within the immediate vicinity, before assuring me that all of the local residents would have automatic membership.

Having presented a perfect *opening speech*, the sales negotiator then spent the next twenty-five minutes meticulously going through various floor plans with me, which included giving me the dimensions of every individual room in each one of the houses, in her attempt at warming-me-up for a possible sale. As soon as she had finished *part two* of her operation, Roman ushered me out through a side door and along a garden pathway, which led to the rear entrance of a four bedroom bungalow, a property which he claimed was "On special offer for this month only" However, he was dealing with an ex-car salesman, and so I knew that if the bungalow was still unsold in six months time, then it would still be the same price as it is today...or maybe even a few thousand bucks cheaper? Set in a quarter of an acre of land, with beautifully manicured lawns to the front, side and rear of this magnificently built property, and encompassed by a small white picket fence, I instantly fell in love with the place, having found my own version of 'The American Dream'.

Constructed with interior block-work and exterior brickwork, it was not unlike a traditionally built British home, except for four lavish pillars, each one of them standing proud on either side of both of the front windows. The house also boasted a double-garage, with a huge door that was almost half the width of the house itself. Inside the house, the living room on its own covered the same surface area as the whole of my house back in Cardiff, and the kitchen was more than twice the size of what Caryl and I had been used to. The four double bedrooms leading off the lounge were all bigger than our main bedroom at home, and the master bedroom had a set of patio doors leading out onto the lawn, which is something that Caryl and I could only ever dream of. Surprisingly, the main bathroom was not a great deal larger than the norm, but then there *was* an én-suite bathroom attached to the master bedroom as well. All four bedrooms had fitted wardrobes and there was plenty of cupboard space everywhere else in the house, although none of the cupboards (or any of the doorways, for that matter) had actually been fitted with doors?

I knew that this was simply a ploy, in order to make the rooms look more spacious, but with a property of this magnitude, well they needn't have bothered. Just beyond the small hallway at the entrance to the bungalow, I opened the only interior door in the place, before standing-back in total amazement! The room it led to was almost as big as the rest of the house put together. Inside, it was as bare as all of the other rooms, furniture-wise, but at least they were carpeted, whereas this 'auditorium' (for want of a better word) proudly showed off its *smooth-as-silk*, creamy-grey, concrete floor. We were now standing inside the double garage, of course, which is an integral part of all American homes, and with each American family owning an average of four cars, I suppose it has to be.

However, in my eyes this represented one of my all-time dreams, and that was to be the proud owner of my own personal snooker room.

Roman said that I would be unable to change the frontage of the house, as the law states that new homes must have a garage "However, there is nothing stopping you from hanging-up a set of full-length drapes inside the garage, in order to close off the doorway", he added. Although this was the first house that I had seen, I couldn't think of anything more that I could wish for, but I still had a look at the two and three bedroom properties, simply to satisfy my curiosity. I asked Roman if timber-framed houses were still being built in Florida, as I had mistakenly thought that these were the only type of houses they built out here, before seeing these brick built ones today. He said that most of the houses currently being constructed are a mixture of brick and block-work, simply because timber-framed houses are no longer viable options. This was due to the severe hurricanes which the state has suffered during the last ten years, their enormous strengths and catastrophic wind speeds demolishing people's homes like they were cardboard boxes.

"Of course this makes the new houses a lot more expensive than the old wooden structures", he added, as the pair of us walked back into the office, and I knew right there and then that the niceties were over, and that we were about to *get down to the nitty-gritty.* I had already estimated that in Cardiff, a four bedroom bungalow such as the one we had just walked around, would cost somewhere between £150,000 and £170,000, depending on which building company had constructed the property along with where it was located, and so I just sat back in the luxurious armchair, while Roman and his learned colleague fiddled around on their calculators, before scribbling-down figures on sheets of paper like they were telephone-numbers. The sales adviser, a tall, attractive brunette, in her early forties I would say, then spent the next ten minutes or so converting the dollars into pounds, before asking me to join her at her desk...or *the negotiating table,* as I preferred to call it.

She began explaining all of the necessary taxes, along with a few additional costs that would be incorporated into the deal, but at this point in time all I really wanted to know was *the bottom-line.*

"Well, in dollars it would look a lot of money to you, and so I have already converted the figures into pounds, in order that you can..."

"How much" I exclaimed, interrupting her waffling, in order to save us both a lot of time.

"£73,000" she blurted-out, her face flinching, as if expecting me to scream out at the ridiculously high price that she had quoted me.

If only the woman knew how unbelievably low that price was to me, although I wasn't about to tell her, of course, and so I simply asked her to write the figures down on a piece of paper for me and slot it inside the folder, which I could see that she had already prepared for me. Roman then joined us at the table, before waffling-on about a new film called "A river runs through it", starring Brad Pitt, and directed by Robert Redford, but I wasn't really listening to him, as I knew that it was all part of a friendly *cornering* pitch, to try and get me to *sign on the old dotted-line* –but I wasn't having any of it. However the pair of them continued doing their uttermost to convince me what an incredible investment it would be for me and my family, and so I decided to play my trump card, by telling them that my wife had insisted on me leaving my cheque book and credit cards at home, so that I wouldn't be tempted into buying anything without her joint approval.

This seemed to do the trick, as the pair of them immediately backed-off, thus releasing the pressure off all of us. As relieved as I now felt, in my heart-of-hearts I would love to have been in a position to say "Go-on then, I'll take it", but sadly nothing could be further from the truth. Bovis Homes were next on our itinerary, but the houses were hardly any different to the ones which they were building back in the UK, except maybe a little larger perhaps. Anyway, they were more expensive than my beloved bungalow, and so they were a *non-starter*. By now it was lunch-time, and so Roman suggested that we go to his local bar for a spot of liquid refreshment and a snack, to whit I whole-heartedly agreed. The bar owners were originally from Nottingham in England, but apparently they had left their homeland a few years ago in order to run this pub in Florida. The husband claimed that he and his wife had moved out here '*lock stock and barrel*' within six days of returning home from a holiday in Florida, which I found somewhat hard to believe, but then one never knows?

I also wondered if introducing me to this couple was another one of Roman's ploys; to show me how straight-forward it is for families to move out here –and also how easy it is to become a success? (*If it was a rouse, then it was working rather well*). Condominiums were last on the list of our show-homes to visit today. Giant circular windows adorned the penthouse-styled apartments, which were quite lavish inside, but they were graced with relatively small balconies, and so they seemed more suitable for the likes of working couples, rather than whole families to live in. However, considering that I had come out here on the pretext of looking at *holiday-homes* these *condos* would have been ideal I suppose. At the rear of the apartment block there is a swimming-pool, complete with a sun terrace for all of the tenant's personal usage. Considering that the temperature was now in the mid-eighties, I have to say that I was sorely-tempted to strip-off my clothes and jump head-long into the pool.

As we walked around the terrace area I began chatting to one of the residents; a woman in her mid-thirties, I would say. She told me that she was originally from California, but had moved to Florida to study at a university in Orlando when she was in her late teens. "And I never went back", she added jovially. By four-thirty, our day together was at an end, and so Roman kindly dropped me back to the 'Days-Inn Hotel' (my hotel) to enjoy the last of the sun's rays by the pool. I was feeling quite satisfied with the days' events, as I got changed into my swimming trunks, but just as I was about to leave my hotel room, I suddenly realised that I had left my camera in Roman's car! Immediately I contacted his office (Roman had already given me the number, of course) and I asked the receptionist to contact Roman, and ask him if he would be kind-enough to drop it off at the reception desk in my hotel, either this evening or before 10.30am in the morning, as this was the time that I would be checking-out of the hotel?

"I wonder if a day will pass in my life when I manage to do everything right", I said, sarcastically, scorning myself as I walked over to my sun-bed, to catch the last rays of sunshine. After enjoying a rather sedate meal at 'Shoney's', a restaurant, which is situated just across the road from the hotel, I sat around the swimming-pool, peacefully writing postcards to Caryl and the children -with a special one for Liam, my big son, of course. I had just seen a young mother walking hand-in-hand with her two sons towards the children's play-area, and a big lump had immediately surfaced in my throat, as I thought of my two little boys back home in Cardiff. Florida is such a wonderful place, and I would dearly love to live here with my perfect little family, but I desperately needed to get a job first, and so next week could not come quick-enough. Sentimentality had to pass quickly, for it would be another twelve days before I would see any of them again, and I had so much to accomplish before then.

As I didn't know the area very well, coupled with the fact that I had no transportation to travel further-afield anyway, I decided to visit the same bar that I had frequented last night. After a few games on the pinball machine I was invited by one of the local guys to join an 'electronic darts' tournament. The rules are generally the same as a normal darts game, except the *arrows* haven't got spikes, but small electronic pads at the end of them, which simply connect with the electronic board, at which point a light flashes-up, letting you know what denomination you have hit... simplicity itself. My darting prowess was up to its normal standard, and so as usual, I got knocked-out in the first round.

CHAPTER 36

Tampa And Clearwater

After a sound nights' sleep I surfaced, ready to face another day. It wasn't even seven o'clock, and yet I felt like a lion. After duplicating yesterday's *lung-clear-out* on the balcony, I returned to my bedroom, to see what my 'twenty-six channel' television had to offer me this morning? Every second channel was advertising *special offers* for people to phone in with their credit card numbers, in order to confirm their 'once-in-a-lifetime' purchase, and so I instantly skipped on them, my credit cards having already taken a severe bashing. Cartoon channels for children were also in abundance, along with news and sports programmes from all over the world, but sadly none of these really interested me, and so onward I pressed, before finally reaching what I had been looking for –a movie channel. Unfortunately the only thing on offer was a gruesome, blood-thirsty horror film, which I managed to watch for about ten minutes, before pressing the 'Off' switch on the remote control.

If those type of *video-nasty's* were being shown before breakfast, then lord-knows what kind of films they would have on offer for their *late viewing* public? Talking about breakfast, it was now time for me to go for mine, and so I popped down the road to the local Burger-King fast food restaurant to have a bite to eat. Having stuffed my face, I then prostrated my body on the nearest sun-bed, to enjoy another blast of rays from that great fireball in the sky. As I lay on my back, listening to a number of my favourite songs being played in the background, I suddenly heard the sound of an engine –so rudely interrupting my little musical interlude, and so I opened my eyes to see what was causing the noise? Flying over the hotel was a light aircraft, and trailing behind it was a huge banner, advertising some kind of bar, or show -or something of interest to its prospective punters? *(Sorry, but I couldn't read it properly, because the sun was in my eyes!)* "What will they think of next", I asked myself, before turning-over onto my belly.

At ten-fifteen Roman turned up with my camera, and so I tried to scrounge a lift with him to the local Greyhound Bus terminal, but he said that he was going the other way, before adding that I would have to get a taxi, as there was no local bus service that could take me there. *(I'll bet a pound-to-a-penny that had I signed on the old-dotted-line yesterday, then he would have been more-than-happy to give me a lift to the bus terminal, but that is the way the cookie-crumbles I suppose!)* My next destination is a place called 'Newport-Richey', which, according to my Florida map, is around one hundred miles from here -right over on the Gulf coast, and so I had better get my skates on if I want to get there in daylight, as I have no idea what times the busses are running, or how long they will take to get there? I rang Reggie James, the gentleman who would be picking me up at the Greyhound Terminal in Newport Richey, and he said that I would have to take a bus to Tampa first, where I would then have to catch a connecting bus to Clearwater, before taking a third and final bus to Newport Richey.

"It's going to take you a few hours to get here Shaun, but don't worry because I will check out the arrival times and be waiting for you when your bus pulls in at the terminal", he added in a positive manner, thus boosting my confidence in him. Whilst waiting for my taxi (or should I say 'cab') to arrive, I walked over to the end of the reception counter, where a huge box, containing sixty-four pigeon holes, was resting against an adjacent wall. Each segment was full of individual leaflets advertising just about anything and everything that one could ever wish to see and do in the *Sunshine State.* From bird sanctuaries to alligator farms and wildlife parks to sea-life aquariums, there was simply no animal on this planet that one couldn't go and see, whether in captivity or running wild -the choice was yours? Alternatively there were numerous theme-parks on offer, the crème-de-la-crème being Disneyworld, of course, with its world-renowned Magic Kingdom, MGM Studios, Animal Kingdom and Epcot Centre standing right at the top of the list.

"What-about popping over to Universal Studios, to relive some of the most bizarre and exciting –and even the most frightening moments in cinematic history, by taking one of our daily tours of the studio-sets" said one of the leaflets. Water-parks were also in abundance, of course, with Disney's 'Blizzard Beach', 'Typhoon Lagoon' and 'River Country' once again standing out a clear mile in front of all the rest, with their death-defying slides, manmade beaches, and incredible wave-machines. Cruises to the Bahamas, along with various other Caribbean islands started at only $99, which was ridiculously cheap, of course, and there were also daily excursions to the Kennedy Space Centre, to see a Space Shuttle being launched, which is something I would have loved to witness whilst I was here. Sadly all of those wonderful excursions would be out of the question on this trip, but it had certainly given me *food-for-thought* for the future, and that was exactly what I had come out here to do –to plan for my family's 'future'.

The taxi arrived within minutes, its driver immediately offering to take me all the way to Clearwater "Only $140 to you, my friend", he said, ever-so persuasively, but there was simply no-way that I could afford it. (*It was bad-enough having to pay twenty bucks, which I could ill-afford, for a fifteen-minute taxi-ride into town!*) The Greyhound Bus terminal was actually larger inside than the parking lot was outside, and the place certainly wasn't as packed as I had expected it to be. In fact there were only two people standing in front of me at the ticket counter, and so I knew that I shouldn't have to wait long to be served, although I still kept a watchful eye on my bag, which was now resting on one of the chairs, about ten feet away from me. As the woman in front of me walked off to get her baggage tagged by one of the conductors, I began my *pretty-please* enquiry.

"Excuse me, but do you have a ticket that would cover me to travel anywhere in Florida for the next eleven days, and if so then can you tell me how much would it cost me please?"

"We only have a monthly ticket, which is $297 sir", the teller replied, her tone telling me that it was simply a case of "Take-it-or-leave-it sunshine?" £200 was way out of the question, and the woman could see it in my face.

"Why don't you just tell me exactly where you want to go sir, and let me work out a price for you", she then added, in a far-more sympathetic tone.

"Well, I have to get to Newport-Richey today, and then in a few days time I will be travelling south to Fort Myers for another two day stop-over, before heading due east to Fort Lauderdale and Miami. From here I will be travelling up the east coast to West Palm Beach, before returning to Orlando, in order to catch my flight back to the UK in eleven days time. If I have a few days free, in-between all of my prearranged visits, then it would also be nice to be able to spend a couple of days down in Key West, of course?"

"Leave it to me sir; I won't be a moment", my *learned friend* replied with such confidence, that I think that I would have entrusted my life to her, right there and then.

After scrutinizing several timetables on her computer, the woman finally turned back to face me: "I can put a special package together for you that will cost you $128 in total -if you would like me to do that for you sir?" £87 for a 1,000-mile round trip of Florida seemed like an absolute bargain to me, but it was still way out of my price-range, and so I figured that I had better forego the trip down to the Florida Keys, and then see what figure the woman could come up with? However, before I had the chance to reply, the lady delicately slipped a form underneath the slatted window, before telling me to fill in all of the relative details, but to leave the date blank, as she would fill that part in on my behalf. The document she had given me was called a 'Privileged Travel Offer Form', which was only to be given to people who were able to book their travel arrangements at least three weeks in advance. A fifty percent reduction in fare prices would be given to these 'special' customers.

The speed of my biro made the woman grin like a Cheshire cat, while her eyes surveyed her colleagues and the customers around her, in order to make sure that no-one was aware of what both she and I were doing.

"That will be $64 then sir —and I hope that you have a pleasant stay in Florida".

"I am sure that I will, and thank you for everything; you are most kind", I replied, giving the young lady a quick wink, as I shuffled off to get my bag tagged. Unfortunately I had just missed the 1pm bus to Tampa, and as they run every hour -on the hour, it meant that I would have to hang around the depot for another fifty-odd minutes —and that is if the next bus goes out on time, of course? However, the good news is that I am now the proud owner of a 1,000-mile round-trip ticket, which has cost me around

£43, and I also have in my possession a timetable of every Greyhound Bus that travels throughout the state of Florida –whoopee. After purchasing an ice-cold can of Coca-Cola from one of the many vending machines, I went outside the terminal, to sit on the grass and to *study-form* with my brand new *booklet-of-freedom*.

By now the sun was blazing, and the temperature must have been approaching the ninety degree mark. One of the Greyhound porters was sitting opposite me on the grass, and so I asked him if the weather was always like this in Florida? "Hell no, last month it was really HOT", was his unexpected –and very sincere answer. I dreaded to think what the weather must be like in July and August, but then if I wanted to move out here than I had better be willing to take the rough with the smooth I suppose? In 1994 the World Cup is being hosted by the United States, and if Wales get through to the qualifying rounds then I would love to come over here in the summertime and watch the games, but at present the chances of me being able to do that are about one-in-a-million I would say. Having finished my Coca-Cola within ten minutes I began debating whether to have a second can, when suddenly I remembered that I still had a can of Budweiser in my bag, and so I decided that, even-though it might be quite tepid, it was wet, and therefore it would have the desired effect.

By now a group of fellow workers had joined their *buddy* on the grass, and as I began slurping-away at my can, I could see a handful of them staring at me in disbelief –or maybe it was just jealousy, because I was enjoying a can of beer and they were unable to do so, because *effectively* they were still on duty?

"If I were you my friend I'd put that can of beer out of sight, just in case the local sheriff drives by in his car. In case you didn't know, it is illegal to drink liquor on the sidewalk, and so if he sees you guzzling a can of Bud' like that, he's likely to pull-over, and then he's gonna haul your ass off to jail", said one of the guys, with even more sincerity than his workmate. "Enough said my friend", I answered, before slipping off to a more secluded spot around the back of the building. Sitting next to an enormous skip -or should I say 'hiding behind it' and contentedly minding my own business, I am suddenly accosted by this young lad, who cannot be more than ten years old. He politely asks me if I can please 'loan him' 25 cents to phone his mother.

Naturally I assume that this is just a typical ploy to extract a small amount of cash out of a gullible tourist, but because he asked me so nicely, I decided to give him the required amount, simply writing it off as part of everyday life, as the boy scurried-off into the distance. However, ten minutes later the lad reappeared, and so this time I have decided not to give him a *brass farthing* –no matter how much he asks for? "Back for more eh", I quipped sarcastically, waiting to see his outstretched arm, as he begins begging me for more money.

"Oh no sir, my mother wasn't in, and so I came back to return your money to you, but thank you anyway", he replied, making me feel like the lowest of the low. "You keep the money son; you might need it to call your mother later-on", I insisted, my heart not wanting the money back for any reason whatsoever. I then praised him for his honesty, adding that his mother must be very proud to have a son like him. My bus was over half an hour late arriving at the depot, and so unless the time could be made up on our journey to Tampa, then I knew that I would miss my connecting bus to Clearwater –shit!

While my luggage was being loaded I dutifully rang Reggie, so that he would not be hanging around in Newport Richey for me, because if I did miss my connecting bus to Clearwater, then it means that I would obviously have to catch a later bus to Newport Richey –double-shit! Thankfully Reggie accepted my apologies, along with my explanation, before offering to come and collect me from the Greyhound terminal at Clearwater, rather than us both having to wait ages for the third bus to arrive. "I'll be finished work by then Shaun, and as it is only a twenty minute drive down the coast to Clearwater, I should get there roughly the same time as you. However, if I'm not around when you get there, then just wait for me inside the terminal, and I'll be right behind you", he added. Reggie seemed such an organised person, and so knew that I was going to like him –in fact I already did!

Thankfully, I managed to commandeer the seat directly behind the driver, as I do like to have panoramic views of the road ahead, rather than just staring out of a side window and watching the places pass me by. Unfortunately, sitting directly opposite me is the local loud-mouth; a tall, thin guy, dressed in a tatty pair of jeans, a suede waistcoat and a kind of *Crocodile Dundee* hat, with matching boots. He was by no means a trouble-maker, just the proverbial 'know-it-all', who liked to make-out that he knew everything about everything, and he couldn't wait to give his *captive audience* the benefit of his worldly knowledge. By the time he had finished his speech on pollution, I am sure that half of the passengers were too afraid to vacate the bus and breathe in the fresh air outside –I know that I was...well, not really. Within the first hour of being on the road, it was patently obvious to me that we weren't going to make up any time on the run to Tampa –in fact I was expecting to lose a few more valuable minutes, because our driver was going so god-damn slow!

Tampa is a major city that sits on the west coast of Florida, on Tampa Bay, near the Gulf of Mexico. Tampa is the county seat of Hillsborough County, and it is the largest city in the Tampa Bay area. People have lived on the shores of Tampa Bay for thousands of years the Seminole people being forced from their lands in the mid 18th century, before returning to Central Florida after the US took control of Florida in 1821. Prior to this there were only the native fishermen, along with a handful of Cuban residents living in the area. In the late 1800's illegal lotteries began springing-up, which were very popular with the working classes, and by 1920, Charlie Wall, who was the son of a wealthy Tampa family, had taken over the operation, expanding it to massive proportions. This he was able to do without retribution, due to him making substantial pay-offs to a handful of corrupt government officials and also by paying regular *kickbacks* to a selection of *bent* law enforcement officers. However, this era of blatant corruption all ended in the 1950's with various 'Organized Crime Hearings', and the subsequent misconduct trials of several local officials.

Regardless of our 'disciple of doom and gloom', we all survived the trip in one piece, and the good news is that my connecting bus to Clearwater was still standing on the tarmac, even though it should have left the station over half an hour ago. At first I simply assumed that it must also be running late, but then a fellow passenger informed me that Greyhound have a system where if one bus is running late, then they will hold up all of the connecting busses (within reason) until the first bus arrives at its destination. Actually, I was still trying to fathom-out whether this was a good system or not, when our bus arrived in Clearwater about an hour later.

Clearwater, which gets its name from the many freshwater springs that are found both in the city and also around the harbour area, is a city located in Pinellas County, Florida, Cleveland Street being one of its major avenues. Clearwater is also the worldwide headquarters of the Church of Scientology. During the American Civil War (12ᵗʰ April 1861 -9ᵗʰ April 1865) when most of the able-bodied men were away fighting for the Confederate Army the community's supplies were continually being ransacked by Union soldiers, after sailing ashore in their gunboats. Clearwater Beach, which sits on its own barrier island, used to be separated from the city, but in 1915 a bridge was built across the harbour, joining the two areas together. During World War II, Clearwater was a major training base for US troops, almost every hotel being used to house the continual influx of new recruits. 'Dan's Island', as it was known at the time, was used by the US Army Air Corps fighter bombers for bombing practice, but it has long-since been transformed into a densely populated area with lots of high-rise buildings, and its name has officially been changed to 'Sand Key'.

The Clearwater Marine Aquarium, which is located on Clearwater Beach, opened in 1972, and since then its entire workforce have concentrated their efforts on the rescue and rehabilitation of sick and injured marine animals, before releasing them (whenever possible) back into their natural habitats and environments. Its inhabitants includes nurse sharks, sea turtles, river otters and stingrays, many of which are permanent residents, due to them having sustained substantial injuries that prevent them from being able to be released back into the wild. Best known of all of these residents is a bottlenose dolphin called *'Winter'* that was rescued in 2005, after getting her tail entangled in a crab trap? Unfortunately, the extent of her injuries required the amputation of her tail, which was then replaced with a prosthetic tail, an operation which would unexpectedly give *Winter* (and the facility) worldwide acclaim, as *Winter* soon became a movie-star, after the release of the film 'Dolphin Tale' in 2011, which was subsequently followed by its sequel 'Dolphin Tale 2' a few years later.

In years to come (eight years later in fact) I would enjoy a lifetime's worth of experiences on Greyhound Busses, as I circumnavigated the whole of the United States on dozens of their 'doggy' coaches, but that is another story –and another book! The bus stop in Clearwater was a far cry from the terminals in Tampa and Orlando. Set back off the road, amid a small bunch of shops, it looked more like a mini-travel agency, rather than a coach terminal. There was also no sign of Reggie, and so I tried to ring his office, but it was now after five o'clock, and so I ended-up listening to an answer-phone message. The only thing I could do was to hang around the terminal and hope that he would turn up sooner or later?

Half an hour went by and dusk began to settle in, as the horizon darkened by the minute. The area was quite secluded, and only a small tobacconist shop remained open, and so I decided to pass the time away by talking to the owner. As the clock on the wall struck six o'clock, the owner said that he was closing-up for the night, which was certainly *not* what I wanted to hear right now, but I was powerless to do anything about it –and I certainly didn't want the owner to think that I was concerned about being out here all on my own. Just as my companion was putting the bolt on the door of his shop, a large, white Mercedes convertible came tearing down the street, before coming to a screeching halt at the side of the road. Immediately the driver's door opened, and a stout man, probably in his late fifties, with a grey head of hair, and sporting a perfectly-trimmed beard and moustache, stepped out of the car.

He obviously knew who I was, because I had this huge holdall at my feet, and so he immediately introduced himself as "Reggie James", before profusely apologising for being so late, although he tendered no explanation, as we walked to the car together. Tonight I would be staying at the Sheraton Hotel in Newport Richey, and then tomorrow morning we would rendezvous at Reggie's office, which is situated opposite the reception desk in the same hotel "At 9am sharp", Reggie stated, in a polite, but regimental manner, and so I immediately knew that I was dealing with an ex serviceman here.

CHAPTER 37

Newport Richey

Newport Richey is a relatively small city in Pasco County, Florida. At the beginning of the 20th century, the area around Orange Lake was initially called 'New' Port Richey, whereas the older part of the city was dubbed as 'Old' Port Richey. However, it wasn't until a new Post Office was built for the residents of the south of Port Richey in 1915, which was called 'New Port Richey' that the name of Newport Richey effectively became official.

In 1928, a famous actor of the silent screen called Thomas Meighan, built a large house on the river, and it was envisaged that Hollywood stars would follow his lead, by building similar properties in the area, before making Newport Richey their winter retreat. Gene Sarazen, one of the top golfers of that era, who was also the man who invented the *sand wedge*, also built a home in Newport Richey. The famous bandleader, Paul Whiteman, along with the legendary music composer, Irving Berlin, both placed deposits on individual houses, but neither of them completed on their purchases. Sadly, the Florida Land Boom, along with the Great Depression, was catastrophic for Newport Richey, and so dreams of it becoming a millionaire's retreat for the rich and famous never reached fruition.

The day had been a long one, but America wouldn't be America without eating at a MacDonald's —and so I duly indulged myself in a *Quarter-pounder* for my tea, before slipping off to the local watering-hole for a couple of pints of the amber nectar. Yesterday must have taken more out of me than I thought (*and before you ask, it had nothing to do with the two litre pitcher of alcohol I consumed last night*) because I have overslept this morning, and so without running around like a fool, and skipping on breakfast, I would never have made it to my meeting with Reggie at nine o'clock.

Over several cups of percolated coffee Reggie and I discussed the many benefits of living on the northwest coast of Florida, along with one or two of the pitfalls, which seemed quite negligible really. Weather-wise it is actually somewhat different to the southern half of Florida, and according to Reggie the coast is normally cooler than the inland areas, because of the coastal breezes. (*Apparently there is an invisible dividing line between the two halves of Florida and sometimes the difference in temperature between the top half and the bottom half can be as much as twenty degrees Fahrenheit*). "That doesn't mean to say that it gets cold up here Shaun" Reggie added, tickling his moustache "I almost wish it would sometimes", he then said jovially. Reggie continued his speech by openly telling me that Newport Richie predominantly catered for people of retirement age, rather than young couples, who were wishing to bring-up families here, and I have to say that I appreciated his honesty.

However, I was already well-aware that Florida is renowned as being *the wrinkly state,* because of the continuous influx of pensioners, who invade the place every year, to avoid the bitter-cold winters up north, including the severe medical problems that the cold invariably brings with it! As for me personally, well I don't want to suffer chronic arthritis in my bones, or die of hyperthermia when I am old and grey –and I certainly want to enjoy everything that the warmth and sunshine of Florida can offer me and my family, rather than seeing us all suffer from the cold and dampness of the long British winters.

One of the biggest attractions that the west coast of Florida holds for me, is the fact that it is surrounded by the warm, placid waters (well, *mostly placid* anyway), of the Gulf of Mexico, whereas the east coast has to deal with unpredictable Atlantic currents, that can be positively deadly at times. However, I have no doubts that the water on this side of the Atlantic Ocean is several degrees warmer than its counterpart over in Europe, but until such time as I am able to test it for myself I will just keep an open mind on this whole issue.

The Gulf of Mexico, which formed around 300 million years ago, incorporates the states of Alabama, Mississippi, Louisiana, Texas and Florida on its northern shores. These five states make up the 'Gulf Coast of the United States', although it is often referred to as 'The Third Coast'. The south and southwest of the Gulf is bounded by Mexico, and the southeast by Cuba. The Gulf of Mexico Basin is over nine hundred miles wide, and one part of it is connected to the Atlantic Ocean via the Florida Straits which separate the US mainland from Cuba. The Gulf of Mexico is also connected to the Caribbean Sea via the Yucatan Channel, which flows between Cuba and Mexico. This area is officially known as the 'American Mediterranean Sea'. Although Christopher Columbus was credited with discovering the Americas back in September, 1492, he actually sailed into the Caribbean Sea, and so his ships never actually reached the shores of the Gulf of Mexico.

The offshore oil platforms in the Gulf produce around sixteen percent of the petroleum output for the whole of the United States. In 2010, one of the oil platforms off the coast of Louisiana exploded, before sinking to the bottom of the Gulf. It is estimated that around five thousand barrels of oil spilled into the Gulf every day, until they were finally able to stop it. The Gulf is home to around fifty species of shark, including bull sharks, tiger sharks and lemon sharks, along with five species of endangered sea turtles, including leatherhead, loggerhead and hawksbill. A selection of sperm, minke and humpback whales, and all kinds of loveable dolphins and manatees can also be found swimming around in the Gulf. The Gulf of Mexico is also a major crossing for migratory birds, which includes two and a half million that land in Louisiana during the migratory season. The Gulf also has its own share of shipwrecks, including the Madi-Gras, which foundered around the beginning of the 19th century, and lay dormant at the bottom of the Gulf, until it was discovered by an oilfield inspection crew in 2002 –which was ten years *after* my visit to the Gulf.

By ten o'clock Reggie, like the true professional that he is, had qualified all of my wants and needs, and so it was now up to him to satisfy those requirements as best as he could, and as we vacated his office I had this gut-feeling that that was just what he was going to do. Walking out into the parking lot together, Reggie humbly suggested that before taking me to see any of the houses on offer, he would take me to the beach first...

with a view to getting my adrenaline rushing, before going in for the kill, no-doubt. Cruising along the coastal highway with the wind in my hair and the sun up above our heads, its powerful rays glinting off the cars' bonnet and windscreen, I felt as if the pair of us were re-enacting a fifties movie scene, and so I looked behind me, just to see if there was a film crew following us, but of course there was nothing. I was in my own little dream world right now, reality having been blown into oblivion, and I was enjoying every minute of it.

I was so relaxed and contented with life that I literally could have stayed in Reggie's car all day, driving around in circles for all I cared, but sadly it had to come to an end, of course, as Reggie finally brought the car to a halt in this rather secluded parking lot. After a short walk, the pair of us is now ambling along this truly glorious beach, and to my right I can see the Gulf of Mexico, its turquoise waters beckoning me to sample its delights of pure, unadulterated luxury. It wouldn't have long to wait, as I jogged across the bleached white sands, tearing-off my shoes and socks as fast as I could, as I made my way to the water's edge. Reggie could see that my enthusiasm had gotten the better of me, and so he just stood there, watching me in complete and utter silence, as I waded up and down the shoreline, my feet gently kicking and splashing the lukewarm, crystal-clear waters that surrounded me. A hazy mist hovered above the waterline, blotting-out the distant horizon and giving this idyllic of settings a somewhat eerie atmosphere.

Thoughts of the Bermuda Triangle ran through my mind, as I walked further and further out to sea, as if expecting to stumble-upon a long lost galleon, or possibly come face-to-face with a ghostly pirate ship, its tattered flag flickering in the wind, with not a soul on board to sail her. By now my imagination was running wild, and even though the water level had still not reached my knees, the density of the fog had caused me to almost lose sight of Reggie at the shoreline, and so I decided to retreat back to terra-firma, before *Long-John-Silver*, or *Blackbeard's ghost* appeared on the scene and I was inevitably forced to 'walk-the-plank'! As Reggie and I strolled along the palm-fringed shores I knew that my days as a 'city boy' were definitely numbered. No-longer would a bustling metropolis like Cardiff, with its cramped buildings, congested traffic and hectic lifestyle dominate my world and the way I wanted to live? No-longer would I be a prisoner of the inner-city and all its pollutants, but a free man, living close to nature, and enjoying the delights of a healthy climate and a happier environment. Next on the agenda was the local park, which was about four miles along the coast.

It is a good job that I was hoping to work in the car industry, and therefore have a company car, as there is absolutely no-way that one can survive in America without having some form of transportation.

Like every house I had visited in Florida so far, the lawns had been kept in pristine condition, and the landscaping was simply immaculate. Being surrounded by so much greenery, one could easily believe that they were back in *merry-old England* in the middle of the summertime, rather than being in a sub-tropical climate in late-autumn. Either Florida has a very wet rainy season, or their irrigation system is second-to-none. After walking around the park for about twenty minutes, chatting about anything and everything, we came upon the picnic area, which consisted of around twenty slatted picnic benches, half of which were sitting out in the open, while the other half were situated under a huge timber canopy, which was primarily supported by a dozen brick pillars. Apart from being available to the general public during the parks' normal opening hours, Reggie told me that this picnic area was also available to book for children's birthday parties, or any kind of family functions, even after opening hours, if necessary.

Every conceivable *want and need* in order to enjoy one's leisure time was completely catered for in this Utopian paradise, and I was simply 'in love' with the place. Driving around to an adjacent car-park, it was patently obvious that our car was the odd-one-out, and so I questioned Reggie about being able to park here, but he assured me that it would be fine, as we would only be staying for a few minutes. This parking lot was chock-a-block with pick-up trucks, each one of them brandishing individual trailers of various shapes and sizes —and all of which were currently empty, I might add. Reggie confirmed my thoughts that we were near to the water-sports bay, adding that the majority of these trailers were used to carry all different kinds of boats and jet-skis, along with just-about anything-else that travelled on water, down to the waterfront for a day out *on the ocean wave.* I suppose that being a water-sport enthusiast is a must when one lives in Florida —indeed, to the majority of people living here, I daresay that it has become a way of life.

However, enough of these frivolities, and learning all about the lifestyles of the idle-rich; I had come here to look at houses, had I not, and so let us get on with the show. Even in this department Reggie had excelled himself, taking me to the 'best on offer' first, safe in the knowledge, no doubt, that it would surpass anything that I had ever seen in my life before. From the outside the front of the house looked absolutely enormous, with its 'six-peaked' roof and triple garage, but little did I realise then that I was about to be completely 'blown-away' with what was inside the place. As soon as the door opened my chin hit the floor. Unlike my last 'dream-house' which was totally barren on the inside, this place had been kitted-out with the most luxurious furniture. Plush curtains, shag-pile carpets and expensive window-blinds also adorned every room –the place was simply a palace. After a quick show-around of the four adjoining bedrooms and the two bathrooms (one of which was an en-suite to the master bedroom) Reggie and I returned to the living room, where he said that he would now show me the highlight of this "Dream home".

One of the walls of the main living area had been completely obscured by a set of full-length drapes, and so Reggie walked over to one corner of the room, before gently pulling on a set of cords that were hanging on the wall. As the curtains drifted apart, sliding majestically across the top of the carpet, I stood back in total disbelief of what my eyes beheld? To say that I was completely and utterly mesmerized would be a gross understatement of the truth, because I was simply in awe of its majesty. A giant conservatory, much larger than anything I had ever seen in my life before (and no-doubt the biggest that I am ever likely to *clap-eyes on* in the future) stood there in front of me, its grandeur notably enhanced by what it encompassed in its gargantuan shell. A huge kidney-shaped swimming pool, complete with a small diving board and two climbing ladders, was just begging me to take a dip in its refreshingly-cool waters –and Reggie knew it.

Sitting on one side of the ornately tiled patio area was a beautiful wooden dining table, complete with six matching chairs (including their respective cushions, of course), a multi-coloured table-cloth, and even a sun umbrella, though lord-knows when that would be needed, as the whole area was encased in what looked like a shell of darkly-tinted glass. To compliment the patio-set there were also two hardwood sun-loungers, with matching cushions, sitting next to the pool, along with a free-standing barbeque, standing next to a drinks trolley, in one corner of the room. Dotted around the perimeter of the conservatory was a handful of cane tables, each one of them supporting a variety of ornate statuettes, and standing proudly in-between each of these tables were pairs of elegantly-shaped vases, each one containing a myriad of multi-coloured flowers –just to give the place that added touch of splendour. Upon closer inspection of the conservatory walls and ceiling, I discovered that they were not made up of tinted glass at all, but a very fine mesh; a kind of darkened gauze, very similar to what is used in your average tea-strainer. Apparently the holes are large enough to allow the air to flow through them freely, and yet small enough to stop the mosquito's getting in, which is of paramount importance when one is living in a sub-tropical climate. This epitome of luxury would set me back $140,000 (around £97,000), which was way out of my price-range, of course, but considering that our house was valued at £82,000 two years ago, and it doesn't even come close to this place, I suppose that Caryl and I might not be so far away from obtaining our dream home in the sun after-all! The triple garage was not only big-enough to hold a full-size (12' x 6') snooker table, but also a ping-pong table and a pool table as well –with plenty of room to spare. Lord-knows what else I could build in the huge back garden that came with the property –a sauna-room and a gymnasium perhaps? It was no good -I would have to stop day-dreaming, before I ended-up putting pen-to-paper on something that I could ill-afford –and I knew it.

Thankfully, Reggie has just insisted upon taking me to one of the local beach bars for lunch, which will hopefully take my mind off this *world of make-believe* that I was currently living in. About fifteen minutes later, we are pulling-up at this ornate little restaurant overlooking the coast.

If truth be known, I couldn't confirm how long it took, or in which direction Reggie had driven to get us here, because my mind was still entranced with that wondrous house –and I was completely oblivious to everything else that was going-on around me.

As the pair of us vacated the car, Reggie literally *slammed* his driver's door –in an attempt to snap me out of my trance, no-doubt, and it must have worked, because I remember walking across a courtyard of pebbled stones before reaching the doorway of this enchanting eatery. The owner of the restaurant obviously had a somewhat macabre sense of humour, because outside the main entrance there were three piles of earth, each one stretching the length of a (dead) man's body. At the head of each mound stood a small headstone, with an epitaph to the deceased individual, engraved upon it, all of which were quite cynical about the way that each person had *supposedly* croaked-it! At the base of each mound a pair of winkle-picker-type shoes stood upright, their toecaps pointing away to the stars, as if proclaiming that each man died with his boots on...except the middle one. This one had a pair of boots pointing downwards, with a single high-heeled shoe pointing upwards on either side of the boots, and so it didn't take much imagination to guess what this guy was supposedly doing when he went to meet his maker!

As Reggie and I sat outside on one of the picnic benches, a rather voluptuous young waitress sauntered over to our table. Donning a pair of very tight shorts and a low-cut skimpy vest, she then proceeded to lean precariously over our table, whilst Reggie and I stuttered-out our requirements to her. "Florida is looking more attractive with each passing minute", I joked to Reggie, as the young lady wiggled-away with our order. "Why do you think I brought you here Shaun", Reggie replied. Reggie was not a lecherous old man, but a very cunning and shrewd salesman, who would stop at nothing to entice a customer into buying one of his properties –and so I had to admire his tactics –however sleazy they might seem.

If Caryl had been accompanying me today then I am positive that Reggie *would have taken us to an entirely different restaurant.*

Having enjoyed a plateful of spicy chicken drumsticks, along with a huge bowl of mixed rice and salad, Reggie and I headed-off to a nearby construction site, where suntanned builders worked at a painfully slow pace in the blistering heat –and who could blame them? Reggie had brought a small camera with him, and so he proceeded to take several pictures of one of the half-built homes, before asking me to stay put for a few minutes, while he disappeared out of site, around the back of the building.

True to his word, Reggie returned within minutes, before crouching-down on one knee and taking a few more photographs of the front of the house –and then he was done. Feeling slightly curious by now, I asked Reggie why he had taken so many photographs of a house that was only half-finished, and he told me that it was company policy to take pictures of all of the properties that had been sold, *'before'* (meaning the bare plot of land, I presume), *during construction* and *after completion,* so that any agencies involved with a purchase on behalf of a client could verify that all of the structural work was conforming to the customer's specific requirements, and that everything was in accordance with Florida's building regulation standards. I gathered that these homes must be for V.I.P. customers, and it wasn't long before Reggie confirmed my theory. "This house has been bought by the drummer of Merillion", Reggie said, as we drove past one of the completed *mansions* on our way to the sales office.

As soon as we entered the office I could see a large photograph on one of the walls of Ronnie Briggs (from Coronation Street) standing outside his new home, whilst accepting his keys from one of the sales advisers. It was all very impressive stuff, and the show-homes were once again out-of-this-world, but I had already set my heart on that first house, and so nothing, or nobody was going to change my mind –no matter how hard they tried. However, tomorrow I would be meeting up with my third and final housing representative in Fort Myers, which is 150 miles south of Newport Richey, and so I couldn't wait to see what he or she would have on offer for me? Reggie very kindly offered to give me a lift into Tampa tomorrow morning, in order for me to catch my bus, as he said that he would be going that way anyway, and so, after arranging a suitable time with him, and then checking out my bus timetable, I duly rang the office in Fort Myers to inform them of my ETA (estimated time of arrival).

Things were going along swimmingly, and I only wished that I had the money to ring Caryl and let her know how prosperous and enlightening my visits had been, but I just daren't risk running-out of money.

CHAPTER 38

Across The Great Divide

By the time Reggie turned up the next day I had less than twenty minutes to catch my bus, and so he had to drive *hell-for-leather* to get me to the terminal on time. Thankfully, my Greyhound Bus was running about half an hour behind schedule, which seems to be the norm, and so I need not have panicked after-all. Once again I managed to get a front seat, only this time I didn't have it all to myself, for sitting right next to me is a chap called Samuel, who tells me (in a very broad Irish accent), that he is from Belfast in Northern Ireland, and that he is travelling all the way to Fort Lauderdale on Florida's south-eastern coast. He already has a few friends who are on a working holiday there, and so he is planning to meet up with them this evening, and then stay with the guys for about a week or so, before moving on down to Miami and the Florida Keys. Samuel is great company to be with, and so it wasn't long before we had exchanged telephone numbers and addresses, including where he would be staying in Fort Lauderdale, just in case we both had time to meet-up again.

Another friend had been made, and another column in my address book had been filled, which is half the fun of travelling, especially when one is travelling alone. By two o'clock our bus was pulling up in Fort Myers, and so Samuel and I said our fond farewells to each-other, the pair of us sincerely hoping that we will meet-up again real soon.

Fort Myers, which was named after Colonel Abraham Myers, has also been dubbed as the 'Gateway to southwest Florida'. It is also affectionately known as the 'City of Palms', due to the scores of beautiful palm trees that were imported and planted along what was once Riverside Avenue, but has since been renamed McGregor Boulevard. Fort Myers, which is the county seat and commercial centre of Lee County, Florida, was built in 1830, around the time of the American Indian Wars, its fort being used as a base for operations during the Seminole Wars and the Indian Removal period. Cattle ranchers and Confederate blockade runners were also based in Fort Myers during the American Civil War. In 1898 the magnificent 'Royal Palm Hotel' was constructed, immediately putting Fort Myers on the map as a major winter tourist destination, and six years later, access to the city was greatly improved by the opening of the Atlantic Coast Line Railroad.

In 1924 construction began on the Edison Bridge, and seven years later, on February 11th 1931, the bridge was officially opened. As this day was also Edison's 84th birthday, it was quite fitting that he would not only be the man to dedicate the bridge, but he would also be the first commuter to drive across it. Today, the city of Fort Myers is very popular for both foreign tourists, as well as the residents of Florida itself, all of whom flock to the area in their droves primarily to enjoy its sun-kissed shores and fabulous beaches. Two of the most famous attractions are the winter estates of Henry Ford (The Mangoes) and Thomas Edison (Seminole Lodge). Thomas Edison and Henry Ford rank among the most famous men in America history, along with Harvey Firestone, and all three of these gentlemen were once members of an elite group entitled 'The Millionaires Club'. In Fort Myers' Centennial Park, there are three individual statues dedicated to each of these great figureheads, along with a copper plaque that reads: Uncommon Friends, Edison, Ford and Firestone 'Fathers of the American Industrial twentieth century'.

As luck would have it, there were plenty of taxis at the terminal, and so it wasn't long before I was *winging my way* to 'Tarawoods', my next port-of-call, and hopefully my home for the next two nights. The lady I spoke with on the telephone yesterday had told me that the site was only two miles off the main highway, and that the Greyhound Bus would actually pass the main entrance on its way into the city-entre, and so I knew that I did not have far to go, which was a good thing, as I only had a small amount of dollars to my name right now. However, *sod's law* managed to show its face once again, as I find that I am twenty-five cents short of the fare, and a small part of me now regrets having told that young lad to keep the money to phone his mother, as my *minuscule gesture* had now left me in this somewhat awkward position.

Thankfully, the driver is a decent chap, and so he duly let me off the twenty-five cents, which is *karma* I suppose *—one good turn and all that stuff.* I am greeted at the reception office by a very attractive, and also quite petite blonde lady, who looks about the same age as me. (*I am 33 years old, by the way*). She introduces herself to me as "Jasmine", before telling me that she had only been informed a few hours ago that I was on my way! "Déjà-vu", I say under my breath. After doing the usual check-in formalities, Jasmine and I, along with my luggage, of course, boarded a golf buggy, before heading-off into the main site. As we cruised along the concrete pathway Jasmine expressed to me that this was primarily a retirement complex for elderly people "And so don't expect to meet up with anyone less than sixty years old Shaun —except me, of course", she added with a roll of her eyes and a hearty chuckle.

Feeling slightly out of place right now, I decided to tell Jasmine a little white-lie, by saying that besides keeping my eyes open to all options for my own immediate family, I was also looking at property on behalf of my in-laws, who were seriously considering retiring to Florida, just so that Jasmine would not think that I was here under any false pretences. The site was immaculately kept, and also beautifully landscaped, with a small lake set right in the centre of several acres of grounds. Sitting on a small hillside adjacent to the lake was an owner's clubhouse, which boasted a large restaurant and bar area, an outdoor (heated) swimming pool, and a huge Jacuzzi that could accommodate around twelve people at any given time. As for the mobile home that I would be staying in (again, I use the term 'mobile' with my tongue locked firmly in my right cheek) well it was simply 'exceptional'. Apart from having a large driveway, which leads to an independent garage, the surrounding area also boasted an enclosed patio area with eight brick pillars and a huge sun canopy. From the outside the place truly looked magnificent.

By now I already knew that the *inside* would not let the *outside* down, as the American standards are obviously way-above anything that we Brits are used to, and so with that thought in mind I duly followed Jasmine into the main living area. It was just like walking into Dexter's place in San Diego all over again, hence it is not worth bothering you with the intricate details of the interior, and so I will simply say that it was "Unbelievable". After showing *yours truly* around the place, Jasmine handed me over a set of keys, before apologising for the lack of entertainment that goes on around this area in the evenings. However, I assured her that everything was just fine, and so she left me to my own devices. Jasmine had not mentioned anything about taking me out on a *full-blooded* sales presentation in the morning, and as she had only been informed of my arrival an hour prior to me turning up on her doorstep, she had obviously not planned anything in advance, and so I got the feeling that on this occasion it would be entirely up to me to make up my own mind on whether or not I wanted to buy a place here?

As this was self catering accommodation, it meant that for the first time I had my own fridge to put food and drinks in, and so I decided to find the nearest shop where I could stock-up on groceries to cover me for the next two days and nights. Just as I was locking my front door, I heard the toot of a horn, and so I turned around, only to see Jasmine beckoning me from the end of the driveway. She said that she and a few of her friends were popping over to a Halloween party in town this evening, and that I was more than welcome to join them. Thanking Jasmine for her kind invitation, I gladly accepted her offer, and so Jasmine said that she would collect me from my place at around seven o'clock this evening. As I was now going out for the evening, I decided to reduce my food intake, as one thing I cannot do is go out drinking on a full stomach. Approximately one hundred yards from the entrance to the Tarawoods site is a 'Circle K' grocery store, and so it didn't take me long to stock up on the minimal amount of provisions that I would need for the next couple of days. However, just as I was about to join the short queue at the counter, I suddenly noticed that the shop also had hot meals on offer.

Two hot-dogs, along with a beaker of percolated coffee was being advertised for $1.99 (around £1.35) which was an absolute bargain in my eyes, and would probably work out as cheap -if not cheaper, than cooking for myself. After returning a few items back to their rightful places on the shelves, I immediately took the store up on their *generous* offer, paying my subsequent dues, before wolfing-down two deliciously tasty hot-dogs –yummy. I then took my coffee with me to the swimming pool, where I spent the next couple of hours sunbathing and relaxing –this really was the life. Feeling sufficiently *bronzed* for one day, I returned to my *palatial* home, where I then decided to have *forty-winks* in one of the four beds that were currently on offer to me. By six-thirty I am ready, and at seven o'clock sharp, Jasmine is knocking at my door, along with her friend, Eleanor, who had apparently driven fifty miles from Sarasota in order to attend the party this evening.

Jasmine then drove the three of us to a bar in the town centre, where I was cordially introduced to Jasmine's boyfriend (whom I shall call 'Alex' from now on –as I cannot recall his real name), along with two of his buddies, one of them having apparently come-along as a blind date for Eleanor. The six of us got on famously together, and we all laughed buckets at some of the other party guests, who had turned up at the pub in the most outlandish costumes. It was after midnight when we finally got to the party, but as it was expected to continue-on well into the early hours of the morning, we were by no-means the last people to arrive. Balloons, pumpkins and multi-coloured decorations hung from every tile in the ceiling, and gruesome creatures were in abundance, predominantly witches and warlocks, who surrounded us in every direction, their long, crooked noses, lime-green faces, and black-pointed hats, making them look like *the real McCoy*. The night was alive, as young *cow-girls,* dressed in chequered blouses and shorts, along with knee-length cowboy boots and huge Stetson hats, posed for photographs with the guests.

A makeshift robot also mingled his way through the crowds, before *strutting his funky stuff* on the dance floor with a mixture of *ghosties* and *ghoulies.* I told my fellow companions how much I loved Florida, and how I couldn't wait to live here, only to be shot-down in flames, as they duly informed me that I would have no chance of living in Florida, unless I was married to an American woman...or man, of course! "I'll marry you", said a female voice behind me, but as I turned to face my willing-fiancée, I not only realised how drunk she was, but also how inebriated I would have to be, before even considering accepting her proposal! As the clock on the wall struck 2am I realised that for me, this was truly 'the witching hour', as my legs couldn't take any more dancing, and my stomach couldn't face any more alcohol. Thankfully, Alex was also feeling slightly worse-for-wear, along with Eleanor, who incidentally, had decided not to take matters further with her blind date, and so Jasmine, who was the only sober one out of the four of us (or 'six' if you count Alex's two friends) offered to drive us all home.

As we vacated the building a black stretched limousine pulled-up into the car park, although no-one got in or out of it during the time it took us to walk to our car, and so lord-knows who it was for, or indeed, if anyone was inside of the vehicle? However, it was the first limousine that I had ever seen *in the flesh*, so to speak, and so just seeing the *beast* had rounded-off my night perfectly. Today is the first day of November, and last night Alex had kindly offered to take me out on his boat around lunchtime today, to a small island off the coast, where we will apparently do a spot of fishing together, and so I am really looking forward to this latest adventure, albeit only a small one, of course. It has just turned ten o'clock, and although I am not suffering too badly from last-nights' party, I still need to wash the *sleep* out of my eyes and the cobwebs off my face, and there is no better way of doing that than diving headlong into a swimming pool. Relaxing on my sun-bed and drying off the residue from my *dip in the drink* I am suddenly accosted by an elderly lady who, up until now had been having a leisurely nap on one of the sun-beds nearby. (*I know this, because I had heard her snoring when I first arrived at the pool*). She politely asks me what I am doing in an establishment like this, before jokingly adding: "I think that you are about thirty year's too soon young man?"

Feeling obliged to explain myself to the lady, I tell her the whole story, including the fact that I am hoping to pay a visit to Key West whilst I am in Florida. However, no-sooner had I said this than the lady immediately extracted a pen and a piece of plain paper from her handbag, before scribbling-down a few notes. "Now this is my daughters' telephone number and also her address; she lives in Sugarloaf Shores, on Sugarloaf Key", the lady said, whilst handing me over the piece of paper. "If you do make it to The Keys, then please feel free to pop in and see her. Just give her a call first, to let her know when you are coming and she will be happy to put you up for a couple of nights. Simply tell her that you met 'Molly' in Tarawoods and everything will be fine, I promise you young man". I was simply overwhelmed by her trust in me, and also by the friendliness which she had shown to a complete stranger. We had known each other for less than half an hour, and yet I already felt as though I was a part of her family —what a woman!

It is Jasmine's day off today, and so I sat talking to her colleague in the office, whilst waiting for Alex to pick me up at 1 pm, which we had agreed upon last night. Unfortunately, he never showed up, but then considering the amount of alcohol that he had consumed when the arrangement was made, I cannot say that I was overly surprised. Jasmine's secretary kindly rang Jasmine's home number, to see if Alex had stayed there the night, but there was no answer from Jasmine's phone, and so that was the end of that. As I began walking back to my humble abode, I soon realised the mistake I had made by not putting socks on my feet this morning, because my training shoes had managed to create a small blister on my left heel that was really painful to walk on, and so I ended-up trudging barefoot all the way back to my temporary home. With nothing really better to do, I decided to sit back on the giant sofa and relax in front of the television set. The comedy film 'Down and out in Beverly Hills', starring Nick Nolte, Bette Midler and Richard Dreyfuss, had just started, and so I spent the next ninety minutes laughing at the hilarious antics in the movie, before retiring to the hot-tub for a long, lingering bath.

Tonight it would be up to me to find my own entertainment, and so after doing my usual wash and brush-up routine, I sauntered over to the local golf club, where I intended finishing-off my stay in Fort Myers with nothing more than a couple of quiet beers at the bar. However, my idea of a sedate night, chatting peacefully with the resident bartender, soon turned sour, as the security guard on duty refused to let me into the club without a pass, and so before I knew it, I was back home in my *oversized caravan*, making a cup of coffee in the kitchen, before trundling off to bed for an early night. *Check-out Day* has now arrived, and after six nights of complimentary accommodation, the cost of putting a roof over my head will be strictly down to me...well, at least until I get to meet-up with Leo again in West Palm Beach. I have time for one last swim in the pool before breakfast and then I will be off on my travels once more, my latest journey taking me eastwards, across what is affectionately known as 'Alligator Alley', to Florida's southern Atlantic shores.

At the poolside I am accosted by a lady in her late sixties, early seventies perhaps, who says that she has been watching me from her sun-lounger, adding how intrigued she is that I am so young to be living in a *retirement home!* The ladies' name is Maisie, and just like Molly, she begins asking me all the usual questions, like 'Where am I from and where am I going'; 'how long am I staying in America for'...the usual stuff. As she seems a very nice lady, I do not want to disappoint her, and so, just like Molly, I tell her my full story, before politely sending her packing, by saying "Oh well, I must be off now, as I have a bus to catch". However, by the time I have changed into a pair of dry shorts and rolled-up my bathers and towel, Maisie is back at my side, hell-bent it seems on getting to know me more intimately, as she hands me over her business card. On the front of the card are all of her personal details, such as her name and telephone number, along with the name and address of a company, which I gather she owns, as she is a very well-spoken lady.

On the blank side of the card Maisie has written the number of her mobile home in Tarawoods, which she and her husband now live in permanently, and as she pointed-out where it was situated, she made the following speech:

"If you and your lovely family decide to come and live in Florida, all five of you are most welcome to live with my husband and me for as long as it takes for you to secure full time employment, and for you to find a home of your own. My husband and I have no children, and so we would love to help you out –and also to spoil your children rotten...like honorary grandparents, if you like...I hope you are not offended by this".

How could anyone be offended with such a wonderful act of kindness and generosity? I was speechless. All I really wanted to do was to give her a massive hug, but somehow I managed to refrain myself from going all sloppy and gooey with a person I had never met before in my life, and so I politely thanked her for her kind gesture, saying that I would definitely keep in touch with her.

With my bags all packed and ready to go, I made my way over to the office, with the simple intention of asking Jasmine to book a taxi to the Greyhound Bus Terminal in the centre of Fort Myers for me. However, when I got there Jasmine was full of apologies (along with a handful of excuses) for her boyfriend letting me down yesterday, saying that he had to go and collect a car for her that her ex-husband had written-off the night before, before adding that her ex was not insured to drive it, and so it had caused all sorts of problems for her. In recompense for leaving me in the lurch, Jasmine kindly offered to give me a lift to the bus station, which was an offer I gladly accepted, as time was once again running short. By the time we got to the bus depot, we had missed the original departure time by seven minutes, but as ever, the bus schedule was running fifteen minutes late, and so once again I was a very happy man. Thankfully, there were no inquisitive passengers wanting to give me the third degree about my past, present and future life, nor any *know-it-all's* that were ready to spread *doom and gloom*, for the entire world to hear, and so I decided to grab myself a front-line seat, and have a friendly chat with the driver.

As we began crossing 'Alligator Alley', our driver, whom I have simply called 'Fred', decided to enlighten me about these notorious giant lizards that have roamed the Earth for millions of years.

"There are literally thousands of alligator's living in the swamps of Southern Florida, and contrary to popular belief, they are actually more scared of humans than we are of them, and so they will only ever attack a human if he or she stray's too close to their nest, because like all creatures great and small, they will protect their off-springs to the death. Their diet primarily consists of fish and raccoons, although they are partial to the odd dog, and so should any budding angler be fool-hardy enough to go fishing in a small boat with his 'best friend', then he might just be in for a heap of trouble! Alligator's jaws have a tremendous amount of strength, and they can rip a limb from most animals with the greatest of ease. Their bodies are also very versatile, as alligator's are able to spin around in the water at incredible speeds, as they viciously deliver what is commonly known as their 'Death Roll', disorientating their prey, before drowning or choking them, and then taking them to a safe haven to be dismembered and devoured at their leisure".

I was so engrossed with Fred's knowledge of these glorious creatures that I simply sat back and listened intensely, while he carried-on with his *lesson*.

"An alligator's tail is another deadly weapon in this creature's armoury, which he often uses with terrifying force when defending himself from any would be assailants, such as hippos, lion's, or even other gators. On land, alligators can run up to forty miles an hour, and they are often seen taking short-cuts across the highway, in search of food, or a mate. Occasionally an alligator will get knocked down by one of the thousands of vehicles that use this highway every day, more-often after dark, when the drivers are unable to spot them until it is too late, which is a sad reality of life that the animal lovers of Florida have had to come to terms with".

We actually drove passed a dead alligator, and also a dead raccoon on this trip, which was certainly not a pleasant sight, and so I fully understood how the animal lovers of Florida must feel. Alligators are a protected species in Florida, and according to my tutor for the day, they can have anywhere between 100 and 200 mates at any given time. Fred then went on to tell me how much he enjoys fishing in the swamps, with nothing more than a pair of thigh-length waders to protect him from these *amphibious dinosaurs,* which made me realise what a nutcase he must be...sooner you than me, my friend. About halfway across the alley, we stopped at a kind of trading post-cum-diner for a spot of lunch. As I meandered my way along the main food counter, tentatively collecting-up the odd sandwich pack, along with one or two biscuit bars, as a special treat to myself for the day, I am suddenly confronted with this giant alligator's head, its hideous jaws standing to-to-toe with me, as if ready to tear my head off at any given moment –and its beady eyes staring me directly in the face, as if letting me know "You're next buddy".

Thankfully, the animal was long dead, and it had been duly stuffed, but it was enough to give me total respect for not only alligators, but crocodiles as well, both of which are sadly hunted for their skins by poachers all over the world. We were now in the heart of the Seminole Indian Reservation, and so as we re-boarded our bus, I asked Fred if he would be kind enough to enlighten me about the tribes who still lived deep in the swamplands of the Everglades. His rendition of events went something like this:

"The Seminole Indians are a Native American tribe, who originally lived in southern Florida, although the majority of them have now moved to Oklahoma. Back in the 1800's the Seminoles battled with the settlers primarily over land issues, but also over escaped slaves, which the Seminoles gave food and shelter to. The government eventually stepped in, insisting on moving the Seminoles to a chunk of Indian Territory west of the Mississippi, but they refused to go. Over a period of forty years three major wars raged between the Seminoles and the soldiers who were sent to oust them, the Seminole numbers decreasing with every new treaty that was introduced and enforced to remove them from their homeland.

Vastly outnumbered by their enemies, the Seminoles used guerrilla tactics to outsmart the soldiers for many years, as the Indians knew the everglades like the back of their hands, but by 1958 almost all of the Seminoles had finally been relocated. However, around two hundred of them still remain in the *'Land of the Watery Grass'* and it is claimed that to this day the Seminole Indians of Florida are still the only tribe in the US who have never signed a peace treaty with the United States government".

I thoroughly enjoyed my latest history lesson, and looking out of the front and side windows of the bus, at the hundreds of square miles of swampland that surrounded us in all directions, I could easily understand why the soldiers had so much trouble in finding the Seminoles. Having always wanted to tour America on a motorcycle, but trying to imagine what it would be like crossing this stretch of road without any external protection, I duly asked Fred how dangerous it would be for a motorcyclist, should he happen to break down, or simply run out of fuel in this gator-infested area?

"My friend, the mosquito's would get him long before the 'gators did!" was his rather surprising answer. Taking his left hand off the steering wheel, Fred then used his thumb and forefinger to give me an idea of how big they are. It made my flesh creep. Having realised that his *somewhat inquisitive* passenger had obviously had enough of talking about deadly creatures and the like, Fred changed the subject completely, enlightening me instead about some of the weird and wonderful passengers that he had carried on his bus over the last twenty two years, although lord-only knows why I never made any notes on these stories? The mileage that Fred had covered during his time on the Greyhound Busses was overwhelming, and it made my relatively miniscule journey look like *a walk in the park*. As we approached our destination, I asked Fred where I would find the nearest youth hostel accommodation, in case Samuel was unable to put me up at his friends place in Fort Lauderdale.

Fred said that there were two hostels in Fort Lauderdale, one at either end of the city, before adding that I would be able to find the telephone numbers of both of them in the phone booth inside the Greyhound Station.

The city of Fort Lauderdale, which is located twenty eight miles north of Miami, is the county seat of Broward County. Fort Lauderdale boasts an average of 3,000 hours of sunshine per year, with temperatures rarely dropping below seventy five degrees, hence in is a very popular tourist destination. With over five hundred hotels and motels, encompassing well over thirty thousand rooms, and almost 300 parkland campsites, there is certainly no shortage of accommodation in this part of Florida. The city also encompasses around four thousand restaurants, lord knows how many bars —and over 130 nightclubs, and so there is definitely no lack of entertainment for the discerning party-goer. Add to this a plethora of museums, a dozen shopping malls, and over sixty golf courses to keep one busy in the daytime, and it is no wonder that people fly into this city from all over the planet. Visitors also call into the city via the numerous cruise ships that sail into its harbour each year, and with around 100 marinas, housing over 40,000 resident yachts, suffice it to say that there are plenty of millionaires living in Fort Lauderdale too.

Fort Lauderdale takes its name from a series of forts that were built in the name of Major William Lauderdale, a planter-soldier from Virginia, who fought in many conflicts, including the Creek War, The Battle of New Orleans and the Seminole Indian War, although prior to the 20th century, the area was actually known as 'The New River Settlement'. Like so many other areas in the state, Fort Lauderdale flourished during the Florida Land Boom in the 1920's, only to come crashing down during The Great Depression of the 1930's. During World War II, Fort Lauderdale became a major base, encompassing a naval air station that not only trained pilots, but also radar and fire control operators. Just off the coast of Fort Lauderdale is Osborne Reef, which was originally made up of around three quarters of a million disused tyres. Unfortunately, this man-made reef eventually turned into an ecological disaster when the straps and cables that originally secured the tyres back in the 1960's, eventually rusted, before snapping and subsequently releasing the tyres from their moorings. In the years that followed thousands of tyres washed-up on the beaches, especially during hurricane season, causing a major clean-up campaign to be instigated by the local authorities, which involved the US army, Navy and Coast Guard.

CHAPTER 39

Fort Lauderdale

After vacating the bus, I duly rang Samuel from the said booth, but unfortunately he was unable to put me up at his friend's place, and so I rang one of the hostels, but likewise there was *no room at the Inn* -and I was soon to find out that the other one was also full! I then noticed an advertisement that had been taped to the wall alongside the booth, for a nearby hotel, and so I immediately rang the number, but with a price tag of sixty five dollars per night, it was way out of my price range, and so that one also *bit the dust*. This was my first night of having to find my own accommodation, and things were already looking pretty grim.

"Come on Shaun; you're a salesman, for god's sake, so get rid of this defeatist attitude and think of something positive to make this thing work", I scorned, under my breath, of course, as the waiting room was currently packed to bursting with fellow commuters, who would probably think that I am crazy, if I said it out loud. I decided to ring the first hostel again, only this time I really laid it on thick, pouring my heart out to the guy on the other end of the phone about my current situation, before virtually insisting that he sort me out a bed for the night. Trying to give someone a guilty conscience is somewhat unethical, I suppose, but desperate situations often require desperate measures, and I guess it is fair to say that this was one of those instances. If the guy genuinely cared about helping people, and he was in a position to do anything about it, then hopefully he would assist me with my plight? However if he wasn't that type of person, and simply ignored my plea, then all I have lost is a *quarter* for my efforts.

427

"The only thing I can offer you is a mattress on the floor" said the owner of the hostel, adding that it would cost me $12, and that I would have to share a room with several other people. This suited me fine, and so after thanking him profusely, I quickly hailed a taxi to take me to the hostel in question. However, when it came to checking-in for the next two nights, there were a few more additional costs which the guy had omitted to tell me over the telephone. Because I was not a student, or a member of any kind of youth travel organisation, the cost was now $15 per night, along with $2 for the loan of a bed-sheet at $1 per night. I would also have to pay $3 deposit on the aforementioned bed-sheet, along with $5 deposit on the room key, making it a grand total of '$40'. Okay, so I would have $8 returned to me at the end of my stay, but I had only cashed a $50 traveller's cheque today, and now, having also paid the $7 taxi fare, meant that I only had $3 left to my name!

Having to pay the equivalent of about ten pounds per night for a roof over my head, I daresay that I shouldn't really grumble about the cost, but I was tentatively counting every dime that I spent, just in case anything untoward happened during the week remaining, before catching my return flight to the UK. Upon opening the door to our dormitory, I was shocked to see that there were three sets of bunk beds taking up most of the wall space, along with one of those 'put-you-up' beds, which was still in its folded position, standing ominously in front of the only window in the room. It would not take a genius to work out who would be sleeping on that *hunk of junk* for the next two nights...namely yours truly. Initially, I was ever so grateful for the guy putting me up for a couple of nights, but after working out the mathematics, I calculated that if all of the rooms were just like this one, and all of the beds have already been taken up by guests, then the owner was making around $100 per room, per night, which, considering that the place was no bigger than your average double bedroom, was 'extortionate', to say the very least.

As I began unfolding my bed, a group of lads, all in their early twenties, I would say, entered the room. They seemed to be a decent bunch of chaps, all laughing and joking together, as if they had known each other for years, and so with a friendly atmosphere all around, needless to say that it wasn't long before I was introducing myself to the majority of my *fellow in-mates*. The guys then reciprocated, of course, and over the course of the next few minutes, I had become acquainted with Max, Tommy, Lewis, Damon and Jayden. I also met 'Archie', although we were not formally introduced, as he had been sleeping on one of the top bunks when I first came into the room, and by the time the four of us left the dorm, to go down to the local pub around two hours later, he looked more in a state of unconsciousness, rather than simply sleeping to his hearts' content. Apart from a list of rules and regulations, which everyone is given when they are checking into the hostel, all of us had also been handed vouchers for a free drink at one of the local bars, as part of our individual welcome packages, and so without further ado I decided to make full use of mine.

Our friendly barmaid is a lovely Australian lassie called Amelia, who tells us that she has been living in Florida for the past six months, and that she will soon be embarking on a thirteen night Caribbean cruise, which will be setting sail from the shores of Miami. However, apart from this trip being a dream holiday for her, that Amelia was looking forward to immensely, of course, she added that it was also a journey that she had take for legal reasons. Seeing the look of confusion on our faces (but hopefully not the fear in our minds that she might be *doing a drugs run*, in order to support her fabulous lifestyle) Amelia then explained the situation.

"You see, under American law, foreigners –I.E. 'Non-American citizens' who have a temporary working permit, are only allowed to remain in the US for a maximum of six months, before having to leave America when their permit runs out. By going to the Caribbean, a person is actually leaving American soil, and so, apart from filling in a mandatory 'exit' visa their passport is duly stamped, stating that they have left the country altogether. Upon returning to Miami, two weeks later, I will simply apply for a new six months working permit, as there are virtually no time constraints between finishing one permit and starting another one. The only stipulated criteria are that your passport shows that you have left the country before the end of your last permit, before re-entering it again at a later date."

Like the majority of Australians I have met on my travels, Amelia has already travelled around half of the world, and she doesn't intend stopping until she has covered the other half. Apart from admiring this young lady tremendously, for what she has accomplished at such a young age (she was still in her teens) I was also in awe of her tenacity and self-belief that she would one day achieve every one of her goals in life, which included seeing as much of our planet as is humanely possible during her lifetime. I salute you Amelia. Max's story was equally as fascinating as Amelia's, albeit that the funding for his travelling came from the sale of his house in the UK, rather than having to work his way around the world.

Max began his travelling in Australia, where he claims that he lived with an exceedingly rich couple for several months, before flying over to California, to pick up a car, which he then drove 2,500 miles across the southern states, before finally dropping down into Florida. The car in question belonged to an ageing American couple, who had apparently paid an agency around $500 to have someone drive their car across America, after being told that it would cost them over a grand to have their car transported by rail from coast to coast. Max was duly paid $250 for his eight day crossing of the American continent, which included the $120 that he would have to spend on fuel, in order to cover the distance. However he was well happy, having been paid $130 to be a part of the *all-American dream*, even though he had spent the majority of the cash staying in very cheap youth hostels along the way. Max also admitted that he had spent virtually all of the profit that he had originally made from the sale of his house, and so pretty soon he would have to return home, and once again become a member of the ubiquitous working classes, which was something that he was dreading with a passion.

Whilst he was in Australia, Max's parent's had sold their home in Bolton, after his mum had been made redundant, and his father had retired, before emigrating to Alicante, in order to live-out the remainder of their lives basking in the glorious sunshine, on Spain's beautiful Mediterranean coast. Max said that he would probably fly over to their villa for Christmas, before returning to Bolton, in order to join the 'rat-race' of hard grafting workers "With very little reward for one's efforts", he added in a venomous tone. Max was now beginning to sound like me, and yet he had already been lucky-enough to *live the dream*, albeit temporarily, whereas I was still chomping at the bit. Max and I have now been joined by a German chap called Gunter, who is ever so friendly, although there is confusion on all sides, as he tries his uttermost to converse with us in English, even though he barely knows a handful of words. However, with a multitude of hand signals, along with several pitchers of lager, and a stream of dirty jokes, the three of us managed to laugh our way into oblivion. Incidentally, the other two lads had long-since left our company, after chatting-up two lovely young ladies from lord-knows where?

By 8 am the following morning it is scorching outside and the humidity level inside the dorm is stiflingly oppressive, making it impossible for me to sleep on this uncomfortable mattress that is no thicker than your average pin-cushion! Also, having the aroma of seven dehydrated, sweaty carcasses, wafting heavily around the room, is not doing my nostrils any favours right now. After forcing my *crispy eyelids* apart, I staggered wearily to the door, whereupon opening it the blistering sunlight hit me like a blinding flash, causing me to squint intermittently as I meandered my way to the swimming pool that was thankfully only a few steps away. Without a second thought, I simply threw myself into the 'drink', the cooling waters immediately cleansing the sweaty pores in my skin, and refreshing my whole body in a feeling that can only be described as '*Orgasmic*'. Unbeknown to me, but today is 'Election Day' in America, and before I know it I am being handed a sticker that says 'Vote for Perod' by one of the candidates devotees, who is combing the streets of Fort Lauderdale, happily handing-out leaflets, badges, paper hats and stickers, along with anything else that will assist in promoting the campaign for his *great leader*.

I did not envy his job one iota, and even watching him parading up and down the sidewalk at an erratic pace was making me sweat profusely. Having dried myself for the umpteenth time, I begin chatting to a gentleman who has been sitting around the pool since I first left the dorm. In fact he may well have been sitting there all night, because he was still wearing the same red shorts and white tee-shirt that he had on when I first spotted him on my way home from the pub last night –and he is also sitting in exactly the same chair as he was sitting in when I went to bed. During autumn time in Britain, this would be unthinkable, of course, as being dressed in such flimsy attire overnight, one would probably freeze to death, but in Florida's sub-tropical climate, this was quite a feasible assumption to make. Unlike my good self, this man is tall, dark and handsome, and he is donning a thick, bushy moustache, which makes him look much like the actor, Tom Selleck, who is more affectionately known as (Thomas) 'Magnum', from the television series of the same name, which was filmed on the island of Oahu, in Hawaii, and was so popular with viewing audiences throughout the eighties.

The permanent tan that this guy is also donning, tells me from the off that he is definitely 'not' a tourist. After doing the usual introductions, my new found friend (who I shall call 'Mason') and I begin chatting-away, nineteen to the dozen, telling each other our own personal stories, simply to while away our time in the glorious Floridian sunshine. Mason tells me that he is a sea captain, who spends half of his life on land, and the other half sailing up and down the Floridian coast, and around a selection of the Caribbean islands. Although Mason is not the proud owner of a sea-going vessel himself, he is registered with one of the many seafaring agencies in Fort Lauderdale, and so whenever a yacht owner, or simply a party of tourists want to set sail for a sunshine cruise, his job is to take charge of the vessel in question for however long it takes. I asked Mason if his lifestyle was as glamorous as it sounded, or does it ever become repetitive and possibly even boring at times?

"Well, it's funny that you should say that Shaun, because only last year, at the end of a very hectic summer season, I decided that I would take time out from the job, and head off back home to New York for the winter season. Apart from seeing all of my friends and family again, I was also looking forward to a change of climate, as quite often the searing heat and stifling humidity which is ever present throughout the summer months in the southern half of Florida, can really get to you. The blessed hurricanes, that are prevalent throughout September and October, can also be a *pain in the butt* –especially if they are of the 'extreme' variety, because not only do they tear up the town, but they also curtail any chances of work for me, due to the cancellation of all shipping lines. However, when I worked out the cost of buying a whole new set of winter clothes, which I would obviously need for the freezing-cold weather up north, I said "Sod that" and just sent my folks a Christmas post card instead".

I had to chuckle at the irony of Mason's story, for here was I, having no choice but to live through the freezing cold winters that the UK is famous for, while my body and soul simply yearned to live in the warmth and sunshine that he enjoys all year round and here was Mason, happily willing to sacrifice everything that I dreamed of in my life, simply to 'have a change of climate'. Of course, this was not strictly true, because primarily he would be going home to see his beloved family, and I could fully appreciate his sentiments, because without having Caryl and the children to go home to, my life would not be worth living. It also made me wonder whether Mason was truly 'Living the dream', or perhaps deep down inside, he was really a lonely person, who was trying to kid himself into believing that he was extremely happy...who knows? Apart from a complimentary beer, anyone staying at this hostel is also entitled to free coffee and doughnuts for breakfast every morning, and so that was the next item on the agenda for Mason and me.

Although I am trying to watch my weight, especially after seeing the photos from last years' trip to California, my pocket is more important to me right now, and so I was happy to wait in the long queue, before accepting my handout with relish. Our nearest beach is located behind a string of hotels that dominate both the skyline -and also the promenade on the other side of the road from our backpacker's hostel. However, how one is supposed to gain access to any part of the beach is a mystery to me right now? Up and down the avenue I wandered, hoping to find some kind of entrance to the beach, or even a small alleyway that would lead me to those golden shores, but there was nothing? After one final attempt, but again to no avail, I decided that the only way that I was going to get through to the other side of these mammoth hotels, was to walk directly through the middle of one of them, and hope to god that I manage to come out the other side without being confronted by one of the hotel staff...well, here goes.

Tentatively making my way through the reception area, whilst trying my uttermost to look as inconspicuous as possible, I pass a handful of porters, who have just exited the lift, along with a trolley full of luggage that they are now pushing and pulling at great speed, their erratic pace telling me that they are in more of a hurry than I am right now, and so I nonchalantly ignore them, slipping-out through a side door and into a long corridor. At the end of the corridor is the main restaurant, where I can see a volley of waiters, all clearing-up the tables, in readiness to close their doors, probably until lunchtime? Picking up my pace, I quickly slip through the sliding glass doors, scuppering past the last remaining diners, before exiting onto the main patio area, where dozens of people are now relaxing on their sun-beds around the main pool, and a handful of children

433

are playing happily together in the paddling pool. Beyond the terrace is a golden, sandy beach, stretching-out as far as the eye can see, and behind that lies a billion gallons of crystal clear water, just beckoning me to take advantage of its cooling, calming commodities.

Having completed my *five day working week*, I told myself that it is now the weekend, and therefore I am entitled to enjoy myself, and so without further ado, I kicked off my training shoes and ripped off my tee-shirt, before charging across the sand and flinging myself headfirst into the Atlantic Ocean. I am pleasantly surprised at how warm the water is, and so I continue frolicking about for several minutes, before prostrating my carcass on the wet sand at the waters' edge, letting the small, frothy ripples of salty water gently cascade over my body in a soothing rhythm that almost sends me to sleep. Having fully rejuvenated my energy levels, I decide to walk the length of the beach, which is no mean feat, as it literally went on for miles, but in the end I managed to find a slipway leading back to the main road, and so at least I now knew where the opening was! Adjacent to the entrance-cum-exit is a small beach bar, and so I decided to pop in for a cool beer, as I figured that I deserved to *wet my whistle*, after having completed such a long trek.

Sitting opposite me are three lovely young ladies, probably in their early twenties, I would say, although I was never any good at guessing peoples ages. One of the girls is doing most of the talking, regaling her fellow travellers with numerous tales of her *American adventure*, whilst at the same time extracting various road maps, a book on bus and train timetables across the US, and several packs of photographs that she had obviously had developed along the way. Intrigued by her dauntless enthusiasm, I decided to shift my chair around to the other side of my table, in order that I could eavesdrop on her *somewhat one-sided* conversation. "Travelling by bus in the day is fine, but the overnight journeys can be exhaustive, especially when you do twelve in a row, which is what I did, in order to cover as much ground as possible" she bleated, one of the girls telling her that she must 'out of her tiny mind' to travel in this way. I was inclined to agree with her, as our *learned traveller* will have missed-out on all the beautiful countryside that the US has to offer, by travelling mostly at night.

Anyway, it was now time for her next story:

"When I first arrived at the hostel in Miami, there was a massive 'bottle fight' going-on outside the place, and I ended-up having an armed escort into the building...and during the week that I stayed there, some poor guy got shot!"

In contrast to that rather disturbing story, the young lady then went on to say that her main *claim-to-fame* was walking into one of the main streets in Chicago, just as they were about to start filming an episode of the situation comedy, 'Cheers'. While our reverent speaker finally paused for a few seconds, in order to take a well-awaited swig of her drink, one of the other girls decided to jump in and say her piece. "The Greyhound Bus system is good, but I decided to opt for the Delta Airlines '30-day' pass instead, which, for only three hundred dollars allows me to go on standby, to fly to any destination in the US for up to one month". I liked the sound of that deal, and so I thought that it would be well-worth checking this out, in readiness for my next visit to the US...whenever that might be?

Having eavesdropped enough on their conversation, I decided to take leave of these young ladies, regardless of how many more exciting stories I would miss out on, and head off back to my humble abode. When I finally reached the hostel, which was almost an hour after leaving the snack bar, a group of guys were gathered around the pool, and just like the young ladies that I had said my farewells to only an hour ago, there was one major speaker, who was taking full advantage of his captive audience. This rather loud young lad was telling *so-called* true stories about an Irish backpacker who had been staying at the hostel a few weeks prior to my arrival, and some of the antics that this *paddy* had apparently gotten up to sounded absolutely hilarious. Apart from the fact that he had managed to fall straight through the canvas of the retractable sun canopy, after trying to scale the walls of the building whilst being under the influence of alcohol, he had also been barred from the backpackers altogether, after he was seen jumping off one of the balconies into the swimming pool in the middle of the night —once again while he was pissed out of his skull!

Drunken revelry is par for the course when one is young, of course, and being on holiday is just the excuse that anyone needs to go *overboard* in every sense of the word, but so long as no-one is seriously injured, then I guess it is acceptable to a point. Having been on various '*football tours*' to Magaluf in Majorca and San Antonio in Ibiza, with a dozen crazy

Welshmen, I have done my fair share of serious partying over the years, but now, as a respectable family man, a husband and a father of three children, I was more than happy to take a back seat and listen to everyone else's tales. Returning to our dorm, which was like walking into an oven, I finally met Archie, who was awake for the first time since I had arrived here a couple of days ago. He was not looking that healthy, and so I asked him if he thought that it might be a good idea to visit the hospital, to whit he retorted that that was where he had just come from, before explaining to me exactly what he had been doing for the past week or so.

Archie had been released from hospital only three days ago, after volunteering to be a part of a clinical trial, by testing a new drug for the treatment of schizophrenia. For the use of his body he had been paid the princely sum of eight hundred dollars, which he said that he was more than happy with, even though the aftermath of the treatment had left him with 'the shakes', along with bouts of impaired vision that sometimes lasted for great lengths at a time. "There's a special trial coming up in a couple of months time, that apparently lasts for sixty nights, but it pays around five grand, and so I'll definitely be putting my name down for that one" Archie added, before rubbing his two hands together...in unison with the rest of his vibrating body, I might add. "You want to try it Shaun" Archie exclaimed, as he waddled out of the dorm. "Not for five fucking million buddy" I shouted back, before shaking my head in sheer disbelief at what some people will do for money. Tonight would be pretty-much the same as last night, only this time it would be 'my last night', for I knew that I must be on my way, if I intended getting to Key West on this trip.

In less than thirty hours I had struck up numerous friendships for life, and I had also added a dozen or more tales to add to my ever-increasing portfolio of true stories —stories that I would tell to my friends and family, work colleagues and total strangers in years to come —and now here I am, writing about them for all the world to read. Perhaps the friends that I have met along the way, and the people that I will no-doubt meet in the future, will one day be telling my stories around the globe, in the years that follow...who knows?

CHAPTER 40

Key West

Having consumed every one of the *farewell cocktails* that the lads had kindly treated me to last night, I was feeling slightly *worse for wear* this morning, and so I was pleased to see that the sky was overcast, as I don't think that my brain could have coped with blazing sunshine beaming down on my aching head all day. The humidity levels alone in southern Florida are enough to drain a person of their energy but couple this with a searing hangover along with a serious lack of sleeping hours over a period of several days, and total lifelessness is inevitable. Sitting outside the backpackers, happily minding my own business, whilst trying to get as much fresh air into my lungs as was humanely possible, I was also quietly contemplating whether to simply catch the bus into Miami, or to splash out for once, and treat myself to a taxi? Suddenly there is a tap on my left shoulder, and so I turn around to face Archie, who, for the first time, is actually looking a lot healthier than I am feeling right now.

Archie tells me that he is popping into Miami for a shopping trip, in order to spend a wad of his *ill-gotten gains*, before adding that I am welcome to hitch a ride in his car, so long as I am packed and ready to go within the next half an hour. I did not need to be asked a second time, and so I immediately began packing-up my gear, whilst wolfing-down my complimentary coffee and doughnut, as the first part of my rehydration programme that would probably take me the best part of my day, before getting anywhere near normality again? Next in line was to return my bed sheet and door key, in order to collect my eight dollars refund, as agreed when I checked-in. However, the guy on reception was hell-bent on selling me a tee-shirt, complete with their back-packers logo on the chest, as a memento of my stay, rather than returning my hard earned money, but after several minutes of trying he finally gave up the ghost, before handing over my refund. Under normal circumstances I wouldn't have thought twice about buying a tee-shirt, but as tight as things were right now, those eight dollars could be the difference between me getting a meal, and going hungry for the day.

The journey into *downtown* Miami only took around half an hour to complete, and Archie was kind enough to drop me right outside the Greyhound Bus Station, which was a major bonus, as there was absolutely no doubt that I would get hopelessly lost in a city this big. There are actually three bus terminals in Miami, one at 'Miami Beach', a second one in 'North Miami', and this one in 'Downtown Miami', which is where I will need to catch my connecting bus to Key West. By midday the clouds had all-but disappeared from the skies above, and now the sun was beating down mercilessly upon us all once again. My bus was not due to leave for another hour or so, and so I decided to go looking for somewhere that I could buy a packed lunch, in readiness for the long journey ahead. According to the desk clerk, the local shop should have been open by now, but for some reason its steel window shutters were still down, and the huge chain of rusted shackles that secured the main entrance, looked as though they had not been removed in years!

For over half an hour I waited patiently for someone to turn up, before succumbing to buying a handful of chocolate bars, along with a family bag of crisps from one of the vending machines that were situated outside the terminal. Finally my bus arrived, and this big, black driver emerged from the cab, the stern look on his face telling all of the passengers who were hoping to ride his bus, that their tickets had better be in order...or else! The sheer bulk of this man was also more than enough to ward off any would-be trouble-makers, and so I knew that I was going to be on my best behaviour today. About a dozen people emerged from the bus, with a view to collecting their respective cases from out of the hold, but not a single package was being handed over to any of the passengers until each individual ticket had been thoroughly checked-over by our man in charge. Whilst all of this was going on, I decided to make my way to the front of the bus, in readiness to board it, as and when we are ready to leave, when suddenly I heard this very deep, *Paul Robeson* type, voice bellowing from behind me "And where do you think you are going?"

Immediately I stopped in my tracks, as it was patently obvious who the voice belonged to, and no-way was I going to cross him. Suddenly, there was this huge hand, about the size of your average American baseball glove, looming in front of my face, as he boomed-out his next statement: "Coz you ain't goin' nowhere until I see your ticket, my friend". I could not get my travel pass out of my wallet quick enough, just to prove to him that I was in possession of a bona fide multi-ride ticket that entitled me to travel on his bus. As the driver checked over my papers, I noticed that the shop across the way was now open, and so I politely asked if I would have enough time to pop across and purchase a cup of coffee. "That is your choice entirely, my friend", he growled "Because this bus leaves in exactly five minutes –with or without you on board", he added, letting me know by the tone of his voice, that he was deadly serious. No way could I risk missing this bus, and so I decided that on this occasion, a soft drink from the vending machine would have to suffice in quenching my thirst.

A part of me was saying that this was one of those chance meetings that maybe I could have done without, whereas another part of me was saying that if this was the harshest person that I was going to have to deal with on my travels, then I could easily live with that. Working with the general public is never an easy task for anyone, and knowing the reputation that Miami has for more than its fair share of 'bad guys', I guess that this man has to put up with a volley of troublemakers almost every day of his working week, and so perhaps I shouldn't be so harsh with my judgment of him? By eleven-thirty we are winging our way to Key West, which is 164 miles south-west of Miami –and 370 miles south of Orlando –the city where my journey had begun just over a week ago. Orlando will also be the final destination of my one thousand mile round trip, before flying back to the UK in five days time.

After forty-five minutes we reached the small town of Homestead –or at least what was left of the place after Hurricane Andrew had demolished the majority of it back in August this year. The whole area had been completely flattened, and it now resembled a huge rubbish tip, rather than looking like your average American seaside town, where people had lived and worked as a happy community only a few months ago. Bulldozers by the dozen continued the laborious task of collecting-up a million-plus tons of timber and rubble that were once people's homes, unceremoniously dumping the shattered debris into the backs of awaiting trucks, before being carted away, never to be seen again. The American people are a very resilient nation, and so I knew that they would soon begin restoring their beloved town and that within a few years it would be like new once again. However, looking around me at the complete and utter devastation that this horrendous hurricane had caused, I could not help but shed a few tears for the thousands of people whose homes and lives had been completely shattered by what many would call *'An act of God'*, and so sitting silently in my seat, I decided to say a small prayer for them.

After a quick stop to pick up a handful of passengers in Homestead, we continued on through Florida City on the mainland, before finally driving out onto the skeletal line of islands that are more affectionately known as the Florida Keys. The Florida Keys ('Key' having been taken from the Spanish word 'Cayo', which means 'Small Island') is a coral cay archipelago, which is located off the southernmost tip of Florida. The climate in the Florida Keys is similar to the Caribbean, and for many years the Keys were only accessible by water, but all of that changed with the building of the Overseas Railway, which, despite suffering three hurricanes during construction, was finally completed in the early 1900's. However, nothing could compare to the tragedy that struck the Keys on September the second, 1935, when what became known as the 'Labour Day Hurricane' made landfall near Islamorada in the Upper Keys. 200 mph winds were said to have gusted across the islands, raising a storm surge of over five metres above sea level. At the time of this disaster, a handful of bridges were being constructed, in order to build a highway all the way down from Florida's mainland to island of Key West.

Hundreds of war veterans who were working on these bridges at the time lost their lives, after the evacuation train failed to reach their three construction camps in time. Somewhere between 400 and 600 lives were lost in the hurricane, although the actual figure is unsure. Great portions of the railway tracks were damaged so badly by the hurricane that they were never replaced, and thus ended the railway's twenty-three year reign as the main source of transportation throughout the Keys, 'Highway 1' which runs for over 500 miles, from Jacksonville to Key West becoming its successor. Just offshore from the Florida Keys, is the 270 Km (167 miles) long 'Florida Reef', which is actually the third largest barrier reef system on the planet. Following the Cuban Revolution, a huge number of Cubans fled to Florida, many of them risking life and limb, as they sailed the ninety miles of open sea in makeshift rafts, before finally reaching the shores of Key West.

On we travelled, passing through Key Largo, to Sunset Point and Tavernier, before crossing Plantation Key, and onto Islamorada, which is the halfway stage of our journey. We have now been on the bus for just over one and a half hours, and so when our driver finally pulled over outside a Burger King restaurant, before announcing that we would be stopping here for a twenty minute break, a large cheer rang out from all of the passengers on board, including me. Unlike your average *Burger King* fast food outlet, this place was built in the shape of a large cabin cruiser, complete with several portholes, a steering wheel –and even a complex looking instrument panel. Also, printed in bold letters on the front of the *ship-cum-diner*, was the name 'Kings Pride'.

The inside of the diner was made out of beautifully carved timbers, varnished and lacquered to a perfectly smooth finish, and small, rectangular blocks covered the entire floor, from stem to stern. Two wooden poles, with ropes tightly fastened around the tops, stood either side of the entrance to the diner, each bearing a small plaque, cordially inviting guests to step down inside the boat. After a short stop, we were on our way again. Upper and Lower Matecumbe Key, Long Key, Duck and Vaca Key were traversed in minutes, and by 3 pm we are pulling up in Marathon Key, which is about halfway between Islamorada and Key West. To my great surprise, we then passed a place called 'No Name Key', before our final run through Big Pine Key, Summerland Key and Sugarloaf Key, where Molly's daughter, Clara, lives with her husband, Blake, before finally touching down in Key West.

The island and city of Key West, which is only four miles long, by one mile wide, is located at the southern end of the Florida Keys, and around ninety miles from the island of Cuba. Its main street is Duval Street, which is made up of only fourteen blocks, and is just over a mile in length. Key West has been dubbed as 'The Gibraltar of the west', because of its location, and the people of Key West have a motto, which is

"One human family". For many years, Key West was the largest town in America, and it prospered greatly, not only from trading with Cuba and the Bahamas, but also because it was situated on the main trade route from New Orleans. Key West also made small fortunes from various shipwreck salvage operations in the early years, but as navigation systems improved, their plundering went into decline, and the people had to rely on other sources of income in order to keep their thriving economy afloat. Up until the beginning of the 20th century, Key West was a somewhat remote island, but all of this changed with the building of the Overseas Railway in 1912, and the subsequent extension of Highway 1 from Miami to Key West.

Prior to the Cuban Revolution, which was led by Fidel Castro between July 1953 and 31st December 1958, and its culmination on 26th July 1959, which is now celebrated as the 'Day of the Revolution', there were regular ferries running to and from Havana and Key West. Marathon swims from Cuba to Key West, some with the assistance of a shark cage, and some without, have also been completed by various 'brave' people in recent years. The 'Old Town' of Key West, which is more affectionately known as 'The Key West Historic District' primarily consists of the earliest neighbourhoods on the island to be assembled, along with two of its most famous tourist attractions, which are Ernest Hemmingway's House, where he lived for eight years, between 1931 and 1939, and also Harry S. Truman's 'Little White House' which the president occupied for around six months of his time in office.

As soon as our bus pulled in at the Greyhound terminal, I immediately called Clara, in order to inform her of my safe arrival in Key West. A man called 'Albie' answered the phone, saying that he was a close friend of Clara and Blake's, before adding that he was currently staying at their home for a few weeks. Albie then informed me that both Clara and Blake were out of the house, at present, before adding that I would find Blake at 'Sunset' on the 'Havana' Platform, which is right at the end of the pier. "Blake is one half of a duo that will be singing on stage, and his singing partner's name is Ellis, who is black, by the way," Albie added –just to make sure that I would accost the right people when I get there. The streets were packed with tourists from all over the world, and one could easily see why Key West has an aura about it that is so majestic and vibrant, that it just makes you want to stay here forever. Swish cafe-bars, plush restaurants, and ornate little coffee shops, adorn the main street, along with a plethora of trendy gift shops and bustling souvenir shacks, all packed to the gills with a multitude of hand-crafted ornaments, along with an abundance of dainty little trinkets, that would put *Aladdin's Cave* to shame.

As I approached the end of the pier, I could see the main platform, and standing in the centre of it were two black musicians, one of them playing a keyboard, while the other one gently strummed his guitar. The pair of them was singing a harmonious ballad that seemed to touch the hearts of their audience, as the majority of the people swayed in unison with the music. Unfortunately, with all the noise going on around me, I was too far away to hear the lyrics of the song, but their words had obviously struck a chord with their listeners. Having already deduced that these guys were obviously *not* Ellis and Blake, I decided to pull up a chair at one of the nearby bars and treat myself to a cool, refreshing beer, whilst watching a myriad of amazing yachts and unbelievable cruise ships go sailing by in the distance. This place was truly amazing, and it was certainly worth the long journey down from Fort Lauderdale, even though I knew that I would soon be doing the trip all over again, only in reverse this time, of course.

To top everything else, the sunset was just about the most amazing spectacle that I have ever seen in my life, and as I sat there watching the sun gently disappear over that crimson horizon, I fully understood why many thousands of people, over several generations, have been flocking in their droves to visit this utterly beautiful place. Having quenched the proverbial thirst, I continued-on, ambling my way through the many streets that adorned the main square, simply enjoying the warmth of the evening, whilst soaking-up the atmosphere all around me of people laughing and joking, and generally enjoying themselves, as if they didn't have a single care in the world. It was truly magical. To the right of me I can see a rather large building, with a canopy running the length of one side of it, and underneath that canopy are two guys, one black and one white, who are likewise singing a duet in harmony, and so I decide to gently manoeuvre my way through the madding crowd, in order to get a closer look at the people whom I believe to be Blake and Ellis.

After a few minutes of pushing and shoving, I am finally standing at the front of the stage, and apart from having a great view of the guys I can also hear them singing their beautiful song. "*So hold me close, don't let me go, so hold me close, don't let me go*", was the words of the main chorus line, which they sang ever so softly three or four times throughout this beautifully romantic song, which I later found out had been written by Blake only a few weeks earlier. Raptures of applause followed their last song of the evening, and I found myself joining in with the chanting of this frenzied crowd as they continued yelling "More" at least a dozen times. As the guys returned for a final encore, this tiny arena simply erupted into a mass of cheers, screams and wolf-whistles, before the guys finally took their last bows and waved their goodbye's to their adoring fans. Complimentary plate-loads of boiled rice smothered in Bolognese sauce were now being handed out in the main lounge of this bar-cum-restaurant, and it wasn't until I enquired about the price of a drink, that I realised how easily they could afford to *dish out the freebies* to their customers.

After mingling amongst the packed crowd for several minutes, I finally managed to shuffle my way through to the lads, who were currently being surrounded by a volley of besotted fans and autograph seekers, the majority of them being young and female, of course, as is inevitably the case when a band consists of two (or more) virile young men, who, in this instance, look not only handsome, but they also have fine physiques as well. After introducing myself to the lads, I ask them if they could possibly give me directions to the nearest backpackers establishment, as I currently had nowhere to stay for the night, to whit Blake immediately said that he would have been happy to put me up, had Albie not been occupying the one and only spare bedroom that they had available in their home. My subtle hint had duly taken the nose-dive that it deserved, although Ellis kindly offered to drop me off at the nearest youth hostel, as soon as they had packed all of their gear into the back of the van, which the three of us soon managed to accomplish between us.

Before Blake left us to head for home, he told me to give Clara a call at the house tomorrow, as he would be out working for most of the day, before adding that he would do his best to try and meet up with me for a couple of beers before I left Key West. As luck would have it the hostel was only a mile away from the pier, and so in no time at all we were pulling-up outside the place. Ellis also kindly waited in the van, until I was able to confirm with him that they had a vacancy for me to stay the night, which thankfully, they did.

However, not only would the cost be $16 per night, which was somewhat dearer than in Fort Lauderdale, but the ground rules would also be a lot stiffer. No guests or any alcohol whatsoever would be allowed in the rooms at any time during the day or the night, and all guests must vacate their rooms by 11 am on the day that they leave the hostel. I also had to hand over my passport to the receptionist, who said that I could collect it as and when I checked out of the hostel –but not before. After paying my dues, and going through all the usual checking-in procedures, I stopped to have a chat with the manager, who seemed a really decent chap. Apart from welcoming me to his hostel, he duly warned me to be very careful with my property, as theft had been rife in the area recently, and so everyone was on their guard right now. He then assured me that the area was not dangerous at all, so long as people used their common sense, especially when going out alone after dark.

After thanking him for his somewhat ambiguous advice, I made my way to my room, hoping and praying as I crossed the main courtyard, that I wouldn't get mugged before reaching the other side? There was no swimming pool this time, just a battered old pool table, seeking refuge under a ripped sheet of tarpaulin that had obviously seen better days. This mangy sheet was currently being supported by half a dozen rusty poles that were falling in all different directions, and so it looked more like the sheet was supporting them! Sitting in a circle under this *sorry canopy* were a bunch of teenage lads, not one of them donning a smile on his face, and all of them sitting in total silence, as if some sort of unspoken curfew had forbidden them to talk after dark? As I passed-by their table, one or two of them glared up at 'the old man', making me feel somewhat out of place for being amongst a much younger generation –and I could also sense that this group of guys did not appreciate my company whatsoever. In fact, I felt like I was entering some kind of borstal, rather than a hostel, and I was so glad that I had only booked to stay for one night.

No-one was in the dorm when I entered the sleeping quarters, and so I immediately hid my luggage under one of the bunk beds, making sure that I still had my important documents, along with my camera, and anything else that an opportunist thief might want to steal, still about my person, before exiting the room. It was a very sticky night, and my clothes clung to me, as I walked along Duval Street, before stopping at Rick's

Bar for a beer. Just across the way I could see the infamous 'Sloppy Joe's Bar', buzzing with activity, and packed to capacity with night revellers, who would no-doubt be singing and dancing the night away, well into the early hours of the morning. Had I been a single, much younger man, then I would have been more than happy to pay the two dollars entrance fee, before joining-in with the endless frivolities of the drunken party-goers, but as happily married man, with three wonderful children, I was quite content just sitting peacefully at my table, daydreaming about a deliriously happy —and indeed a more prosperous future, that hopefully lies in store for me and my family.

The singer in 'Rick's Bar' is very entertaining, and he is currently singing songs that are being chosen by individual members of his audience. However, at one dollar, fifty cents per half pint of beer, I will only be staying for one drink, before moving on down the road to 'Captain Toni's Bar', where the drinks are considerably cheaper. Although there are no singers to entertain the guests at this rather quirky establishment, there are thousands upon thousands of small business cards from all over the world encompassing the walls and ceiling of this cavernous bar, and so one simply makes one's own entertainment, by reading as many of the interesting cards as possible, before either leaving this place, or falling over drunk with dizziness and dehydration. Hundreds of the cards are really funny, dozens of them are outrageous, and the rest have simply been stapled, taped or glued to the walls and ceiling by the thousands of businessmen who have frequented this outlandish bar over the years, and simply want to leave proof of their visit.

After a couple of drinks, and having read but a tiny proportion of the commemorative cards, I took my leave of the place, heading in the direction of my latest temporary home. Within minutes I am accosted by a frail old Chinese gentleman, who is trying his damndest to encourage me to eat in his humble restaurant, including offering me 10% discount off every meal on the menu. However, enticing as it was, I duly declined his generous offer, even though I was absolutely starving by now. Further down the road I stopped outside a Mexican restaurant, slowly pondering over the mouth-watering menu that was being advertised in the window, but again their prices were way too high for my rapidly diminishing budget, and so I swiftly moved on. Then, just when I had accepted the fact that I would be going to bed hungry this evening, I spotted a small hotdog stand, which had been set up halfway down one of the small side streets. At first I was a little dubious about venturing down a dark alleyway on my own, but the aroma of those frying onions, sizzling-away on that big, black griddle, soon made me forget all my fears, and so off I went, into the *black hole*.

Unfortunately, or should I say 'luckily', I have no story to tell about my venture into the darkness, except to say that the hot dog was "Bloody lovely", and *just what the doctor had ordered*, as far as I was concerned, before crashing-out in my *pit* for the night. To my utter dismay, there are no freebies for breakfast at this hostel, although for only two dollars I am at liberty to help myself to a choice of cereals, a selection of fruits and yoghurts, and also as much coffee as I care to drink, which sounded rather good to me, and so I duly indulged myself in the healthiest breakfast that I had eaten since I can ever remember? Having finished my morning *pig-out*, I duly ventured outside the hostel to bask in the warm morning sunshine, which was an enjoyable pleasure that would soon be curtailed when I returned home to *wet and windy Wales* next week. Standing outside the hostel are a selection of pushbikes, all of which have been interlocked with a huge chunk of chain, before being padlocked to a long metal handrail.

Thoughts of the petty thievery that was supposed to be in abundance in Key West –well, according to the hostel manager anyway, ran through my brain, as I began checking-out the prices for hire, which included a substantial security deposit, just in case the bike failed to be returned, of course. And so yet another small treat unceremoniously bit the dust, as I continued counting the pennies in my pocket with great dexterity, so as not to miscount my stash, or worse-still, drop one of my beloved coins into the gutter, only to watch it roll gently away, before slipping down into an open drain, disappearing under the murky water, never to be seen again. Once again it would be *Shanks's Pony* for me, not only because my money was dwindling fast, but also because I needed the exercise, or so I kept telling myself in a kind of compensatory way, as I tried desperately not to dwell on the fact that once all of my money had finally been spent, I could end up in all kinds of trouble?

Onward I trundled, heading for the shoreline, as the day really began warming-up, and I could feel the suns' powerful rays burning away at any patch of my skin that I had dared to leave uncovered. I passed a small motel that had a big wooden sign in its garden, proclaiming it to be 'The Southernmost motel in the U.S.A.', which did have a certain ring to it, I suppose. I then stumbled upon 'The southernmost Bar in the U.S.A.', before coming face to face with 'The southernmost Restaurant in the U.S.A.' and so it went on. Whether these *claims to fame* were true or not were immaterial to me, as I had no intentions of finding out the cost of eating, drinking or sleeping in any of these 'prestigious places' –well not on this particular visit to the Keys anyway. Finally I reached the golden, sandy beach area, and as I walked peacefully along the coast, breathing in deeply, to suck in all of that fresh Floridian air, I suddenly spotted what looked like a giant pearl from a distance. Only the top half of it was showing, as it lay there, embedded in the sand, and surrounded by a small cluster of black seaweed, like a crystal ball being held in the hands of a witch-like clairvoyant, the swirling mists inside its opaque shell becoming clearer by the second, as if in readiness to reveal all of one's darkest secrets.

Upon closer inspection, this unidentified object that had temporarily transfixed my mind, now resembled a lump of translucent plastic, or putting it more bluntly, a disused breast implant? If that was the case, then where was the other one –and who do they belong to? By now, curiosity had now gotten the better of me, and so I reached down to pick up *the offending item*, when suddenly my fingers sank deeply into this mass of slimy jelly...'jelly' being the operative word, for only now did I realize that I was holding in my hand some kind of jellyfish. Unluckily for him, but luckily for me, he was *as dead as a Dodo*, and so I escaped the wrath of his almighty sting, which is something that I will be eternally grateful for. However, this brief encounter had convinced me not to go for a dip in the Atlantic, just in case a handful of his friends or relatives were out there in the ocean, watching me from a distance, and simply waiting for this poor, unsuspecting tourist, to throw his carcass into their lair.

On my return journey to the hostel, I passed a 'Moped Hospital', complete with the usual Red Cross insignia on the metal sign that was standing proud outside the main entrance. This was definitely a first for me, and I couldn't help but chuckle as I wondered whether the bikes came under any kind of *National Health Scheme* –or perhaps they would have to belong to B.U.P.A. before being allowed any specialist treatment? As it turned out, this was simply a rental agency for motorbikes, mopeds and scooters that is still going strong twenty-five years after I happened to come across it on one of my many walks around the world. As I turned the last corner, where my hostel was located, I caught sight of the local Conch Train, which carries hundreds of tourists a day on several guided tours around Key West, and I really wanted to jump on board, as it pulled up at the next stop. However, time was not on my side, as officially I had to be out of my room be eleven, and it was almost that time now.

After checking out of the hostel, I duly rang Clara, who told me to take the Greyhound Bus to Sugarloaf Key, which is a small island that sits adjacent to Summerland Key. Thankfully, the hostel has a complimentary shuttle bus service that picks up and drops off backpackers at the Greyhound depot whenever the bus is due in, and so my transportation to the terminal was soon sorted. As we drove passed the 'Purple Porpoise Pub', my driver, who happens to be from Sussex, tells me how he intends moving out to Florida permanently in the very near future. He says that he has been working in *the sunshine state* for the past nine months, during which time he has made several friends and acquaintances, many of whom have offered him a *permanent job* "Without upsetting the authorities too much" he adds, smiling and winking at me, and generally admitting that he will be working illegally. This is fine for a young, single man, of course, as the worst thing that can happen to him, is for the authorities to deport him, which will also save him the cost of his return fare to the UK, of course.

However, in my case, where I have a wife and three children to consider, I dare not risk any illegal tactics, and so I openly tell him that I will be doing everything legitimately, and *by the book*, as they say. At this point, my friend just burst out laughing, before turning around to face me:

"Well, unless you are a brain surgeon, or an astronaut, then I would forget it, my friend, because as far as I am aware there is at least a five year waiting list for working visas, and they are apparently only allocating them to people who work in specialist fields. According to the newspapers, the last person to be accepted was a specialist engraver, about three years ago, and the only reason that he was given a visa is because there was no-one else in the country who was doing the same job".

His words tore deep into my soul, and I cursed myself for not doing more research on the subject, before embarking on this journey of *hope* –a journey that was sadly looking more *hopeless* with each day that passed!

449

Caryl had repeatedly told me to make further enquiries with the American Embassy in London, before embarking on this *flight of fancy*, but I was so afraid that with everything closing-in all around us, our world would collapse long before we had the chance to sort out all of the bureaucracy that would be encompassed in the move, and so I had insisted on doing it sooner, rather than later. Now I was regressing myself bitterly. Had I really made such a grave error of judgment, and am I to be made a laughing stock once again? Is everyone still laughing behind my back, saying what an absolute idiot I am, and are they chastising me for wasting the very last of our money on one big fools' errand? My mind was now racked with guilt. At least last years' venture to California had been sponsored by someone else, and so it hadn't cost us a penny, but this time Caryl and I had used up the very last of our resources to get me here, and already the fear of failure was setting in. My only hope now, is that Leo can sort me out with a permanent job, so that somehow; someway, I can legally work here, and my family can likewise live here, no matter how long it takes. This was now all I had left to cling on to.

CHAPTER 41

Sugarloaf Shores

The bus leaving late was a foregone conclusion, but what I had not anticipated was us running into a massive traffic jam along the way, which was something I had not experienced as yet. The combination of these two factors ended-up delaying our arrival time by almost half an hour, and so when we finally arrived at Sugar Loaf Shores, there was no sign of Clara, which is what I had half-expected really. The island of Sugarloaf Key is technically split into two halves 'Sugar Loaf Shores' being located on the coast of 'Lower Sugarloaf Key', and sitting directly opposite the ghost town of Perky, which has the historic 'Bat Tower'. The island is shaped like the letter 'U', with Lower Sugarloaf Key being situated between 'Park Key' and 'Saddle-bunch Keys', and Upper Sugarloaf Key nestling between Park key and Cudjoe Key. The name 'Sugarloaf' was either attributed to an Indian mound on Upper Sugarloaf Key, which was said to resemble an old fashioned loaf of sugar, or from the 'sugarloaf' pineapple, which used to be harvested in the area.

A small corner of the island, which has the wonderful name of 'Pirates Cove', used to house one of the 'Overseas Railroad' stations, back in the early 1900's. Although Lower Sugarloaf Key is somewhat smaller in stature than its counterpart, it has a larger population. A phone call to the lady, in order to express my apologies was certainly in order, even though it was not my fault, of course, but there are certain situations in life which, when they do occur, someone must be held accountable for the consequences, regardless of whose fault it is —and this was one of them. Thankfully, Clara had obviously inherited her mothers' wonderful, friendly personality, telling me not to worry about it at all, adding that she would be with me in around ten minutes time. Although Sugar Loaf Shores is not an official stop for the bus, and therefore I had to ask the driver to stop there, as I was the only person getting off the bus, I have to say that it turned-out to be one of the most picturesque places that I have ever had the pleasure of visiting in my life.

This small cove only had a handful of boats moored alongside its quay, and apart from a tiny water-sports shack, that housed mostly fishing equipment, along with a couple of food stores, that were set back off the road, there was very little else going on in this quaint little piece of heaven on Earth. My waiting was short-lived, as within minutes my host appeared on the scene, gently rolling into the car park in her brand spanking new open-topped sports car..."Wow". Donning a mass of long blonde hair, that went all the way down to her tiny waist, and wearing a pair of large, oval sunglasses, Clara looked like a professional model, and when she got out of the car to greet me, thus revealing her long, slender legs, that seemed to keep on going, before finally disappearing into a pair of very tight shorts, I had no doubts whatsoever that I was right. However, having only just met the lady for the very first time, I was not about to ask her. Clara's welcoming smile was identical to the initial smile that her mum had given me back at the poolside in Tarawoods, and I could see that it was not only her wonderful personality that Clara had inherited from Molly, but also her facial features, as well.

After a relatively short drive we were back at the house, where I was formally introduced to Albie, the guy whom I had spoken to on the telephone yesterday. Unlike Clara, who was immaculately attired from head to toe, Albie was quite the opposite. Donning a scruffy tee-shirt, and a pair of knee-length shorts that had obviously seen better days, he looked somewhat worse for wear. Albie was also sporting a rough-cut goatee beard, along with a mass of un-brushed hair, and so I suppose that these didn't help matters either. In my eyes, Albie looked like your archetypal hippy. Mind you, I hadn't looked in the mirror for a long time, and so I daresay that with my pot belly now hanging over the top of my shorts, and my hair receding to the crown of my head, I guess I wasn't looking that good to him either? Despite his *dressed-down* appearance, Albie was actually a really nice chap, of course, and so the three of us got on famously together.

Clara disappeared into the kitchen for a few minutes, before returning with a plateful of sliced melon for the three of us to share. If truth be known, I could have eaten a horse by now, but with the other two looking so slim and healthy, I decided to make the most of my calorie-free brunch, and just pretend to myself that I was really scoffing a 'Full English!' Clara then went into the master bedroom downstairs, in order to get changed, and the second that she disappeared from sight, Blake came walking in through the front door. Just like Clara, Blake is a very pleasant person, warmly welcoming me into his home, before giving me a show-around of the place. Inside, the bungalow looked much like any other two bedroom property, but when we ventured out onto the patio, well that was a different story altogether. Huge trees, surrounded by a myriad of tropical plants engulfed the whole garden area, making it look more like a South American rainforest, rather than someone's back garden, and the sheer density of the overhanging branches meant that nearly all of the garden was completely enshrouded in shade from the scorching heat of the day.

As we reached the bottom of the garden Blake took me on a mini-tour of his music studio, which from the outside, looked like nothing more than a garden workshop. Made entirely out of solid timber it was probably robust enough to withstand a minor earthquake, but on the inside there were thousands of pounds worth of delicate instruments and audio equipment that I was scared to even go near, for fear of knocking something over? Thick metal cables intertwined with endless streams of plastic coated wiring adorned the walls and ceiling of this amazing place, and a volley of junction boxes and multi-sockets hung from every rafter, like a stream of ornate Christmas decorations that had been smothered in a mixture of black, white and chrome coatings. After clicking-on a handful of power switches, Blake began playing one of the many keyboards in the room, his head shaking from side to side, like Stevie Wonder at his beloved piano, as he began singing one of the songs from his latest album, which had just been released by a major record company. It was heavenly.

After Blake had finished singing –and I had finished clapping, he reached over a pile of speakers to a small shelf, where an open box was sitting all alone, and after fiddling around inside the box for a few seconds, he produced a brand new cassette tape of his latest album. "For you, my friend, as a memento of your visit, and a token of our friendship –I hope that it gives you many hours of pleasure". What a wonderful gesture, from a wonderful man, and I could not thank him enough for both his kindness and also his hospitality. Blake is truly one of a kind. Apart from four loveable cats that roamed the house at their leisure, Blake also owned a rather noisy dog, which he had lashed to one of the trees in the garden, prior to my outdoor tour. From the moment that I stepped out onto the terrace, the blessed thing barked and growled at me for all he was worth, jumping up in the air and gnashing his teeth together, as Blake and I walked through the garden, Blake assuring me that his bark was a lot worse than his bite. However, I still kept my distance from this half-breed mongrel, having no intentions of giving him the opportunity to prove to me what a lovely doggy he really is?

After an hour or so Blake set off for Key West again, as he had to do an afternoon gig at one of the bars on the pier, and as we shook hands and said our farewells to each other, I felt truly privileged to have had my own personal audience with this extremely talented young man. After he left, I said to Clara that I was sure that his 'big day' would come very soon, to whit she replied "I think that it already has Shaun, because Blake has just finished writing part of the soundtrack for the latest James Bond movie". Clara then added that she had also played a small role in the film, by being a stand-in for one of the Bond girls, before adding that it was "No big deal", as film moguls were often seen filming in the Florida Keys, and that they were always looking for extras to play bit-parts in their productions. "I guess I just happened to be in the right place at the right time", Clara said modestly, although I have no doubts that she did a great job –and that she looked every bit as stunning in the movie, as any one of the other Bond Girls.

The pair of us then took off along the coast in her super sports car, Clara driving quite cautiously considering the power that she had under her bonnet, but I was happy to just sit back and enjoy the ride, whilst feeling the gentle breeze blowing through the little hair that I have left on my head. After a couple of miles, we turned off the main highway, into a narrow street, and about halfway along the length of the road Clara pulled over next to a small opening in the hedge. As I got out of the car, I could see a large lake at the end of a muddy walkway, and jutting out into the water was this small wooden jetty, about fifteen feet long and approximately three feet wide. "I hope that you are good at fishing", Clara shouted, whilst taking two halves of a fishing rod from out of the boot of her car. "I don't know how good I am Clara, simply because I've never done any fishing before", I jokingly replied. Well, whether I liked it or not, I was just about to learn what the sport of angling was all about. To my great surprise, and with a lot of help and encouragement from my tutor, I might add, I eventually managed to catch two little *tiddlers*.

How disappointed was I, when my mentor insisted that I throw them back into the water, because they were too small to keep! Having enjoyed my first ever fishing challenge, Clara and I then sat on the pier together, exchanging pleasantries, and having a general *chitty-chat* about our individual lives. I told Clara all about my life in Cardiff with my lovely wife, and my three wonderful children, and how I hoped and prayed that one day we would all be living in this beautifully warm and wonderful part of the world, which I openly admitted would be a 'dream come true' for me. Having poured my heart out to a woman whom I barely knew, but I still trusted enough to tell her my life story, Clara reciprocated, by telling me that she would soon be giving up her job, as she had recently found out that she was expecting Blake's child, and that she intended being a full-time mum from the start. "Talking of work Shaun, I have to be on duty at four o'clock, and so we are going to have to make a move, I'm afraid, otherwise I am going to be late", she added, somewhat reluctantly.

For some reason I had not asked Clara what she did for a living, although I gathered she would have a high profile job, as she was such a well-spoken, professional individual, with a great personality and an abundance of confidence. However, it wouldn't be long before I found out exactly what this young lady did to make ends meet. As soon as we arrived back at the house, Clara disappeared into her bedroom, in order to get suitably attired for work, while I sat on the sofa in the living room, watching a string of advertisements on TV, none of which interested me in the slightest, of course. I then heard Clara's bedroom door shut downstairs, and so after hearing her footsteps coming up the staircase, I turned around, expecting to see her dressed in a lovely summer outfit, or maybe even wearing a posh suit, like all top professional business-ladies wear, but when I saw her standing proud in her immaculate uniform, I almost fell off my chair. "You're a policewoman?" I gasped, staring ominously at the leather gun holster that was now straddling her right hip, whilst trying my utmost to come to terms with my biggest shock of the week.

"Sorry, didn't I mention that to you during our conversation Shaun –I hope I didn't shock you too much?" Clara answered, her face beaming with a huge smile, as if she had planned to scare the living daylights out of me all along. If that was the case, well it had certainly worked a treat, although 'Flabbergast' would probably have been a more appropriate word, for I was still totally lost for words. How on earth could this delicate, sweet-looking housewife, who looked as though she wouldn't *say boo to a goose*, suddenly transform herself into a gun-toting law-enforcement officer, who was willing to risk life and limb on the drug-ridden streets of Southern Florida? This was completely beyond my comprehension. Thoughts flashed back to me of the policewoman in California who had scared me half to death, after pulling me over on the motorway for wearing headphones. In total contrast to Clara's angelic personality, that rather stone-faced lady had me firmly believing that if I made one false move, then she would think nothing of blowing my brains out!

At first, Clara kindly offered to drop me off at Marathon Key, but after checking the timetable, she said that it would be after dark by the time the bus arrived, and the depot had a reputation for being unsafe after sundown, and so she said that she would drop me off at Big Pine Key instead. If truth be known, I had always assumed –wrongly, it seems, that the Florida Keys were just about the safest part of the sunshine state, but Clara soon put me right on that score, as she duly enlightened me about what happened to a colleague of hers.

"Back last year, a fellow policewoman was severely beaten-up although she openly admits that she made three vital mistakes that could have prevented the beating from happening in the first place. Firstly, she omitted to radio-in to the main office with the number of the licence plate, after pulling-in a man for driving at seventy miles per hour in a thirty-five mile limit zone. Secondly, she should have immediately called for an ambulance when she noticed an injured man sitting in the back of the car, but she failed to do that, as well And last, but certainly not least, she then broke the most important rule in the book, which is never to turn one's back on the offender...something that is absolutely taboo –especially when you are on your own, which unfortunately she was at the time. What happened next is too heinous to talk about, but what I can tell you is that a lorry driver discovered her lying in a ditch in the woods a few hours later, soaked in her own blood, but lucky to be alive.

Mind you, the poor woman lost all of her teeth, and she had to have an ear stitched back on, before spending three months in the county hospital, with a steel plate firmly attached to her head, and several pins holding the rest of her body together. Thankfully, they caught the bastard who did this to her, and personally I hope that he spends the rest of his life rotting away in some godforsaken jail, and that one of the inmates will do to him exactly what he did to her...or hopefully worse. Surprisingly, the woman is back on the beat, would you believe Shaun, and so I have to admire her spunk, but she is one of the main reasons why I am quitting this job, because pretty soon I will have my own child to take care of, and no-way would I want to risk being unable to look after my own kids –not for all the money in the world."

I am one hundred percent behind Clara, as providing for one's children, and looking after them every step of the way, is something I swore I would do from the moment that I became a father. Holding my son, Liam, for the first time in my hands, made me realise how fragile babies really are, and how much they will rely on you, not only when they are young, but for the best part of their lives. Clara's next story was also quite gruesome although nowhere near as brutal as the last one, thank goodness. She told me about a gentleman who had been bitten by a poisonous spider, before describing in graphic detail how the poor man's leg had swollen dramatically, as the deadly venom continued eating-away at his muscles, before finally settling down, after a long course of anti-venom and antibiotics finally saved his leg from being amputated at the knee. When we finally reached Pine Key, dusk was settling in, and thanks to Clara's stories my eyes were now everywhere, as I crossed the small car park, before finding sanctuary inside the terminal.

Sitting alone in one of the chairs, I thought about Caryl and the children, the tears welling up in my eyes, as I realised that Florida might not be such a heavenly place after all, and that perhaps home is really where the heart should be. Years ago I remember a comedian saying that New York was the only place where one could go out in the evening to buy a newspaper –and the next day they were in it! However, having seen and experienced first-hand, what goes on in Florida, perhaps the same could be said about this place too?

CHAPTER 42

Miami

"Miami –and don't spare the horses", I say to the driver, after carefully depositing my rucksack into the hold. As our bus was already half an hour behind schedule, it looks like my latest host for the evening had already decided to make up time, because, even though it was pitch black outside, we flew along that highway like there was no tomorrow. After a very long drive, we finally approached Key Largo, and I was so tired by now that I considered getting off the bus at this point, and staying overnight, before completing the journey to Miami in the morning, but then, like all naughty girls, I decided to *go all the way*. By the time we finally arrived at the 'Bayside' terminal in Miami, I was completely and utterly shattered. It was nine o'clock.

The city of Miami, which has been ranked as the richest city in America, along with being nicknamed 'The Capital of Latin America', is the cultural, economic and financial centre of Southern Florida. Miami also holds the title of being 'The Cleanest City in the United States', thanks to its year-round air quality and vast greenbelt area, along with its clean streets, fresh drinking water and extensive recycling programmes. Metropolitan Miami is also second only to New York in the tourist stakes, and the Port of Miami is the number one cruise ship passenger port on the planet. The area of Miami has been inhabited for thousands of years, with Indian tribes dating as far back as 500 - 600 B.C. At the beginning of the twentieth century, two-fifths of the population in Miami were made up of migrants from the Bahamas and black African Americans. Like many other officers of the law, Miami's ex-Chief of Police, H. Leslie Quigg, was proud to say that he was a member of the infamous Ku Klux Klan.

In fact, he once publicly flogged a black boy to death for simply talking directly to a Caucasian woman. During World War II, Miami became a base for the US defence against German submarines, and in 1959, when Fidel Castro came to power in Cuba, many wealthy Cubans fled to Miami. In just over a century, Miami's population exploded from just over one thousand residents, to around five million, thus earning it the nickname of 'The Magic City', as the rapid growth of its inhabitants was deemed to be "Like magic". Miami is split into several areas, 'Downtown Miami', which is effectively situated on the eastern side of the city, being classed as 'The heart of the city', probably because it has more international banks in the area that anywhere else in the United States. The west side of Miami was once a predominantly Jewish community, but in recent years, it has been flooded with immigrants from Cuba and Central America, thus earning it the title of 'Little Havana'.

As I enter the main terminal building, I begin searching around for any kind of tourist information, such as posters, leaflets –or even anyone dressed in a Greyhound uniform that might be able to help me, but at every turn I drew a blank? There is not even a telephone number of the nearest youth hostel, which is supposedly only a couple of miles away, and no sign of any taxi that could take me there anyway! According to various people whom I have spoken to along the way, I am now in the centre of one of the most notorious areas, in one of the top ten most dangerous cities in the world. Even in the daytime, Miami is not a safe place to be, I've been told on several occasions, but here I am, in the pitch darkness of night, travelling all alone, with nothing but my fists to protect me. Thoughts of being robbed, beaten-up, or even murdered, are now running through my head at a rapid pace, as I try to figure out what the hell to do next? Suddenly I see a set of headlights flickering in the distance, and as the vehicle approaches the entrance to the terminal, I can see a 'Taxi' sign on its roof.

"Hallelujah", I shout, before grabbing my belongings and running towards the small taxi rank sign, where he has now pulled-up. However, just as I am about to accost the driver, I see one of the back doors open on the opposite side of the cab, and before I know it, a young backpacker has tossed his bag onto the back seat, before jumping aboard, and hollering to the driver to take him "Downtown". I have barely a moment to think straight, and so I decide to throw all caution to the wind, and simply jump out in front of the taxi, before the driver has time to take off. Waving my arms frantically in front of his windscreen, I say that I want to share the taxi *downtown* with his passenger, to whit he duly turns around to face the young lad for his permission, and he thankfully confirms that that will be fine. My latest companion is called Jenson, and he tells me that he is originally from Denmark, although he has been working in a restaurant in Key West for just over a year. Jenson says that he intends in staying in one of the many youth hostels in town for the night, but first he wants to go *downtown* for something to eat, as he is absolutely starving.

Apart from feeling quite peckish myself the thought of having the company (and the backup) of a strapping young lad for the evening was music to my ears, and so I was more than happy to go with him. Jenson tells the taxi driver which restaurant to take us to, before assuring me that the food is excellent there, adding that he knows downtown Miami like the back of his hand. After paying the taxi fare between us, Jenson and I ventured inside the restaurant, which was small and quite quaint, and the menu prices were not extortionate, which was more than a relief to me. After ordering a couple of beers, Jenson tells me that he will be meeting his Nicaraguan girlfriend at Miami Airport tomorrow. "She has been on holiday in Athens for two weeks, staying with a couple of Greek girls that she met in college a few years ago", he added, the excitement in his voice telling me that he obviously couldn't wait to see her. When the waitress arrived to take our meal orders, Jenson immediately asked her if there were any vacancies for kitchen workers.

No sooner had she left us than, the manager was joining us at our table, chatting to the pair of us like we were old friends, before asking Jenson to call in and see him again tomorrow afternoon. I had to admire Jenson for his openness, for I would never dream of being as forthright as that when it came to looking for work, and yet, in my own little way, I suppose that not many people would have dared to do what I've done this past year, and so perhaps I am not so shy after all? After we had finished our meals, Jenson suggested that we book a twin room at the hotel which is directly opposite the restaurant, saying that as it is only $34 a night, it is well worth paying the extra couple of bucks each, rather than having to share a dormitory with about six other guys for around $15 each. I was a little hesitant at first, having only known Jenson for a couple of hours, but during that time I had also felt quite safe with him, as he was a very switched-on guy, who seemed to be nobody's fool. The lad was also very open and honest about everything, not pulling any punches about whatever he felt deeply about, regardless of whether the person on the receiving end liked it or not.

Jenson could also be feeling the same way about me, of course, and yet this man was willing to take his chances with a guy that he didn't know from Adam? Anyway, it was me who had accosted him in the first place, and not the other way around, and so why shouldn't I trust him? Unfortunately, the room turned-out to be $59 for the night as it was now high season, but having already told Jensen what my current financial situation was, he kindly offered to pay the difference. The standard of the hotel was designated as 'three stars', but I have to say that whoever gave it that rating at the time, must have been in a very good mood that day, for I would be more than *pushing the boat out* to give it even two stars! Having unceremoniously dumped our bags in one corner of the room, Jenson and I once again ventured-out into downtown Miami, to see what the place really did have to offer? "Every night is a Saturday night in Miami Shaun", Jenson quipped enthusiastically, as we made our way into one of the many themed bars that lined the avenue. Once inside, Jenson tells me that although he has primarily worked in bars and restaurants for the last ten years, he would love to pursue a career in 'Musical Engineering'.

Unfortunately, Jenson can only afford to do this on a part time basis at present, and he knows deep down inside that without getting the proverbial 'break' in his life, then it is very unlikely that he will achieve his goal. Upon listening to this, I decided to give Jenson, Blake's telephone number, in the hope that he might possibly be able to help Jenson in one way or another. Wouldn't it be uncanny if my whole trip to Florida, coupled with my chance meeting with Jenson, had actually been part of my destiny from the start, and that regardless of whatever happens with me, I may well have unwittingly been the break that Jenson had been waiting for all along? Selfishly, I hoped that this was not the case, as I desperately wanted my dreams to reach fruition, of course, but having done my good deed for the day, I contentedly rested back on my laurels, leaving everything else to lie within the lap of the gods. By 2 am the pair of us was just about bushed, and so we left the Miami revellers to continue-on with their partying, while we shuffled-off back to our hotel.

As we are walking along the main corridor leading to our room, Jenson suddenly gave a loud 'sniff', before stopping dead in his tracks. "I know that smell very well", he said, before broadening the prominent grin that never left his face. I could also detect a strange aroma, like burning incense, in the air, but I had no idea what it was? Just ahead of us we could see a door on the left hand side of the corridor that was partially open, and it was obvious to both of us that this was where the smell was emanating from. Jenson's pace quickened, as if the smell was drawing him to it like a magnet, and so, not wanting to be left out, I quickened my pace too. As we approached the half-open door, the smell became quite intoxicating, and we could see two sets of legs hanging over the edge of a double bed. Jenson gently pushed-open the door, revealing a couple of teenage lads, the pair of them lying spread-eagled across the bed. The guys were still dressed in their daytime attire, even though both of them looked completely spaced-out, as though they had been sleeping like babies, before the creaking of the opening door had awakened them from their slumber.

Jenson tapped his knuckles on the inside of the door, before uttering softly "I don't suppose you would have a little bit of that to spare, would you guys?" One of the lads glanced over at us for a second, before slurring "Sure thing –help yourself, my friend." As Jenson and I entered the room, that same guy got up off the bed, before handing Jenson what looked to me like a rolled-up cigarette, but, of course it was a spliff. In fact it was the first spliff that I had ever encountered, and having never even smoked a cigarette in my life beforehand, my ignorance was probably bliss. In all fairness to the chap, he kindly offered me the same delight, but I politely declined his offer, saying that it was "Not my scene", before quietly exiting the room and leaving the three of them to get *as high as a kite.* Being in one of the drug capitals of the world, I guess I should have expected nothing less, but the mere thought of what could have happened, had we interrupted some kind of drugs deal, was enough to send shivers down my spine. Still, it was yet another situation in my life that I might never have encountered in the UK, and so I decided to put it all down to experience.

I woke early the next morning, only to find Jenson crashed-out on the other single bed, his loud snoring telling me that he would not be waking up for quite some time. Wearing only a pair of shorts, and having nothing more than a thin cotton sheet covering my body, I shivered in the air-conditioned room, where the temperature must have been around sixty degrees, whereas outside it was already high in the seventies. By the time I had finished in the bathroom, Jenson was beginning to stir in his slumber, and so I gently tapped his shoulder, telling him that I was going down for breakfast, to whit he mumbled that he would be right behind me, before turning over and drifting off into the land of nod once again. A pot of freshly percolated coffee, along with a tray of Danish pastries was included in the cost of our stay, and although I am not a lover of sticky buns at any time of the day, let-alone first thing in the morning, I tucked into my latest continental breakfast without a second thought, primarily because the funds situation was beginning to look somewhat dire, and so any freebie food would be devoured without question.

By the time I had finished my *Scandinavian scavenge,* Jenson had joined me at the table, his eyes looking somewhat worse for wear, which was no-doubt attributed to his extra-curricular nocturnal activities with our neighbours. We chatted for a short while, and then Jenson went back to the room to get washed and changed, before popping over to the restaurant for his interview with the manager, while I had my first stroll along Miami Beach itself. The fine white-powdered sand burned the soles of my feet, and so I made a b-line for the waters' edge, where I paddled my feet in the warm ocean, while the sun above my head slowly, but surely melted the rest of my torso, until I was as brown as a berry. Scantily-clad women paraded up and down the beach like fashion models on a Parisian catwalk, their tight-fitting bikinis and tantalising g-strings turning every male head in the vicinity –and the ladies knew it! The expression "If you've got it, then flaunt it" certainly sprung to my mind.

Further along the beach it was time for retribution, as the male body-builders were out in full force today, their huge, muscular torso's bursting-out of their t-shirts and vests, like the Incredible Hulk on a bad day, as their pipe-sized veins continued pumping blood at a horrendous rate, while they bench-pressed ridiculous amounts of weights -primarily to impress their entourage of female admirers, of course. Surrounded by dozens of happy people, who were enjoying life to the full, I found it hard to comprehend that by this time tomorrow someone will have probably been brutally attacked, or even murdered in the streets that surrounded me, but then I knew this long before I left the green, green, grass of home to go in search of paradise. However, as I have no intentions of living anywhere near Miami, I would just make sure that I left this place totally unscathed by whatever dangers were lurking out there in the dark shadows, and hopefully bring back many happy memories of this most beautiful part of the world.

After an hour or so, I returned to the hotel, where Jenson was in the middle of packing-up his stuff, in readiness to check out of our room. As my things were already packed, I had a quick shower, primarily to wash-off all of the sticky sand that had attached itself to my carcass, and also to cool-down my sweaty –and by now, severely-tanned body. After dropping off our bags in the luggage room, Jenson and I sauntered over to one of the beach bars, in order to celebrate the new job that Jenson had just been offered by the bar owner from last night. "I'll have to buy a new pair of bloody shoes for work now" Jenson exclaimed, before pulling back the front half of his right boot, thus revealing his five crooked toes, which almost caused me to fall off the picnic-styled bench with laughter at the state of his *somewhat dilapidated* footwear. By three o'clock the sun had gotten the best of my tender shoulders, and the Budweiser was certainly getting the best of Jenson, and so we decided to retreat inside the restaurant area and have a late lunch, before collecting our bags from the hotel.

Jenson was heading in the same direction as me, and so we stood on the bus stop together for forty-five minutes, the pair of us melting in the sweltering heat, before our transportation finally arrived. As our bus begins making its way through the dense traffic, I politely ask the driver how long it will take us to get to 'Bayside', adding that I have to catch a connecting bus in less than twenty minutes. "You're going to have a very long wait, my friend, as this bus isn't going to Bayside –you should have boarded the number seven bus to 411th Street". This was all I needed to hear, and so after saying my farewells to Jenson, I duly jumped off the bus (courtesy of my sympathetic driver) at the next set of traffic lights. I then spent the next ten minutes trying to flag down a taxi, in the vain hope that the driver can whizz me to the Greyhound terminal before my bus for West Palm Beach leaves Miami. Finally a cab pulls over to the side of the road, but after jumping onto the back seat, the driver duly informs me that it is around eight miles to the Greyhound Bus depot, and that at this time of the day, it is going to take us at least twenty minutes to get there.

I already know that If I miss this bus, then the next one won't be leaving Miami until eleven-forty-five this evening, and so my only hope now is that the bus is running true to form –in other words about half an hour behind schedule? Once again my heart is pounding, and my head is throbbing with nervous tension, as my driver dodges in and out of the many lanes, taking full advantage of every gap in the traffic, as he races through the city on his errand of mercy. When we finally pull up outside the terminal, I can see that the fare is $11.20, but as I have no time to count out what little change is left in my pocket, I quickly hand him over a $10 and a $5 note, thanking him for his efforts and telling him to keep the change. This was something I could ill-afford to do, of course, but I was in so much of a hurry that nothing else really mattered right now, as I simply had to get to West Palm Beach this evening. When I enter the terminal there is only one solitary bus on the starting blocks in the coach park, and I can see that the driver is about to reverse it out of its parking space.

Running to the 'exit' door as fast as my little legs will carry me, I can now see that it says 'West Palm Beach' at the top of the window, and so I begin waving frantically at the black, female driver. At first she doesn't see me, as she is too busy checking in her rear-view mirrors on both sides of the bus, but as I get nearer to the glass-panelled door, she suddenly spots me, my arms now flailing like a fool, and the next thing I can hear is the sound of air-brakes, as the bus comes to a grinding halt. By the time the main door opens I am at the front of the coach, expressing my profuse apologies to the driver, who is thankfully a really lovely woman, and so she simply checks my ticket before putting my rucksack into the hold. I've made it! The traffic congestion is appalling, and even though there are six lanes going in either direction, every lane on our side of the freeway is chock-a-bloc with vehicles. This very steady flow of traffic continued almost all the way to Fort Lauderdale, where the bus stopped to pick up another load of passengers, and due to the volume of people boarding, I was *given the okay* by our driver, to make a 'very quick' phone call from inside the terminal.

I rang Leo's number, in order to confirm my arrival time in West Palm Beach this evening, (today is the date that that we had prearranged to meet up) but even-though it was now 6 pm, and therefore he should be at home, there was no answer from his number? As we make our way back out onto the freeway, the driving conditions have changed dramatically, thanks to a tropical storm that has been raging for the last twenty minutes, or so. The torrential downpour has literally flooded the roads completely, thus bringing everything on the road to a complete standstill. The windscreen wipers on our bus are now going at full pelt, in order to try and clear the massive deluge of water that is hitting the front windows by the bucket-load, thus obscuring the driver's view of the road ahead. As I am only dressed in a pair of shorts and a tee-shirt, I am once again shivering, as the air-conditioning has now kicked-in, big-style. All of my clothes are safely tucked-away under the bus, and so I have no option but to shake like a fool. In Boca Raton we stop at a level crossing, in order to let a train go whizzing through, but long after the light has changed to green, the barrier still refuses to lift up and let us through, and the traffic is now seriously building-up behind us.

As the honking begins, a man who is probably from Texas, as he is donning a ten-gallon hat, on his head, suddenly exclaims "Hey driver, we can't stop here all day; open the blessed door and I'll get this problem sorted". On his way down the steps, the guy says that he needs a volunteer to help him, and so as I am nearest to the door, I duly offer my services. Manually lifting the barrier was much easier than I thought it would be, and my friend and I were able to hold it open while the bus crossed the railway line, where, incidentally, the barrier on the opposite side of the track was already in its vertical position. Being out in the warm air again was simply wonderful, and a thousand goose-pimples covered every inch of my naked flesh, as the blood began flowing properly through my veins once more. Unfortunately, this was not to last, and very soon I was back to shivering and shaking like a washing machine on its final spin, as we sped off down the highway once again. However, it seems that not only had a barrier been 'lifted' outside on the road, but it also looked as though a few barriers had been 'brought down' inside the bus, for everyone was now coming alive, especially the driver, who was continually making jokes about how "Hopeless" her husband was -and also how petrified he was of her.

Her light-hearted insight into the trials and tribulations of her marriage to this *poor* man, had everyone in fits of laughter, but when one of the passengers dared to make a derogative quip about him, albeit jokingly, of course, she duly defended his honour, before ripping into the perpetrator for daring to interrupt her when she was in full swing...it was hilarious. Apart from being a great stand-up comedienne, who was also blessed with a belly-laugh that would put a hysterical hyena to shame, this lady could also sing –and sing she did, belting-out some of the best Tamla-Motown hits from the comfort of her driver's seat, as we trundled happily along the road, her captive audience sitting in total silence throughout each wonderful song, before erupting in raptures of applause each time the lady stopped to take a breath. At 8 pm we finally pulled into the terminal at West Palm Beach, after what had unequivocally been the most memorable journey of the whole trip.

West Palm Beach, which is separated from its neighbouring Palm Beach by the Lake Worth Lagoon, is the county seat of Palm Beach County, Florida. In the late 1800's, a few hundred settlers from all parts of the world moved into what was then called 'Lake Worth County', and on November 5th 1894, the town of West Palm Beach was officially incorporated into what was then known as 'Dade County'. Like most other areas in the state, West Palm Beach flourished in the Florida Land Boom of the 1920's, its population increasing by four hundred percent within a seven year period. Just prior to the onslaught of World War II the Palm Beach Air Force Base was built, bringing thousands of military personnel into the city. A decade later hundreds of these servicemen and ex-servicemen returned to the area, where they immediately set up permanent homes for them and their families to live in. In the 1960's the 'Palm Beach Mall' was opened, along with an indoor arena, but in the decades that followed the crime rate became so bad that many people, along with several major companies and hoards of local businesses, moved away from the downtown area, leaving a mass of urban decay in its wake.

CHAPTER 43

West Palm Beach

By now thunder and lightning had joined forces with the rain, making me feel right at home with the weather. Almost reluctantly, I disembarked the bus with the last of the remaining passengers, as dozens of people who had initially been sitting patiently in the waiting room, began swarming around the poor driver, a handful of them rudely shoving their tickets in her face, as she tried desperately to read the printed manifest that she had just attached to her clipboard. It was obvious to everyone that the amount of people wanting to travel northwards far outnumbered the capacity of seats available on this bus, but the good news is that another bus was on its way. However, that still didn't deter people from pushing and shoving their way to the front of the queue, for fear of being left behind, and having to wait until the next bus arrived, which would probably have been no more than ten to fifteen minutes. As soon as I had retrieved my bag, I rang Leo's home, but again I could only leave a message on his answer-phone, which was now starting to worry me, as I currently had nowhere to sleep for the night?

My only recourse was to sit in one of the chairs in the waiting room, and watch the pandemonium that was now going on all around me. Finally, the second bus arrived, and within minutes the chaotic commotion had subdued into nothing more than a few raised voices, as order was duly restored by the driver of the second bus, who remained calm, cool and collected throughout this ridiculous ordeal. It was almost nine o'clock, and so I figured that Leo must surely be at home by now, and so I called him again, but there was no answer, and so all I could do was leave yet another message on his answer-phone. When we met up in the UK, he told me that he would only be home at weekends, and so I had given him the dates of my flights, along with a general overview of my itinerary, hence we had agreed to meet up in West Palm Beach on my last weekend in Florida, and so what could be simpler? At 9.30 I tried again –and once again at 10 pm, but there was no human response, just that damned answering machine, asking me ever so politely to leave Leo a message, before assuring me that he will get back to me as soon as he has retrieved my message.

I could only hope and pray that this would be the case, because I had no money left to pay for any kind of accommodation, and barely enough dollars in my pocket to cover the cost of my food over the next three days, let-alone anything else that might crop up in the meantime? After a final call at 11 pm, I succumbed to the fact that I would not be hearing anything from Leo this evening, and so I went over to the ticket counter and asked the young lady what time the depot closed, as I prepared myself for my first night of sleeping it rough on the streets? Alternatively, I could end-up sharing a park bench with a *wino,* or if I struck it really lucky, I might find myself spending the night in a *down-and-out's* refuge, snuggling-up to a drug addict on one side of me, and an alcoholic on the other? My mind was awash with every worse-case scenario that I could think of, and as the realisation of my situation finally hit home, I have to admit that I began to panic somewhat.

To my pleasant surprise, I was duly informed that West Palm Beach was one of only two Greyhound terminals in Florida that remained open around the clock, and so there would be no problem with me sleeping in the depot overnight, if I wished to do so. Spending the night in a *glorified bus shelter* was a daunting thought in itself, but this was unequivocally a case of beggars being unable to be choosers, and so I duly accepted the cashiers open invitation with relish. The young lady also informed me that there was a local bar just a few blocks away, where I could enjoy a couple of beers, before crashing out for the night, so long as I was willing to get a soaking on the way, as the rain was now lashing-down outside. Looking around at my humble abode for the evening, I decided that it would be worth getting drenched, just to get out of here for an hour or two, and so I slipped into the gentlemen's toilets, in order to get cleaned-up, and to treat myself to a change of clothes, as my current attire was beginning to whiff a bit!

I then hired one of the small lockers for a nominal sum, before squeezing my rucksack inside its chambers for the night. At $2.75 for a bottle of Budweiser, it was certainly not going to be a boozy night for me, but at least the place was warm and cosy, and a live band played excellent music into the small hours, thus keeping me and the five other customers that had braved the elements to make it here on this damp and dreary night, well happy. The resident barmaid was also a lovely lady, and having enlightened her with my sob story during the course of the evening, she kindly treated me to a complimentary beer "On the house". By 2 am I ready for my bed, and as I made my way back to the terminal, I was feeling quite confident that I would have no problems in sleeping throughout the night -with a little help from the alcohol, of course. As I entered the waiting room, I could see three black guys sitting in individual seats, in the cinema-styled plastic chairs that were situated at the back of the room. They all seemed to be sleeping to their hearts content, and so I decided to try and follow their lead, by plonking my butt in one of the chairs in the front row of seats.

Unfortunately, this never worked for me, as no way could I get anywhere near comfortable, trying to sleep in an upright position, and so, after rearranging a handful of single seats nearby, I proceeded to prostrate myself across the four of them, in readiness for a good nights' sleep. However, this turned-out to be even worse, as the ridges on the edges of the chairs soon began cutting into my bones, and no matter how many times I rolled over onto my left side, my right side, my front and my back, I simply could not get into a comfortable position for more than a few minutes at a time? There is only one thing for it -I must follow the lead of my fellow (female) inmate, who has crashed-out on the tiled floor in one corner of the room. Unfortunately, this was not going to be half as easy as it looked, as I have nothing to rest my head on, and being someone who always likes to sleep on his side, it means that my head is now on an angle that will make it impossible for me to fall sleep. To combat this I decided to retrieve my rucksack from the locker, even though it means that I will have to forfeit my small deposit, but without having any support for my head, I will never get any sleep at all, and so out comes the bag.

Finally I drop off to sleep, but within a few hours I am woken by the sounds of several loud voices and the shuffling of numerous bodies around me, as one by one they drop to the floor, either very drunk, or spaced-out on drugs. This motley crew, I later found out, were from Cuba, and I have to say that I have never seen such a mean looking bunch of deadbeat's in all my life before. However, having joined the ranks of the homeless and impoverished, albeit temporarily (or so I hoped) I simply took it in my stride as one more unexpected experience to add to my ever-growing catalogue of weird and wonderful encounters. The most uncomfortable nights' rest I can ever recall lasted for only four and a half hours, but during that short space of time I still managed to stiffen every bone in my body. I could not even hazard a guess at how many times I flipped my body into a different position on that sheet of hardened ceramic, but every muscle and joint in my body was aching so much that I reckon I would have had a more comfortable sleep on a bed of frigging nails.

Sitting upright was a major upheaval, as was opening my eyes, but in the end I managed to do both, only to discover that I now had a great lump of gooey chewing gum sticking to the bottom of my right trouser leg...Fuck! For the next twenty minutes or so I tried everything to remove this sticky, dirty and dusty piece of crap from my clothes, but I was failing miserably, and so I headed off to the bathroom once more, where after scrubbing and scratching the offending *shit* of my trousers as best as I could, I duly swopped them for a pair of clean shorts, before shoving the trousers in a plastic bag, and hiding them away in the bottom of my kit-bag. When we arrived last night, I could have sworn that the terminal was in the middle of nowhere, but now, in the cold light of day, I can see a marina in the distance, and so it was obvious that the coastline, and hopefully a nice, sandy beach, would not be too far away either? The young lady who is on duty in the terminal this morning, kindly gives me directions on how to get to the nicest part of the coastline, but as usual, I ended-up getting completely lost, before finding myself wandering aimlessly in the back streets of West Palm Beach.

Burned-out cars, disused parking lots and makeshift rubbish dumps surround me, as I swiftly made my way through the dirty, smelly streets and alleyways, in an attempt to get out of this godforsaken place as soon as possible. A Rastafarian chap, with dreadlocks down to his knees, is sitting on the edge of the pavement, smoking what looks like a joint from where I am now standing. The guy is staring at me with daggers, as if warning me that I don't belong in this neck of the woods, or at least that it how it looks to me, although I could be totally wrong, of course? Speeding up my pace, I turn into the next street, where an elderly couple is walking their dog, and after politely introducing myself to them, they are happy to give me a set of directions to the marina. When I finally reach the waterfront, I can see two men fishing on the bridge that I am about to cross, and so as I pass by them I say "Good morning", to each gentleman in turn, purely as a friendly gesture, now that I am feeling safe again. Both of them return the compliment with a smile and a gentle nod, before getting back to their simple worlds of utter peace and tranquillity.

As soon as I spotted the ocean, I immediately made a b-line for the beach, but before I knew it I was being escorted off the sand by a burly security guard, who had just informed me that this was a private beach, before kindly pointing me in the direction of the public area...oops! The beach is completely deserted, and although the sky is awash with clouds, and the air temperature is quite cool, I am determined to dip my feet in the water. Running across the sand like a man possessed, I go full pelt into the Atlantic Ocean, but within half a dozen strides I have gone from ankle-deep in water, up to my neck in the stuff...I am not impressed. It was only later on in the day, after chatting to a few of the local people, that one of them informed me that this was a man-made beach. The water was not as clean, nor as warm as it was in Miami or Fort Lauderdale which rather surprised me, seeing at it was the same ocean? Maybe Palm Beach is not renowned for its beach, but it certainly lives up to its name of 'Palm' Beach, as there were palm trees everywhere.

Along the promenade, down the side roads, on central reservations, and just about anywhere else where there was room to put one, a tree had been planted. Set out at equidistant intervals, these wonderful trees gave Palm Beach a magnificent touch of class, the likes of which I had not seen anywhere else on my travels. Along the coast and for several blocks inland, the streets were absolutely immaculate, and it looked like all of the major buildings had been strategically spaced-out, so as not to spoil the picturesque waterfront with a mass of skyscrapers, tower blocks and multi-storey hotels. Looking at this haven of sheer beauty, I found it hard to believe that less than half a mile inland the place looked like a tip! The main row of shops that adorned the seafront were primarily made up of designer clothes boutiques and branded jewellery stores, their top quality merchandise displaying prices that looked more like long distance telephone numbers. This place was definitely a playground for the upper-classes, and I could sense an air of snobbery all around me.

Trying to find a cheap meal around here was certainly not going to be easy, as the mere suggestion of someone opening a roadside cafe along the seafront would no-doubt have the locals up in arms. An elderly man, who is sitting near to the seafront, contentedly reading his newspaper (or at least he was, before I so rudely interrupted his day) advises me to go to 'Greens Pharmacy', saying that it is only a few hundred metres up the road, and that they have an excellent cafeteria there, which serves top quality food at very reasonable prices. Thanking the gentleman for his time and advice, I swiftly move on, leaving him to return to his peaceful world, hopefully without any further interruptions from annoying tourists? From the outside, the cafeteria looks quite plush, and looking through the glass windows, I can see that it is also very busy, which tells me immediately that the food must be of a decent standard, and so inside I go. Apart from the normal seating arrangements that one would expect to see in your average cafe, there is also a line of bar stools running the length of the main serving counter, which I am now standing at.

Most of the stools are already occupied, primarily by men and women in their working attire, who have obviously popped into the cafe for their morning cuppa, whereas the majority of the seating areas have been taken-up by families, the parents enjoying the delights of a Saturday morning treat with their young children. Noticing a small gap at the other end of the counter, I move swiftly down the line, subsequently accosting the revolving stool, before anyone else has the chance to *steal* it! All of the waitresses are very chatty, even though they are continually working at a frantic pace, in order to supply the massive demand of the ever-increasing customers, and so within minutes my order is taken. I have to say that the food was very good, and it was certainly plentiful, which it needed to be, as this one meal would now have to sustain me for the rest of the day. By the time I return to the terminal, where, incidentally, I have already rented my second locker, in order to leave my bag safely under lock and key, the sun is shining brightly, and so after yet another futile call to Leo, I wander off down to the bay area, with a view to topping-up on my tan.

I now have less than $25 to my name. Only one person of the many I have spoken to in the last 24 hours, have any idea where the address that Leo has given me is situated? According to my source it is about six or seven miles from the Greyhound Bus terminal, and so I ask myself whether I should spend a day trying to find the address on foot, or should I *bite the bullet* and use up the last of my resources on a taxi journey? And what if the house is empty when I get there? Should I then stake out the place for the next 48 hours, in the hope that Leo returns home during that time, or should I simply *give up the ghost* altogether, and hope that I can find my way back to the terminal? I cannot afford a two-way taxi-ride, as I have to keep $13 for the taxi journey from the Greyhound Bus terminal in Orlando to the Airport, and as for feeding myself for the next two days, well it looks like my long-awaited diet was just about to begin. The only logical answer, it seems, is to keep on ringing Leo, in the vain hope that at one time he might just answer the damn call? I have left him numerous messages telling him that I am at the Greyhound Bus Station, and so if he is home, then he must have heard at least one of them?

I knew only too well that time was rapidly running out, but what could I do? I was now trapped in a catch-22 situation, and worse than that, I had also left myself with no other choice but to succumb to doing something that I hate doing more than anything else in this world, and that is 'relying on someone else'? Once again I had fulfilled all of my obligations, and given one hundred percent of myself to the task in question, and now it seemed that once again I was going to be let down badly by the people whom I truly believed in? The hours drifted by, and although my body was now fully relaxed, and the stiffness in my limbs had long-since subsided into nothing more than a small ache in my side, my mind was in turmoil, as I contemplated the dark future that lies ahead, if I don't manage to contact Leo before I leave Florida? What I wouldn't give for a good snooze right now, as the lack of sleep from last night's floor-bound fiasco, had certainly taken its toll on my energy levels, but with everything that was currently going on in my mind, there was just no way that I would be able to drift off into any kind of slumber.

To make matters worse, my hunger pangs had now revived themselves, and even though only four hours had passed since I eat that hearty breakfast, I was now feeling hungry for something to nibble. Walking past a *Woolworth's* store, which is one of the most famous brand names in the UK, I am surprised to see one of their outlets in America –and also pleasantly surprised to see that they are selling meals even cheaper than they are at Greens Pharmacy? Unfortunately, my joy was short-lived, as the waitress informs me, somewhat apologetically, that I have missed the opening hours of the cafeteria, and so they are no-longer serving food today, which is just the kind of luck that I didn't need right now. Looking to while-away another half an hour of the day, I decide to wander back over to Greens Pharmacy, as it seems to be the only other place in this area that serves cheap food? On the way over, I am accosted by a beggar, who asks me if I can spare him a couple of bucks for a meal, and my only recourse is to smile sadly at him, before walking away.

"If only you knew, my friend", I say under my breath, knowing full well that the way things are panning out, tomorrow morning I could well be standing alongside him with my begging bowl clasped firmly in hand. No sooner had the beggar disappeared out of sight, than another man approached me, only this time he us a British tourist, from Devon, asking me if I knew the whereabouts of a new building development, adding that he was hoping to invest a few thousand pounds in one of their condominium options? Again I am unable to assist this man with his endeavours although I secretly wished that he would channel just a small proportion of what was obviously 'spare cash' in my direction? For the second time today I enjoyed a slap-up meal, only this time I asked the waitress to put a sizeable chunk of it in a doggy bag for me, just in case I became hungry again later on in the evening. On my return journey to the terminal I pass a small art gallery, where a number of canvas paintings have already been set up on various sized easels, on the small pavement area outside the shop.

Standing next to them is a large a-board, advertising further displays of arts and crafts inside the shop, along with offering free glasses of champagne, complimented with a selection of hors d'oeuvres, for anyone wishing to have a stroll around the gallery. However, judging by the immaculately attired clientele who are currently taking full advantage of their offer, I am feeling slightly underdressed, standing here in my skimpy shorts, sleeveless vest and a pair of flip-flops that have long-since seen better days! By five o'clock the rain is pouring down once again, but it is of no real consequence to me, for I am sitting safely inside the Greyhound terminal, contentedly reading a book to while-away the time. Over the next three hours I ring Leo on the hour, every hour, but I already know what the result of each call will be, long before the answer-machine kicks into play. My philosophy of refusing to allow myself to get bored is seriously waning by now, and I have already given up hope of ever seeing Leo again in this life. Disappointment combined with frustration and fatigue, has made me extremely irritable, and I am simply dreading my second night of sleeping it rough on the floor of the terminal.

They say that alcohol is a depressant, rather than a stimulant, and yet whenever I am feeling a bit down, like I am experiencing right now, I find that a couple of pints of beer can relieve all of that built-up tension inside of me, thus releasing whatever pressures of life that are currently surrounding me, and so with that thought in mind, I decided to speak to my learned friend at the ticket kiosk. The same gentleman, who told me about 'Bibinis' bar last night, has now advised me to pop over to 'Roxy's Bar' this evening, not only because there is usually a great atmosphere in the place, but also because it stays open until quite late on the weekends. However, more importantly from my point of view, is the fact that its drinks prices are a lot cheaper than they are in Bibinis. Tomorrow is Sunday, and I know that without meeting up with Leo, me staying in West Palm Beach for one more day is going to be a total waste of time. I will call him for one last time in the morning, but if I end up listening to that damn piece of celluloid again, then I am going to be on the next bus to Orlando.

Roxy's Bar is a lot smaller than Bibinis, and it is the type of bar that regular drinker's frequent on a nightly basis, and so the atmosphere is very pleasant, and because of its quaintness, it also seems a lot more homely than its somewhat larger, and therefore less personal, counterpart. I can just about afford one pitcher of lager, which holds around three pints of beer, and so after buying my *jug* for the evening, I headed for the darkest corner of the bar, a place where I can wallow in self pity and personal blame for everything going wrong once again. Three middle-aged men, all of them dressed in virtually identical pairs of white tennis shorts and matching polo shirts, are propping-up the main bar. One of the guys has a cricket-styled V-neck pullover hanging around his shoulders, though lord knows why, as the temperature outside is far too warm to even consider putting it on. For mature men, they are quite rowdy, in a harmless kind of way, and so nobody is paying too much attention to their permanently raised voices, and sporadic bursts of laughter, which is nearly always followed by a great chinking of glasses.

Unfortunately, I am unable to join in with their somewhat frivolous behaviour, as I am in a melancholy mood right now. I am simply dreading the journey home, and I am also worrying myself sick about what Caryl is going to say about me failing miserably for a second time. God love her; she has put up with a lot of crap this past fourteen months, and I know in my heart that I am unworthy of her. However, when I get home I will be relentless in pursuing my efforts to try and live in Florida, if that is what Caryl has truly set her heart on, although something deep down inside of me is saying that my beloved wife is already regretting the day that she made that initial statement. I begin scribbling a few notes in my diary, but I am soon interrupted by one of our tennis trio, who has sauntered-on over to my table, and before I know it, he has introduced himself to me as "Benjamin".

There is no doubt in my mind that he will now want to find out all there is to know about *the stranger from out of town,* but I am simply not in the mood for *'Twenty Questions'* time, and so at first, I try to fob him off, by answering his initial inquisitive questions with one word answers. However the guy is hell-bent on knowing more about my personal life, it seems, for he has now pulled up a chair, thus cordially inviting himself to my table, whether I like it or not? It seems like I have no other choice but to enlighten him on my story from beginning to end, and so off I go, telling him all my private business, to the point where I feel like I am pouring my heart out to my closest friend, rather than just having a friendly conversation with someone whom I know absolutely nothing about? After ten minutes or so, his two mates announce that they are leaving the bar, but Benjamin is adamant on hearing the rest of my story, and so he tells the guys that he will catch up with them later on in the evening.

Off I go again, spilling all the beans on my deepest, darkest secrets, while he just sits there, lapping it all up, like a newspaper reporter who is about to write the article to end all articles. "And that is where I am today Benjamin", I sigh, having finally come to the end of my tale —and almost to the end of my tether. Benjamin thanks me for letting him in on my personal life, before handing me one of his business cards. Printed in bold ink, above his name and telephone number, are the words 'Tennis Professional', a title which Benjamin is obviously very proud of, and so not to shatter his illusions, I give a small gasp as I read out the words, as if I am in awe of his standing.

"I teach a lot of professional people —and also quite a few famous celebrities Shaun, and so if you don't get any luck with your book over in England, then please send me a copy of the completed manuscript, and I am sure that I will be able to do something with it over here", Benjamin said in all sincerity, before taking his leave of me. Perhaps I should learn not to be so flippant about people whom I know nothing about, I thought to myself, having openly prejudged this guy as an *all mouth and no trousers* type of person.

And so once again I have yet another kind gesture from yet another willing participant in my ventures of glory. Unfortunately, failures and let-downs have happened so often now, that I tend to take any positive statements I am given with a huge pinch of salt. I think that it is about time I came back down to this Earth, before something tragic happens with all our lives.

"Why can't you just live your life like any normal human being Shaun —and why can't you just forget all about dreams and ambitions, and just take each day as it comes?" I scoffed, before continuing-on with my self-indulgent rant "Why can't you be thankful for the things that you possess in life right now, and stop planning things for the future —and Why don't you just sit back and let the world continue running your life for you?" I bleated again, before answering my own questions: "Because I am 'Me', and I have to continue believing that someone, somewhere out there in this big, bad old world of ours, is just waiting to stamp my passport to freedom and to prove to all the non-believers that I am a 'Winner'... that's why".

CHAPTER 44

Home Sweet Home

Tomorrow is always another day, and although I am even stiffer now than I was yesterday, after suffering another appalling nights' sleep on the floor, I have shaken-off last nights' miserable attitude towards life, and even Leo's monotonous tape cannot hinder my passion for another adventurous day ahead. I am going home to my loved ones, and so nothing else matters in life, and no-one is going to stop me. I've got no spare cash, no food or drink, I feel like a sack of shit tied up ugly...and I probably look like one too! The only thing in my possession is a rucksack full of dirty washing, which is beginning to whiff like a pile of tramps *hand-me-downs*, and in my jacket pocket is my final ticket to Orlando, along with my flight ticket to Gatwick Airport, and my train ticket to Cardiff, and so those last three things are all that really matter to me in my life right now. My bus to Orlando is due out at nine o'clock, and I have to say that I will not be sorry to leave this place at all, although lord knows what I would have done without it, of course?

Apart from spotting a burned-out car on the freeway, my bus journey to Orlando was quite uneventful, and so I have nothing to write about it, except to say that my tummy was rumbling for food all the way. Within three hours I was back where it had all begun, just twelve days ago. According to my driver, just around the corner from the Greyhound Bus terminal is a bus stop for a local bus service that goes directly to Orlando Airport. Unfortunately, he is not sure if the bus runs on a Sunday, but being in the position that I am right now, I am more than willing to take my chances. For over an hour I wait, and for the first time since arriving in Florida, the air is actually feeling quite cold on my skin, and so I have to put a sweater on in order to keep warm.

A young couple approaches the bus stop, and so I ask them if they know how much it will cost to take a taxi to the airport? "About ten bucks, my friend", says the gentleman "But a bus will be coming along shortly, and that will only cost you seventy five cents", he adds, probably not realising for one moment the importance of his statement. This would now give me twelve dollars to spend on food when I get to the airport, which was a lifesaver. My official flight to the UK is not due to take off for another thirty hours, and so now I must resort to a rather unethical tactic that my friend used in Spain many years ago, which, I am pleased to say, worked a treat for him. With a very concerned look on my face, I tell the Thomson representative that I have had a tragic bereavement in my family, and that I received a phone call from my wife last night, saying that I need to return home immediately, and so I was wondering if there was any possible way that I could change my flight?

The young lady says that the only flight heading for Gatwick Airport today, will be leaving at five-thirty this afternoon, but she will not be able to confirm a seat for me until half an hour prior to take off – in other words, once everyone else on that flight has checked-in, and they have closed the gates. With a couple of hours to wait for my answer, I pop over to the cafeteria for a cup of coffee, and a plate of well-earned grub, as I am absolutely starving by now. I now have six dollars left to my name, and so if I cannot get on this flight, then it is going to be yet another very long night. At four-thirty on the dot, the Thomson rep gives me the thumbs-up sign, and I cannot run fast enough to collect my boarding card from her, before making my way through the check-in terminals, and into the departure lounge. The couple that I have been chatting with in the cafeteria is not so lucky. They are returning from a two centre holiday in Florida and Jamaica, and they have just arrived at the airport, only to be informed that due to a mix-up with their seat allocations, they may have to wait until tomorrow for their flight to the UK?

The husband is furious, and quite rightly so, I believe, whereas his wife is doing her best to pacify the situation, whilst at the same time trying to keep her two children occupied with things to do. The son and daughter are unanimous that Florida was by far the best part of the holiday, because there was so much for them to see and do, and the parents had to agree. The husband also said that the prices of drinks at their hotel in the Caribbean were a complete and utter rip-off, especially compared to the prices in mainland Florida. His wife then followed-up his statement by talking about the poverty and squalor that the majority of the local people lived in, adding how guilty she had felt about her family languishing in the delights of a fully air-conditioned, luxury five star hotel, while the natives of the land were living in the dirty, smelly slums that effectively surrounded the place. Good news was on the horizon for this lovely family, as one of the holiday representatives informed them that there were four seats available for them on an aircraft that would be taking off for the UK within the next two hours.

However, the bad news is that the aeroplane will be flying into Manchester Airport, instead of Gatwick, which is where the husband had duly parked his company car when they flew out to Florida from London two weeks ago. This would now add several hours onto their journey home, of course, but unfortunately they had no other alternative but to accept the offer. As for me, well they have just announced the final call for all passengers flying to Gatwick Airport, to board the plane, and so off I go, feeling extremely happy that I have managed to secure the last remaining seat on the flight. As our plane takes off into the wide blue yonder, I once again say a fond farewell to the United States of America, only this time my elation at having completed the journey is somewhat overshadowed by my failure to achieve my objectives. Although having any negative thoughts goes totally against the grain with my way of thinking, I have to admit that as Florida's eastern landmass disappeared out of sight, I had serious doubts in my mind that my family and I would ever set foot on American soil as one.

In fact, I was beginning to fear the worst for our family life as a whole in the UK, for without the hope of a much better life than we were presently living, appearing on the horizon pretty soon, then I could not see our wonderful eight year marriage prevailing for very much longer. My thoughts are broken by the sound of the captain's voice, as he greets his passengers for the flight to London. Like his counterpart on the journey from Gatwick twelve days ago, our captain has a volley of information for us, only instead of giving us the *low down* on our route home he is a mine of information regarding various facts and figures about what our flight will entail. His spiel went something like this:

"Our take off speed will be 178 miles per hour, and the total distance that we will be flying today is 3,500 miles. Our route will take us via Amsterdam, and we will have a tail wind behind us of approximately 170mph. This tail wind will actually take 700 miles off our actual distance, and if all goes to plan then we should have a record crossing time of seven hours and twelve minutes. Our plane weight is 150 tons and our cargo weight is forty tons. We are also carrying 32 tons of fuel weight. This aircraft uses one gallon of fuel for every half mile covered, and our tank capacity is 9,000 gallons. Our main cargo today, for those of you who may be interested, is flowers from Columbia."

My mind simply boggled at the mere thought of those statistics, and I'll probably never work out where those 700 miles actually disappeared to, but seeing as Rowan Atkinson has just appeared on my screen, as the incomparable 'Mr. Bean', I have decided to put all of those thoughts to the back of my mind, and just sit back and have a good chuckle at the ridiculous antics of this man. The next comedy caper to follow-on from Mr. Bean was a side-splitting episode of 'One foot in the grave' with *Mr. misery-guts* himself, good old 'Victor Meldrew'. And so the comedy shows continued-on into the evening, while our wonderful stewards and stewardesses continued plying us with food and drinks all the way. Gazing out of my window, I can now see the shores of the British mainland coming into focus, and I am beginning to feel at home again, as our plane starts its descent into Gatwick Airport.

It is crazy to think that just over seven hours after leaving Florida we are landing in the UK, and yet because of the five hour time difference, it is only two hours later in the UK —even though my body clock is telling me different, of course. Shortly after collecting my luggage, I cleared immigration, before heading for the nearest telephone kiosk, in order to call Caryl and inform her of my early arrival in Britain. However, rather than being happy that I would be home a day earlier than expected, Caryl just gave me a severe ticking-off for not phoning her sooner, adding that she had been worried sick about me until she received my post card only two days ago. I tried explaining to her what Roman had warned me regarding the cost of international phone calls, but my beloved wife was having none of it –and when I informed her that my trip had not been as successful as I had hoped it would be, she virtually put the phone down on me.

The ticket collector was the next person to get on my case, as I tried explaining to him that even though I possessed a return ticket to Caterham, I wanted to go directly to Cardiff, and not over to Caterham first? After much deliberation, the conductor had the audacity to charge me an additional three pounds for travelling 50 miles less...what a rip-off. It took another four hours before I finally arrived at Cardiff Railway Station, where Caryl, Carl and Hayley were waiting to greet me. Liam was still in school, and so I insisted on picking-up my boy at 3.30pm. Standing by the main entrance, I can see my son making his way to the cloakroom with the rest of the children in his class, in order to retrieve his coat. Lifting his jacket off the hook which has the 'cherry' picture above it, Liam turned towards the doorway, expecting to see his mum waiting for him in the wings, but instead he saw me, of course. For a moment he just glared at me, as if he was in shock, but then a huge smile ran across his little face, as the realisation that I was home suddenly hit him head on.

"Daddy", he shouted at the top of his voice, before running full pelt through the cloakroom and jumping straight into my open arms, the force of the impact sending me reeling back a few steps, before finally regaining my footing. "I missed you so much Liam", I whispered to my son, before kissing his cheek and ruffling his hair, as he held onto me for dear life, his little arms almost strangling me, as he increased his hold on my neck, before wrapping his long spindly legs around my waist. "I missed you too daddy" he said softly, making me want to burst into floods of tears, and but for the fact that we were in the middle of dozens of people, I probably would have done. As we walked home together, Liam sitting high on my shoulders, we chatted about what he had done in school today, and what he did while I was away. My baby boy obviously couldn't care less about how things had turned out in Florida -and at this precise moment in time, neither could I.

Unfortunately, these precious moments that we have in our lives, rarely last for long, and it wasn't long before reality kicked-in, thus bringing me back down to earth with a tumultuous 'bang'. Caryl truly believed that my trip to Florida was nothing more than a joyride to paradise for me, not only leaving her to cope all alone with our three young children, but also to deal with the onslaught of final demands, threatening letters, and possible eviction orders that were now arriving through our letter box on a daily basis. As the days turned into weeks, and the weeks turned into months, what was once a hairline crack in our relationship, had now turned into a huge, gaping hole -one that would inevitably give way to the huge pressures that were now overshadowing our lives. Being in the depths of winter, with bitter cold days that were desperately short on light and sunshine, and freezing cold nights, that kept reminding us both of how our chances of ever living in a warm and sunny climate, were now nothing more than a hopeless dream, I felt sure that something was bound to give way, and at the end of February my worst fears were finally realised, as Caryl suddenly announced that she wanted us to go our separate ways.

There was no question that Caryl would have custody of the children, of course, although I would have equal rights to see them whenever I wanted to, but I would have no choice but to leave the house and move back in with my wonderful mother. Unfortunately, her house was currently in the middle of a total renovation project, which would not be completed until the end of March, and so that was the date that we set for my *eviction order*. When that day finally arrived, I duly left the house with nothing more than a few black bin liners that were primarily full of clothes, towels, sheets and blankets. A friend had offered to drive me to my mother's house, as Caryl would obviously need the car to transport the children to and from school, and also to do the weekly shopping, or go visiting her parents, and so I would either be walking everywhere, or catching busses from this day forward. Leaving my three lovely children on that fateful afternoon was as heart-breaking for them as it was for me, and I can still hear Hayley screaming for me to come back to her, as she hung on for grim death to the small picket fence that surrounded our front garden.

Liam was also crying, but as it was Friday, Caryl had agreed to drop him off at my mother's house later-on that evening, so that he could spend the weekend with his daddy, and so he was being as brave as he could. Carl was probably in the best situation out of all of us, as he didn't really understand what was going on, and so he just gave me a wave and blew me a kiss, as we took off down the road. Driving through the streets of Cardiff my mind was in turmoil as to what might lie ahead, not just for me, but for my family, as well –and as for my poor dear mum, well her whole life was also about to change completely, for she had lived on her own for over a decade, and so having someone to share her house with again, was certainly going to be a big upheaval in her life. As for the house, well it had been virtually raised to the ground and completely rebuilt from scratch over the past three months, and for the majority of that time it had been open to the elements, with no heating whatsoever, and so the temperature inside the place was barely above freezing point.

With nothing more than two gas fires to warm the whole house, my mum had put them on full power from the moment that she moved back into the place this morning, and so by the time I got there, it was a few degrees warmer, but considerably colder that the lovely warm centrally heated house that I had been used to living in for the past seven years. The plaster on the walls downstairs had been hacked off to a level around three feet above the floor, in order to inject insulating foam, and so the once nicely decorated room, now looked nothing more than a building site. The warm wooden floors had also been ripped-up and replaced with cement floors that were rock hard and cold, and totally uninviting, even under the painfully thin cushion of a threadbare carpet that was almost as old as I was. The kitchen and downstairs bathroom (the first bathroom that we had ever had in the house) were completely brand new, and so these did look very nice, and it was simply wonderful for my mum to finally have a toilet that was 'inside' her house, rather than it being encompassed in a wooden framed conservatory that my father had built over thirty years ago.

The two upstairs bedrooms can only be described as 'Baltic', as there was no upstairs heating, and so my mum immediately made use of her two hot water bottles, in order to warm the bed and blankets for me and Liam to sleep in this evening. Mum had slept on the sofa (not sofa-bed) since the passing of my dear father, eight and a half years ago, having been unable to spend a single night in her double bed without her beloved husband to cwtch into. Seeing Liam again was truly wonderful for me, but watching Caryl drive away with Carl and Hayley in the back of the car, my daughter repeatedly banging on the back window and continually screaming to let her out, simply tore me up inside, although I did not want to show my feelings in front of Liam, as he was as happy as a sand-boy, knowing that he was going to spend the weekend with his daddy. Caryl would be bringing Carl and Hayley back to the house tomorrow afternoon, before collecting them again in the evening, as they were really too young to stay out overnight, and she also wanted to get them into a routine of going to bed without their father.

That night Liam, mum and me, sat in the living room watching television together, and my mother said how lovely it was to have company again in her house. 'Kit' (my pet name for my mum, as her real name is 'Kitty') also cooked Liam's favourite delight of 'Nanny's chips' which he devoured with a passion, along with guzzling two small boxes of Ribena juice, before being ordered to bed by me. As it was our first night away from home, I decided that I would stay with him until he fell asleep, but in the middle of putting his pyjamas on, I suddenly burst into floods of tears. Grabbing Liam quickly, I hugged him tightly, in the hope that he had not seen my outburst, but by now my whole body was trembling, and my son is no fool. "Don't cry daddy –daddies don't cry" he said comfortingly, as I dropped down to my knees, before looking him straight in the eyes and saying softly "Daddies do cry son, but they try not to show it". By now I had regained my composure, and so the pair of us got into bed, before cuddling-up together...and that is all I remember.

The following day my mother cooked Liam and I a full English breakfast, and when Carl and Hayley arrived in the afternoon, I took the three of my children to Victoria Park, where they played happily on the swings and slides together, and I played 'The Monster', growling and grunting and chasing them all over the various climbing frames until all four of us were completely worn out. And so my new life as an *'absent father'*, if you like, had begun, and as much as I hated it with a passion, I knew that I would have to get on with my life, otherwise I would fall by the wayside. The following Wednesday I caught two busses up to Thornhill, where I took the children swimming in the local leisure centre, before taking two busses home again. I still loved my wife as much as I loved my children, and so no longer being a part of her life was literally tearing me up inside, and each time that I saw her I just wanted to hold her tight and beg her for her forgiveness, but I knew deep down inside that that time had long-since passed, and so I tried desperately hard not to show my heartbreak to her whenever she dropped-off or collected our children.

Chapter 45

Words Of Wisdom

For the many thousands of people who have been through a marriage break-up, and especially those of you who have had children involved in the split then you will all know what I am talking about when I say that I had 'good days' and that I also had 'really bad days', the latter being some of the worst days that I have ever encountered —and probably *will ever* encounter in my life. On one of those days, which was not long after our break up, my mother came home early from work and much to my dismay she caught me crying like a baby in the living room. I was inconsolable at the time, but my mother simply sat with me until the worst of it was over, and then she did what she does best, and that is give me one of her 'pep talks' on life, love and living.

"I know how much you are hurting right now Shaun, but there is a reason for everything, and we all know that 'God moves in mysterious ways'. You have finally hit rock bottom in your life, but believe it or not, that is actually a good place to be right now".

I just looked at my mother in total disbelief of what she was saying, before retaliating, by venting my anger to the one person in my life who had never let me down.

"How in God's name can there be any possible reason for me losing my wife and kids mum —and why the hell should I be grateful for being at the lowest ebb in my life", I bleated in a barrage of self-pity.

"Because being on the bones of your ass means that 'the only way is up'".

487

My dear mum; I had heard her use that term of phrase many times when describing her upbringing, and only now did I appreciate what she was talking about. However, I could still not find any possible reason to accept the fact that one day I would look back and say that "Every cloud has a silver lining".

Now that I was classed as a 'single man' again, I duly went to the social services offices in Cardiff, where I explained everything to a lovely young lady who was working behind one of many desks that occupied this huge auditorium. The remortgage that Caryl and I had undertaken a couple of years ago, where our payments were around £500 per month, was just about affordable whilst I was working as a New car salesman, but this was before I was made redundant, of course. However, due to the dramatic increases in mortgage lending rates, from 8% to 16 ¾ % in less than twelve months, it meant that my monthly mortgage was now in excess of £ 1,100 per month! Due to the current economic climate, the highest paid job that was currently on offer was £8,500 per annum, which would not even cover my mortgage payments, let-alone anything else, and so I had no other alternative but to sign on the dole until a job paying a lot more money came along. However, rather than do nothing with my life, I duly signed-up for a ten week computer training course that was being held in Cardiff's city centre.

Unbeknown to me at the time, but a new entity, which was called the C.S.A. (Child Support Agency) had recently been set up by the government, and so the next stage of my *interview* was to fill in the relative questions on the C.S.A. form. I openly said that I was willing to work up to sixteen hours a day, six days a week, as I wanted one day off to see my children, and that I would welcome shift-work, as I knew that that would pay the best money. I had already given the lady a copy of my C.V. and so she knew the kind of work that I specialised in, although in my current situation, I said that I was willing to do any kind of menial tasks in order to *pay my way.* With all the paperwork now completed, I was ready to start life afresh, albeit that I knew I had not even reached the first rung of the proverbial ladder, as yet. However, before letting me go, the young lady decided to put me straight on my situation. Our conversation went something like this:

"Now Mr. Donovan, you do understand that anything you earn over £ 42 will be paid directly to the government in respect of your mortgage, and also any monies that have been paid to your wife for the upkeep of your children".

"Of course, this is why I have offered to work sixteen hours a day, six days a week, in order to keep a roof over my children's heads, put clothes on their backs and food in their bellies...that is what fathers are for. This should then leave me enough money to pay rent and lodgings to my mother and also to facilitate transportation to and from my workplace".

"No Mr. Donovan; you misunderstand what I am telling you; I said that 'anything' over and above £ 42 that you earn will be paid directly to the government, regardless of how many hours or days that you work; that is quite immaterial".

"So what on earth would give me the incentive to work all the hours that God sends, if I am going to see nothing more for my efforts than £ 42 –and how the hell am I supposed to pay my mother for my food and lodgings...please answer me that!"

"Unfortunately, you are living with a family member, and therefore there are no stipulated criteria for paying them any money –well not in the eyes of the C.S.A. anyway".

"And so how long will £ 42 be my maximum earnings, as there has to be an end date to this ludicrous figure at some point in the future, surely to god?"

"Well you said that your daughter was now a year old, and so it will be for the next fifteen years, at least...unless she decides to stay on in school, or go to college, of course, and then it could last until she is eighteen, I'm afraid".

"So what you are basically telling me is that my life is now completely on hold until I am fifty years old...or possibly even older...is that what you are saying?"

Upon seeing my anger and sheer frustration at these unbelievable statistics, the lady leant forward over her desk towards me, before whispering in my face "If you want my own personal advice Mr. Donovan –stay on the dole!"

I was so despondent it was unreal, and I felt as though the whole world was caving-in on me, and there was absolutely nothing that I could do about it. On the way home I called in to see my mother at her workplace, as I desperately needed someone that I could trust to talk to, but I could see through the shop window that she was busy with customers, and so I continued on walking, like a vacant zombie that had no idea of where he was going, or what he was doing? The house was cold and empty, which kind of suited the way that I was feeling inside right now, but rather than screaming-out in pain at my frustrations, I decided to sit down and pour my heart out in two poems; the first one to my beloved children and the second one to my wonderful wife. They are as follows:

HEARTACHE

Oh the pain of a withered soul,
As hurt and anguish take their toll,
A heart in two, no longer in use,
A mind entrapped in a tightening noose,

A world at end with nothing in store,
But a lifetime of sadness left to endure,
Oh why must this turmoil continue each day?
I wish I could make it all go away,

With nothing in life that's worth living for,
I sit and I sob, 'til I can cry no more,
My life has diminished to merely exist,
And thoughts of happiness – completely dismissed,

There's only a fragment of hope that one day,
I'll smile again and maybe say…
"I had my chance and threw it away",
But oh what a price I had to pay,

For assuming that I was both clever and wise,
But in reality a fool in everyone's eyes,
Well now I've lost everything that mattered in life,
My three lovely children; my wonderful wife,

Dreams of watching my babies at play,
And growing-up steadily day-by-day,
Are now in the past, and to my dismay,
There's only the heartache left here to stay,

I'm so sorry Liam; I love you so much,
And Carl and Hayley, so wonderful to touch,
I'm missing you all and terribly so,
But there's only one thing I want you to know,

I'll always be there if you need me at all,
To sort-out your problems, however big or small,
For that is what daddies are sent here to do,
And you know just how much that <u>I love all of you.</u>

God bless my little children, Daddy.

ODE TO CARYL – TEN YEARS TOGETHER

I once met a girl, who captured my heart,
And I knew from that day we would never be apart,
She was everything I dreamed of, all rolled into one,
So beautiful, so elegant, a bundle of fun,

We laughed and we joked, then I gave her a lift home,
But could she fall for a midget, the size of a gnome?
We courted for months and our friendship increased,
So I decided right then, she would never be released,

From my clutches of love, that would change all her life,
Because, like it or not, she would soon be my wife,
But first on the list was romance in the sun,
As we travelled through Europe, together as one,

We then bought a house, as a little 'love-nest',
To begin life together, like all of the rest,
But engagement came first, and the giving of a ring,
To seal our true love, and to happiness bring,

Within seven months, the knot had been tied,
As she swore her betrothal, becoming my bride,
After four years of marriage and a new place to live,
She gave me the best gift a woman could give,

A bundle of love that would thrill everyone,
A baby, a boy, my wonderful son,
I remember our joy when Liam first smiled,
And how playful he was as a two-year-old child,

There was nowhere I could go without him by my side,
On a bus, in a car, or on a donkey, for a ride,
We holidayed in Spain, like most families do,
And we couldn't have been happier, just Liam, me and you,

But then we decided we wanted another,
And into our world, came Liam's little brother,
Together they played and our lives were complete,
And now I had two sons forever round my feet,

But I loved every minute and I wouldn't change a thing,
For they both made me feel as though I were a king,
So proud as a father, I walked them for miles,
My head high in the clouds and my face full of smiles,

Life was so perfect, that I couldn't wish for more,
But little did I know, of what was in store,
For eighteen months later, spinning me in a twirl,
You gave birth once more, to a beautiful girl,

Oh Hayley, my treasure, you're my heart and my soul,
When I hold you in my arms, I lose all my control,
You're as beautiful as your mum, and as sweet as a kitten,
And to look into your eyes, leaves me utterly smitten,

I couldn't think of a man, as lucky as was I,
But if only I knew, how soon I would die,
No longer would life have a meaning for me,
But the cruel awakening of reality,

No birthdays of happiness; no parties for fun,
No Christmas together; no playing in the sun,
No more as a family; no holding of hands,
A broken-down marriage, and that's how it stands,

The first words both Carl and Hayley will say,
Won't be treasured by their daddy, for he's gone far away,
And Liam's first trophy, or winning a prize,
Will be watched not by his, but another man's eyes,

And troubles won't be shared by their mum and their dad,
Nor problems solved, like they should have had,
For no-one's expected to burden their life,
With children's needs, or the troubles and strife,

Of bringing-up a family, not theirs from the start,
And they certainly can't give them, that love from the heart,
But what does it matter; I'm only a voice,
And Caryl, you've already made your choice,

If happiness is what you believe you will get,
Then go for it now that the stage has been set,
It was pressures of life that destroyed what we had,
But as a father and husband, I wasn't all that bad,

Oh Caryl why'd you do it, my only ever love,
I thought that we fitted like a hand in a glove,
But it seems I was wrong, and I'll never love again,
For my heart couldn't take any more of this pain,

Ten years of our lives snuffed-out like a flame,
But I'll have to accept it as part of life's game,
For at least I have memories to contend with the strain,
But my poor little babies have nothing to gain.

Goodbye Caryl; I will always love you.

Shaun.

The Moral Of My Story

Suffice it to say that as the months passed, the 'Good days' became more frequent, and the 'Bad days' were somewhat less painful, although I had several relapses in between, for various reasons, which I am not about to go into. Caryl and I did try to reconcile our relationship by taking the children on another caravan holiday for a week, but unfortunately it didn't work out, and in 1993 our divorce was finally made absolute. In 1995, having completed no less than three computer courses, I began working in the carpet industry, and the following year I started a new relationship with a divorced mother of two children, eventually moving in with them, and seeing my children at weekends. However, after two years or so, the lady in question decided to curtail our relationship, although it was quite amicable on both sides, as even though we had *given it our best shot*, we both knew that it was not going to work.

On 31st August 1998, I flew out to Tenerife, in order to start working in the lucrative world of timeshare, which was something that I knew absolutely nothing about, but I was willing to give it a try. Leaving the children was the hardest thing that I have ever done in my life, but I desperately needed a fresh start, and I was determined to make a success of this new venture, as I so wanted Liam, Carl and Hayley to be proud of their papa. It was the best decision that I had ever made. By the end of the following year the children had enjoyed three great holidays on this paradise island, and by now I was climbing the ladder of success at a rapid pace. In July 2000 I took the kids on a six week summer vacation, as we travelled to all four corners of America, before crossing borders into Canada and sailing under the amazing Niagara Falls. Caryl also spent two weeks with us in Florida and Disneyworld, which included a short cruise over to the Bahamas, before returning to the UK. After several years of

497

waiting, I had finally kept my promise to my children, which meant more to me than anything else on the planet.

In the summer of 2001 I met Sally on the island, and immediately I was smitten by her unadulterated beauty –even though she was married at the time! However, by October that same year, both Sally and her son, Jim, had moved over to the island, to live with me in my two bedroom apartment. After my first journey around Europe on my motorcycle way back in the eighties, my ambitions were to one day live abroad in the sunshine. I had also dreamed of having my own private villa with a swimming pool and also a snooker room, as I have been a snooker fan since I was *knee high to a grasshopper*. As for marrying a beautiful lady, with a heart of gold, well I had already done that, but now Sally was my dream woman, and I was simply besotted with her. After two years together, Sally and I bought our first property; a five bedroom villa, which had been built from scratch and included our very own private swimming pool. We also converted the six car garage into a games room, before importing a regulation size (12' x 6') snooker table from the UK, along with buying a full size ping pong table and a championship dart board.

Relaxing on one of the sun beds next to my pool, I would often compare my current lifestyle to those days of desperation less than a decade ago, and I would always recall those fateful words that my mother had said to me all those years ago: "There's a reason for everything Shaun; God moves in mysterious ways" and a smile would always appear on my face, as I counted my blessings for all that I now had in my life. The children were the happiest they had ever been, and over the next few years they would spend almost all of their school holidays in Tenerife, and one year I even took them on an independent tour around all seven of the major Canary Islands. In 2004 I won the coveted title of the number one salesman in the whole of Europe, and for the next two years I ran my own resort in Cyprus, the children coming over to visit me at every opportunity. During our time in Cyprus, Sally and I enjoyed two fabulous cruises, the first one to the mystical land of the Pharaohs, namely 'Egypt', and the second one touring a handful of glorious Greek islands.

In 2007 Sally and I flew to Dubai for a two night stopover, before spending six weeks circumnavigating the Australian continent –a journey that we will never forget. In 2009, after Sally had returned to the UK, and we had finally gone our separate ways, I worked in Portugal, Spain and Gran Canaria, where the children likewise joined me whenever they were free, although by this time all three of them were in full time employment. In 2011 Hayley and I took off on a half world tour together, the boys having working commitments that could not release them for a four month *sabbatical* holiday. During this period my daughter and I covered over 60,000 Km (40,000 miles) and we began our journey by crossing Europe on a forty-eight hour train ride from London to Moscow, before boarding the Trans-Siberian Train to Beijing in China, which would include crossing the Gobi Desert in Mongolia on our way to China's capital city. After visiting Xian, Shanghai, and Hong Kong over the next three weeks, Hayley and I crossed borders into Vietnam, where we then traversed over one thousand miles from Hanoi in the north, all the way down to Saigon (Ho Chi Minh City) in the south of the country.

Cambodia was next on our list, Phnom Penh being our first port of call, before travelling up to Siem Reap to visit Angkor Wat, the worlds' largest temple. From here we crossed borders once again, only this time into Thailand, where we enjoyed all the delights that the Khoasan Road in Bangkok had to offer us, before moving up to Kanchanaburi, where we crossed the Bridge over the River Kwai, before riding the 'Death Train' on the infamous Burma Railway. From Kanchanaburi we headed south to the island of Koh Tao, where Hayley passed her P.A.D.I. scuba diving award, before visiting the islands of Koh Phangan and Koh Samui. Crossing borders into Malaysia, Hayley and I stayed for four nights in Kuala Lumpur, before taking our final bus ride to the incredible city of Singapore, where we spent two glorious days sightseeing and generally relaxing. From Singapore, my daughter and I flew over to Goa, where we met up with Carl and Liam, and also Hayley's boyfriend at the time, and together the five of us enjoyed nine days relaxing on Goa's paradise beaches, and swimming in the warm, turquoise seas, before the boys flew home, and Hayley and I continued-on with our tour of India.

Mumbai was next on the list, where we also took a boat ride over to the amazing Elephanta Island. From Mumbai, we caught an overnight bus inland to the town of Aurangabad, where we visited the incredible Ajanta and Ellora Caves, before flying off to Delhi, and then taking a taxi to Agra, where we visited the fabulous Agra Fort, before paying homage to the unbelievable Taj Mahal. From Delhi we flew to Abu Dhabi for a one night stopover, before heading south to Cape-Town in South Africa. Apart from taking a cable car to the top of Table Mountain, before having a swim with a handful of sharks in the '2 Oceans Aquarium', Hayley and I also enjoyed a bus tour of this magnificent city, which included a visit to one of the major breweries, where we indulged ourselves in a wonderful wine-tasting session of glorious South African wines. A few days later Hayley and I embarked on a one thousand mile bus ride across the plains of South Africa, before crossing borders into Zimbabwe, where we finally came to a halt at the village of Bulawayo.

From here it was but a five hour bus ride to the stupendous Victoria Falls, where we not only did a zip slide across the incredible ravine that separates Zambia from Zimbabwe, but we also spent a whole day riding the rapids on the mighty Zambezi River. Having visited the David Livingstone Museum, in the town of Livingstone, Hayley and I then traversed the length of Zambia, before crossing borders into Tanzania, where we headed northwards once again, only this time to the town of Arusha. The following day Hayley and I set off on a three day safari, which encompassed the Tarangira National Park, Lake Manyara and the Ngorongoro Crater. We also came face-to-face with the infamous Massai Warriors. For our final journey, we headed to Dar es Salaam on Africa's western coast, where we took a short boat ride to the island of Zanzibar, before flying back to the UK via Doha in Qatar. In 2014, Hayley flew over to Tenerife, to live with me for two and a half years, during which time she also worked in timeshare for a short while, before taking a job as a waitress in a kind of 'Hooters Bar', where she had the time of her life working with a dozen girls that she has remained very close friends with ever since, even though the majority of them are now living in various areas dotted around the UK.

During the last few years I have spent a considerable amount of time back in the UK, even staying with my son, Liam, in the old *matrimonial home* on a few occasions, as my ex-wife has long-since remarried, and she now lives about half a mile down the road. The children and I went to Amsterdam for a long weekend in February 2018, in order to celebrate Liam's 30th birthday, and apart from several one week holidays in various parts of Europe, Hayley also spent three weeks visiting her friend in

Australia last year and this year she enjoyed a five day holiday in Chile. This means that my beloved daughter has now visited all six major continents (North America, South America, Africa, Asia, Australia and Europe) before I have, as I have yet to conquer South America, although it is on my 'bucket list', of course. Since I first visited Ted, way back in 1991, he has completed his second journey around the world on a motorcycle, which took him only two years this time, and his follow-up book 'Dreaming of Jupiter' has pride of place on my bookshelf at home.

Ted was also the inspiration for Ewan McGregor and Charley Boorman to do their 'Long Way Round' journey and subsequent television series, Ted meeting-up with the guys in Mongolia at one stage of their Siberian crossing. Since I first began transposing my diaries into this manuscript, which was almost five years ago, Ted has left his home in the United States, and he is now living in the South of France. Even though he will soon be approaching his 90th year, Ted still spends half of his time riding motorcycles, when he is not writing books, or attending functions and award ceremonies in his honour, of course. As for Meat-Loaf's iconic 'Bat out of Hell' album, well that has now been superseded by 'Bat out of Hell –The Musical', which made its stage debut in March 2017, and to my knowledge it is still going strong today (21.O2. 2019). As for "Bat out of Hell -the Movie", well I guess that I am kind of hoping that the people who read this book will have my way of thinking, and like me they truly believe that a movie depicting Ted's amazing journey around the world, coupled with the songs from the 'Bat out of Hell' album as the soundtrack, would become just as iconic as the movie 'Easy Rider' was back in the sixties and seventies.

However, unlike me, when I tried going it alone all those years ago, someone out there might just be in a position to make it work...who knows? As for reviving classic songs that date back to the mid seventies, well simply take a look at the film 'Mamma-Mia' which is primarily made up of ABBA songs from that same era. The film has been such a tremendous success from day one, that a sequel film, 'Mamma Mia 2', which also stars the amazing singer, 'Cher' (Cher sang 'Dead Ringer for Love' with Meat-Loaf) was released in July 2018, and like its predecessor, it has been a runaway success from the start. Anyway, the moral of my story is a very simple one. Always go for your dreams, because although one never knows what obstacles will appear along the way, or what sacrifices one might have to make in order to overcome those obstacles and achieve one's dreams and ambitions, they will always be worth it in the end. To coin a phrase: "You will always regret what you didn't do in your life, a lot more than what you did".

ACKNOWLEDGEMENTS

I would firstly like to extend my grateful thanks to Ted Simon, for without him writing his amazing book, 'Jupiter's Travels' my dreams would never have begun. The handful of days that we spent together at his homes in California has given me memories to treasure for the rest of my life. Also, Ted's continued support throughout my quest for making the movie 'Freedom Run', and the subsequent documentary series 'Steps of Jupiter -20 years on' was something that always kept me going, no matter how impossible things became towards the end. I really wanted to make you proud Ted, and to give you back just a tiny fragment of what you gave to me, and also to countless other people around the globe, who were likewise inspired by your wonderful writings.

Secondly, I would like to thank Jules Gilmore from the bottom of my heart, for without your belief in me, along with your continuing financial and moral support, I would never have gotten this dream of mine off the ground. The many wonderful times we spent together, such as our day trip to see Ted's bike in the British Transport Museum, our evening's out in the Holiday Inn, in Cardiff, where we chatted freely together about anything and everything, and so many other friendly meetings that we had during that fateful year, I will cherish to the end of my days. It was a shame that the Budget rent-a-car office never opened in Swansea otherwise I might still be working for you, my dear friend.

Thirdly, I would like to express my gratitude to all those of you who believed in me and assisted me with my cause from beginning to end. This includes James Gerwyn, Robert Staines, Josef Zagrodnik, Ryan Willis, Mick Chappell and Jim Howells. You graciously gave me your valuable time, along with a never-ending stream of sound advice, all of which I took on board, as I continued my quest for recognition. However, it was my naivety in the business that eventually let the side down, and for that I can only ask you to please forgive me, and for you also to accept my most humble apologies.

Whilst we are on the subject of apologies, well it may have taken me twenty eight years to say this, but I would like to sincerely apologise to you, Caryl, for letting you down so badly at such a crucial time in our lives, although I can assure you that it was never my intention. You stood by me for as long as you could, I can see that now, and the more that I read my own diaries, the more I wonder how you even lasted as long as you did, before finally breaking. Life has been very hard for both of us at times, but as the years have passed, we have had our rewards, our three wonderful children, Liam, Carl and Hayley, being the greatest of them all, of course.

There are many other people that became involved in my project, including several friends and family members, each of whom made their own contribution in one way or another, from simply giving me sound advice, to putting me up for a few nights. These include my great friends, Daniel and Zara, who looked after me in Irvine, California, Dexter and Sophie, who put me up for the night (and also loaned me their car) in San Diego, and Anthony and his family, who kindly gave me lodgings on two separate occasions at their home in Caterham. I would also like to thank Roman, Reggie and Jasmine for all the assistance they gave me in Florida, along with showing me my dream homes, and taking me to several beautiful places, both in the centre of Florida and along the Gulf of Mexico.

As for the people I met along the way, such as Ronaldo, Shauna, Den and Ariel on Ted's farm in Covelo, my companion in San Francisco and Chinatown, Brynley Graham, my roommates in Fort Lauderdale, Archie, Damon, Jayden, Lewis, Max and Tommy, my bosom buddy in Miami, Jensen, my wonderful friends in the Florida Keys, Clara, Blake and Albie, and last, but certainly not least, those two lovely ladies in Tarawoods, Molly and Maisie, well I would like thank you all from the bottom of my heart for your love and friendship, which means the world to me. As for the companies that kindly sponsored me in one way or another, such as Budget Rent-a-car, Stanfords of London and Prontaprint in Cardiff, I am indebted to you for all your assistance, however big or small it may have been.

Finally, I would like to extend my grateful thanks to the many other people who assisted me along the way, such as Caroline Watts and Derek Bryant, for their sound advice on travelling in Africa, along with Joshua Ball, who was a mine of information. In closing, I would like to thank Carter, Tyler and the guys at the Kawasaki dealership in Malpas, Russell Richards from Action Aid, Matthew Green and Jamie for their spectacular photo and newspaper article, Walter Finn and his associates for giving me a wonderful trip down memory lane and our great friend, Amber, for sorting out flights for me to Florida. I would also like to thank Leo Cummings for the time he gave me in Stoke-on-Trent, but sadly 'not' in Florida, and finally a big "Thank You" to all my family and friends who supported me from day one...I Love you all.

If there is anyone I have omitted to mention, then please forgive me, and as for the majority of people that I have mentioned at any time throughout this manuscript, I would also ask you to forgive me for not using your real names in both the book and also my acknowledgements, but if any of you should read my story then you will know who you are, I am sure.

Lightning Source UK Ltd.
Milton Keynes UK
UKHW040718080519
342312UK00001B/24/P

9 781728 386843